The Earth's Changing Surface

Michael J. Bradshaw

Principal Lecturer in Geography and Geology
College of St. Mark and St. John, Plymouth

A.J. Abbott

Headmaster
Alleyne's School, Stevenage

A.P. Gelsthorpe

Head of Geography
Fearnhill School, Letchworth

HODDER AND STOUGHTON
LONDON SYDNEY AUCKLAND TORONTO

Front cover The sea eroding cliffs on the south coast of Cornwall.

Rear cover A LANDSAT image of the Wyoming mountain ranges in the Rockies of western USA. North is to the left. The Snake now cuts its way through the mountains on the left, flowing through the Palisades reservoir (centre, left) and out onto an irrigated plain in Utah at bottom left. At the top the fans formed by erosion of the snow-capped ranges are dissected by tributaries of the Green River.

Line illustrations by Tony Mould.

British Library Cataloguing in Publication Data

Bradshaw, Michael John
   The earth's changing surface.
   1. Landforms
   I. Title   II. Abbott, Arthur John   III. Gelsthorpe, A P
   551.4   GB401.5

ISBN 0-340-22352-9

First printed 1978

Copyright © 1978 M.J. Bradshaw

Printed in Great Britain for
Hodder and Stoughton Educational,
a division of Hodder and Stoughton Ltd.,
Mill Road, Dunton Green, Sevenoaks, Kent
by Morrison and Gibb Ltd., Tanfield,
Edinburgh EH3 5JT

Contents

Preface   4

Acknowledgments   5

PART I   THE EARTH STORY

    Introduction   6
1   Plate tectonics   10
2   Ocean basins   20
3   Continents and mountains   34
4   Internal Earth energy   62

PART II   SURFACE PROCESSES

    Introduction   81
5   Weathering and the regolith   83
6   Mass movement and slope development   98
7   Running water   115
8   Ice   185
9   Wind   215
10   The sea   230
11   Organic processes   259

PART III   A COMPLEXITY OF LANDSCAPES

    Introduction   275
12   Landscapes of the British Isles   277
13   Arid landscapes   297
14   Geomorphology   316

    Bibliography   330
    Index   333

    Colour section   161-176

Preface

The Earth's landforms have provided a source of fascination for many explorers and scientists. The more they are studied, however, the greater is seen to be the complexity of relationships between the landforms and the processes which might have formed them. Today practical, as well as academic, interests are involved in obtaining satisfactory solutions to the problems of form–process relationships: terrain is regarded as an increasingly valuable resource by planners, and extreme manifestations of processes may cause havoc and the loss of costly buildings or materials. An understanding of the surface processes and their products thus has immense environmental significance.

Students of surface landforms today cannot resort to the oversimplified approaches dominant in the early part of this century, whereby landforms in particular climatic zones were grouped into a partially applicable model of development, of which the Davisian Cycle concept was the main example. The essence of a modern study of landforms is an investigation of the processes observable at present acting on the surface rocks. Horizons of space and time have been pushed back, so that local examples may be compared with those in contrasting climatic zones, and the changing patterns of climate within each area over time may be seen as a continuum in the production of particular features and groups of features. It is now time that a greater emphasis was restored to the influence of internal Earth processes and of the properties of surface rocks. There is also the vital consideration of man's activities in producing his own landforms and modifying the effects of surface processes — a factor which is becoming of paramount importance.

The study of landforms is thus one of complex variables and their interactions, and the systems approach is assisting students in a definition of these relationships. This book does not use such an approach as a complete basis, but demonstrates where it may be useful. The book is occupied largely with studies of surface processes: these include the internal Earth forces, which shape the larger scale features (PART I), and the external, or atmospheric, forces which are responsible for the medium and smaller scale features (PART II). These sections provide a basis for the greater complexity of regional studies, and a fuller consideration of the nature of geomorphology, the science of landform study (PART III). There is much to be said for looking at PART III at an early stage: the last three chapters are placed in that position to form a logical conclusion to the book, but there is no reason why readers should not sample the complexity of interpretation which is involved in understanding the genesis of each local landscape in Britain.

Few students may require to study the whole book, and its size reflects the broadening of the subject, rather than the quantity of information which has to be learnt. Some of the side issues, applications and background information are summarised in 'boxes', separated from the main text so that the progress of argument is not lost.

In a subject where fieldwork is the basic method of collecting data, photographic illustrations provide an all-important resource for the reader. Those included here cover a wide range from ground-based cameras, oblique, vertical and stereoscopic aerial cameras, and those taken from the higher platforms of Gemini, Apollo and Skylab spaceflights, as well as the scanning devices in the LANDSAT satellite.

This volume thus provides the student with a variety of resources. It is hoped that the reader will be stimulated to carry the frontiers of knowledge further by his own observation and analysis, and that he will gain a keener appreciation of the delicately balanced environment inhabited by man.

Michael J. Bradshaw

Acknowledgements

A book of this type could not have been attempted without the consultation of a variety of authorities and wide-ranging discussions. As originally conceived, it was to be written by Mr A.J. Abbott with a co-author, who was later replaced by Mr A.P. Gelsthorpe. The present chief author began as editor, but eventually took over the final writing to emphasise the oneness of treatment, and this book has become very much his own. He could not have managed without the earlier work by the co-authors, although the final version is his responsibility.

Two others have contributed much to this book, and merit a special acknowledgement. Mr Roger Stone has acted as Consulting Editor, and his advice is always stimulating to further development. It is surely a good thing for a geographer to have a physicist breathing down his neck! It was a great blow to many that Mr Cyril Everard, of Queen Mary College London, suffered so much illness through 1973 and 1974, but we are particularly grateful for the fact that he has been able to continue to act as adviser. His suggestions have had a large influence on the final form of the book, and he has taken great pains to improve the manuscript. The book is easier to read, more accurate and broader in vision as a consequence of the efforts of these two friends.

The authors would also like to thank various colleagues, including Mr A.J. Dunk, of the College of St. Mark and St. John, and Mr C.R. Whitaker, now of the Australian National University, for reading chapters over a period of some years, and endless students and pupils at Manchester Grammar School and at their present schools and college.

Many books and papers of a more advanced nature have been consulted in the writing of the book: they are acknowledged by reference in the bibliography.

Yet another major acknowledgement must be made to the United States Geological Survey (USGS) for their assistance with information and illustrative materials: in particular Frank Forrester and Don Kelly of the Information Office and Wil Dooley of the Photographic Division are to be thanked. Other important sources of photographs are the National Aeronautics and Space Administration (NASA) and the US Department of Agriculture Soil Conservation Service (USDASCS) in the USA, and the information offices of Switzerland, Canada, Australia and New Zealand. A number of friends have also assisted us in this field, and they are acknowledged in the captions, along with commercial agencies.

Mrs Mary Jones, Mrs Vera Holmes and Mrs Valerie Bradshaw have typed sections of the manuscript with speed and efficiency, and we thank them, especially the last, for their help and encouragement.

Michael J. Bradshaw

Part I

The Earth Story

Introduction

A first examination of the Earth's surface features leaves the impression that there is a nearly limitless variety of landforms arranged in a seemingly haphazard fashion. The awesome size of many of these features, together with the slow rates at which surface processes like rivers and glaciers work, especially in western Europe, the main centre of learning from the seventeenth century, led most people before 1800 to believe in the permanence, rather than the evolution, of the mountains and valleys around them. The physical landscape was merely an elaborate form of fixed stage scenery on which man acted out his life.

In the mid-eighteenth century a number of engineers realised that there was a relationship between, for instance, the filling of a harbour entrance with silt and the movement of sediment along the coast. Later in that century geologists began to doubt the permanence of the Earth's surface features when faced with fossil shells of marine creatures in the rocks at the summits of Alpine peaks. By the early nineteenth century James Hutton and Charles Lyell were suggesting that many rocks had been formed in past ages during which the land was worn down and the debris deposited in layers on the sea bed. Rates of wearing away and deposition were similar to those of the present—very slow. This may not seem to be a revolutionary idea today, but at the time it conflicted with the idea of a very short life-span for the Earth (c 6000 yr). An enormous and imaginative leap was necessary from understanding the physical world to be a static feature, where natural processes had little effect except in times of catastrophic violence, to a position in which the Earth was seen as subject to gradual but profound changes over hundreds of thousand and even millions of years. There was little point in studying the surface of the Earth whilst the former concept was in vogue, since it was unlikely either that the distributions were the result of any orderly process or that any prediction of future changes could be useful. The new view, however, opened the way to a scientific study of landforms and the processes shaping them.

By the end of the nineteenth century it looked as if an American, W.M. Davis, had discovered the key to the understanding of the Earth's surface features. Each landform of fluvial origin could be related to a series of events, known as a 'cycle', by which a landscape would be worn down to lower and lower levels until it almost reached the 'base-level' of erosion, the sea level. This simple view was attractive because each feature could be treated as part of a single classification of landforms, having a place in a developmental sequence, and so could be introduced at school level. The 'cycle' concept has not, however, been confirmed by subsequent closer examination of the landforms of particular areas, or by studies of the ways in which Earth surface processes act. Nature is far more complex than the Davisian model could suggest, and progress in the study of landforms was delayed by the sense of complacency which the simplistic 'cycle' concept induced. Today many detailed studies of landforms and the surface processes are going ahead without reference to the Davisian scheme.

A modern study of the Earth's surface looks at the form, or shape, of its features, together with the underlying geological structures. The processes acting on the surface are studied, firstly so that their workings may be understood, and secondly so that their relationships to the landforms with which they are now associated may be investigated. Processes include those which stem from internal Earth energy sources, such as mountain-building and volcanic activity, and those which are related to external, or atmospheric, processes, particularly the solar energy powered hydrological cycle (e.g. weathering, mass movement, running water, ice, wind and the sea).

Once form and process have been studied the question as to whether they can be related arises. The observations that streamless channels and valleys are found in deserts, and that wide, U-shaped valleys (without glaciers today) are interpreted as having been excavated by glacier ice, suggest that the processes acting on landforms today are not always those which have been mainly responsible for their formation.

Order	Area	Period of persistence (years)	Characteristic landforms	Basic mechanisms controlling relief (i.e. process)
VII	$10 \, cm^2$	1-10	Microscopic features: detail of solution polishing, soil texture.	Climate-controlled processes; rock and regolith texture.
VI	$1\text{-}10 \, m^2$	10^2	Small landforms: gullies, meanders, stone stripes.	Climate-controlled processes; lithology and regolith character.
V	$100 \, m^2$	10^4	Landforms: terraces, cirques, moraines, escarpments.	Climate-controlled processes; geological structure and lithology.
IV	$10 \, km^2$	$10^6\text{-}10^7$	Tectonic irregularities: anticlines, synclines; also hills, valleys.	Geological structure; climate-controlled processes.
III	$10^2\text{-}10^4$ km^2	10^7	Small structural entities: Lake District, Paris Basin, rift valleys.	Interaction at plate margin; broad differentiation of climate-controlled processes.
II	$10^6 \, km^2$	10^8	Large structural entities: ocean ridges, ocean trenches, fold mountain systems, shields.	Interaction at plate margin; 'stable' areas between with warping and fracturing.
I	$10^7 \, km^2$	10^9	Continents, ocean basins.	Plates develop in lithosphere; differentiation into continent and ocean.

In attempting to sort out which processes are responsible for the origin of a particular landform it is important to adopt a sense of scale in both space and time. The groups in the above table may be divided into three.

The orders VI and VII include the smallest features, which would not show up on a 1:25 000 scale map. These are formed over periods of a few thousand years or less and result from the interaction of atmospheric processes with the details of rock structure and texture.

The orders IV and V are landforms which are the products of the interaction of the internal processes with the surface processes, themselves controlled largely by climatic differences.

The orders I–III are the landforms which can be distinguished on a map of the world, or from a satellite image. They are formed over periods of many millions of years, and are the product mainly of internal Earth processes.

PART I of this book is an account of the largest features and the internal energy sources which give rise to them; the remainder of the book uses this framework in the study of interactions of landform and process — both of internal and external origin — at the smaller scales. The study of the largest features of the Earth surface has recently been provided with a theoretical framework in the idea of plate tectonics, which is explained in outline in chapter 1. Chapters 2 and 3 test this hypothesis by referring to distributions of landforms and of volcano–earthquake activity in the ocean basins and continents respectively. Chapter 4 examines ideas concerning the source of energy for the movements involved in the formation of these phenomena and there is a consideration of the Earth's interior structure.

Remote sensing

A variety of methods of observing the Earth's surface are now being used by scientists. They are known collectively as 'remote sensing' when the observer is not in contact with the surface being examined. In the widest sense these methods may include those which attempt to discover the nature of the Earth's interior from the surface (i.e. geophysical methods including seismic and magnetic studies), but generally they are restricted to studies of the surface from various levels in and above the atmosphere. Remote sensing thus normally refers to aerial and satellite photography and scanning. Photography is a passive process, in that light is reflected from the Earth's surface into a camera held at various heights in aeroplanes, satellites or space stations like Skylab. In addition to the visible light part of the spectrum, photographs may also be taken in the infrared portion, or (less commonly) in the ultraviolet portion (Figure 1). A range of photographic images may be produced in this way in black-and-white, colour, infrared (black-and white or colour), or in false colour (i.e. either extracting a part of the visible spectrum, or in a composite of natural and infrared colour). These enable the scientist to emphasise particular features for study: thus growing vegetation stands out in different hues of red in infrared colour photographs, whilst bare rock and office blocks are coloured blue; submarine features can be detected best when the lower end of the visible light range is emphasised; infrared rays do not penetrate water, so that photographs of this type give a sharp definition of the coastline. Photographs taken at increasing heights cover increasing areas, but the resolution (i.e. the detail of the smallest features shown) decreases. Thus whilst the large 1 km square fields of the prairies show up on satellite images taken at 900 km above the surface, the tiny fields of parts of Spain or east Asia cannot be distinguished (Figure 2). Resolution is also related to the focal length and aperture of the camera used and the quality of the photographic materials.

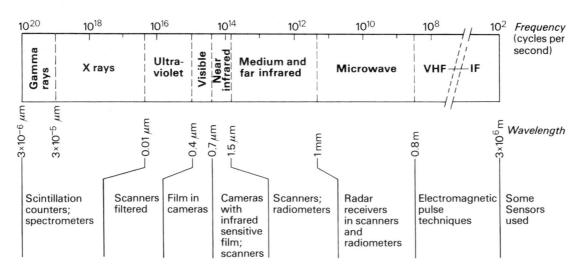

Figure 1 The electromagnetic spectrum and some sensors which image it. (After Parker and Wolf, 1965, in Estes and Senger, 1974)

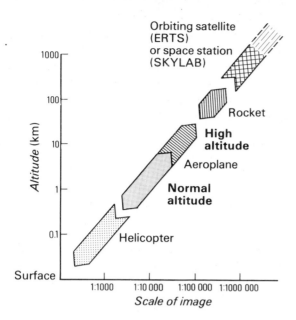

Figure 2 Platforms of observation for Earth sensors and the approximate scales of the resultant images.

In addition to photographs on film a variety of other methods are used, known collectively as scanning. Some of these are also passive, but others are active, emitting rays which bounce back from the surface in ways which can be recorded to give a picture of the surface features. The scanners may act in the visible or infrared wavelengths, but some of their most important uses are in relation to the active radar techniques. These are becoming increasingly important for rapid surveys of large areas of previously unmapped land, especially where cloud cover is common, since radar waves are not obscured by clouds. The Brazilian government has been using this technique in the mapping of the Amazon Basin. Scanners are also useful in satellites, like the Earth Resources Technology Satellite (ERTS or LANDSAT), since the impulses are codified electronically and can be returned to Earth for processing and the production of images. Such images are less detailed than those obtained on film from, for instance, Skylab and the Apollo journeys, but are the only way of obtaining a continuous series of pictures from satellites. Weather satellite scanners operate in a similar way.

The increasing availability of such observational methods has provided new tools for scientists to use. Several pictures of these types occur in this book, and there is a growing science of processing and interpreting the flood of data which is being returned from satellites.

1 Plate tectonics

'The past decade has brought a sweeping revolution in understanding of the Earth, a new basic grasp of the forces that shape it. The change of thought must be compared to the scientific upheavals that occurred when Copernicus showed that our planet was not the centre of the universe, Darwin postulated the slow evolution of living things, Niels Bohr described the atom.' (S.W. Matthews, *National Geographic Magazine*, 1973.)

Zones of Earth activity

People living in the British Isles, western Africa, Australia or the eastern seaboard of North America, have a different view of the Earth from those in Peru, Alaska, Japan or southern Italy. The former areas lie in relatively stable sections of the Earth's crust, whilst the latter are often rent by earthquakes or experience volcanic outbursts. The frequency of earthquake and volcanic activity is related to internal Earth movements (Figure 1.1). Such events are seldom too frequent to affect human settlement, and, indeed, may produce new land from beneath the sea or renew the materials required for the formation of fertile soils.

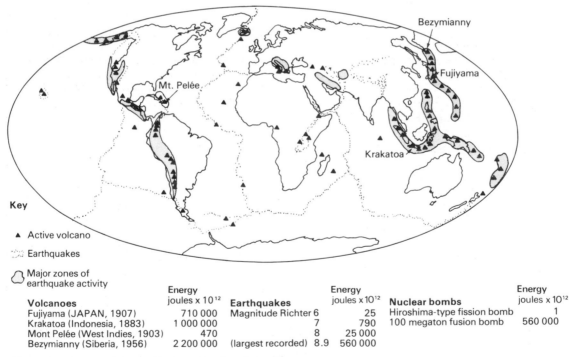

Key

▲ Active volcano

Earthquakes

Major zones of earthquake activity

Volcanoes	Energy joules x 10^{12}	Earthquakes	Energy joules x 10^{12}	Nuclear bombs	Energy joules x 10^{12}
Fujiyama (JAPAN, 1907)	710 000	Magnitude Richter 6	25	Hiroshima-type fission bomb	1
Krakatoa (Indonesia, 1883)	1 000 000	7	790	100 megaton fusion bomb	560 000
Mont Pelée (West Indies, 1903)	470	8	25 000		
Bezymianny (Siberia, 1956)	2 200 000	(largest recorded) 8.9	560 000		

Figure 1.1 The world distribution of volcanic and earthquake activity, together with some estimates of the energy involved.

The San Andreas Fault

The San Andreas Fault of California is perhaps the best known fracture in the Earth's crust. This is largely because of its size (nearly 900 km long), and its closeness to centres of population like San

Francisco and Los Angeles. People who live in California are frightened of the possible results of another earthquake resulting from movement along the fault, like that which destroyed much of San Francisco in 1906.

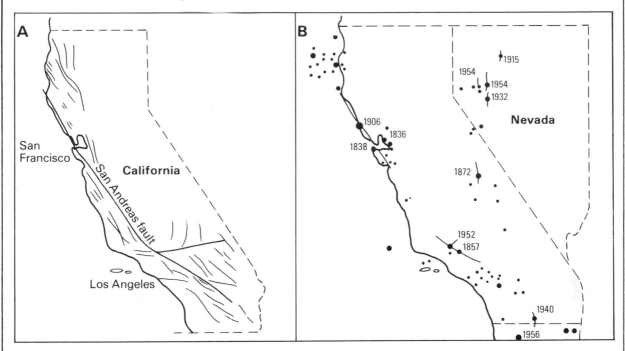

Figure 1 The San Andreas fault and associated fractures in the western USA. (A) The network of faults in coastal California. (B) Epicentres of some large earthquakes in the region. Heavy lines show where ground was broken and the size of dot is proportional to the earthquake's magnitude. (After Crowell, 1962, and Richter, 1955, in USGS pamphlet '*San Andreas Fault*')

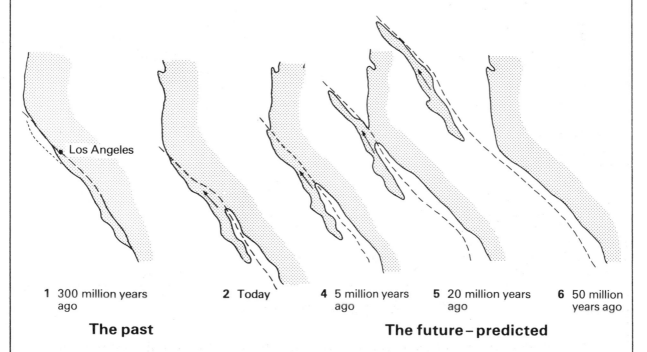

1 300 million years ago

2 Today

4 5 million years ago

5 20 million years ago

6 50 million years ago

The past

The future – predicted

Figure 2 Movement along the San Andreas fault.

> The San Andreas Fault includes a whole system of smaller faults cutting across western California, and there have been a series of associated earthquakes recorded (Figure 1). These are related to the fact that the plate carrying the North American continent has moved over the top of the East Pacific Ridge, and the San Andreas Fault is along the line of a transform fault extending from the ridge (Figure 2). Movement is north-westwards on the Pacific side of the fault: this has already resulted in the opening of the Gulf of California, and could lead to complete separation of this slice of continental crust from the rest of California — in a few million years' time.

The close association of belts of volcanic and earthquake activity has been known for some time, but it is only during the last few years that it has been understood. This association, together with the additional coincidence of young fold mountain ranges on the continents, and trenches and ridges in the ocean basins with the same narrow zones, has led to a new theory of Earth evolution, known as **plate tectonics**. This is an idea which was proposed early in 1968 and has been taken up enthusiastically as more and more evidence has been seen to fit

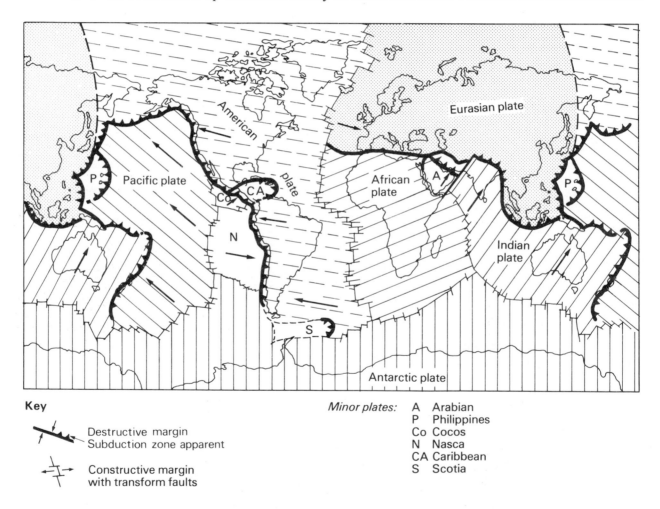

Key

Destructive margin
Subduction zone apparent

Constructive margin
with transform faults

Minor plates: A Arabian
P Philippines
Co Cocos
N Nasca
CA Caribbean
S Scotia

Figure 1.2 The world distribution of plates and types of plate margin. Six major plates are named on the map and six minor plates in the key. Although this map looks clearcut, it must be pointed out that not all the details are finally worked out, and that the polar regions in particular present difficulties of investigation and confirmation. (After Morgan, in Cox, 1973)

the general model. It can now be regarded as a basic hypothesis to account for the evolution of the major Earth structures and morphology. A hypothesis is an idea which has to be tested against observation and experiment. In this case it is necessary to test it by a study of the character and development of ocean basin features (chapter 2) and major continental landforms (chapter 3).

The concept of plate tectonics

The concept of plate tectonics suggests that the Earth's surface layers are divided into large segments, or plates. The plates are approximately 100 km thick, and therefore include the Earth's crust together with part of the upper mantle, but measure several thousand kilometres across. One scheme suggests that there are six major plates (with some smaller ones) covering the entire Earth (Figure 1.2). Plates move over the face of the Earth, each describing a circular path round its pole of rotation (Figure 1.3). In doing so they move away from one adjacent plate, towards another, and between a further pair. These relationships result in three types of plate boundary (Figure 1.4).

a) **A boundary where two plates move apart.** A fissure develops, allowing hot, molten rock to

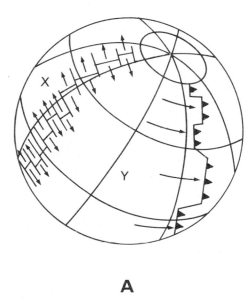

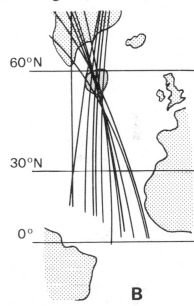

A

Figure 1.3 Plate movement on the spherical Earth. In (A) motion of plate X with respect to Y can be represented by rotation about a pole. In (B) Great Circles have been drawn at right angles to transform faults cutting the Mid-Atlantic ridge: they all

B

pass through the shaded area south-west of Iceland. This point can be thought of as the pole of spreading for the American plate relative to the African. (After Morgan, in Cox, 1973)

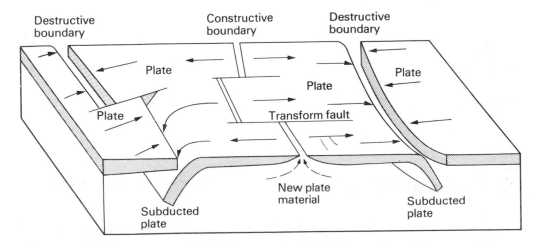

Figure 1.4 Types of plate boundary. (After Isacks, Oliver and Sykes, in Cox, 1973)

well up from the mantle (i.e. the zone in the Earth between the crust and the core, extending from 50 km to 2900 km depth), and to form new plate material as it solidifies. The fissure is called a spreading, or 'pull-apart' centre, and this type of plate boundary is known as a **constructive, or divergent, margin.**

b) **A boundary where two plates collide.** The more dense plate will normally be deflected beneath the less dense, and will be destroyed and absorbed at depth. Such plate destruction takes place at the zone of subduction, and the boundary is known as a **destructive, or convergent, margin.**

c) **Boundaries where two plates slide past one another.** These are shear margins, where little, if any, interaction takes place between the rocks on either side, and are known as **transform fault margins.** They are most in evidence crossing constructive margins, offsetting the trend of an oceanic ridge (Figure 1.2).

The new plate rock, formed by the solidification of molten rock at a constructive margin, then moves away from the constructive margin as more and more new rock forms there. It travels towards the destructive margin where it is subducted and the materials are returned to the mantle. Plates are thus formed of ocean-floor crustal rocks welded to the underlying upper mantle rocks. A plate may have a mass of lower density continental rocks on top (e.g. the western portion of the American plate), and this is carried as a 'superficial passenger'. Whilst the ocean-floor plate material is constantly formed and destroyed, the continents are not consumed at the subduction zone, because their low density provides buoyancy. The continents are subjected to different changes due to erosion and deposition by surface processes, but this has the overall effect of causing rocks to accumulate on these areas. It is significant that the oldest known continental rocks are 3900 million years old, and that the

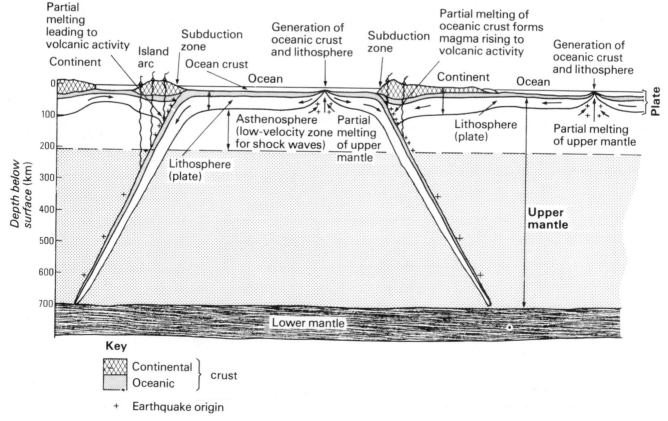

Figure 1.5 The plates and the structure of the outer part of the Earth interior. The lithosphere (oceanic crust plus sections of rigid upper mantle, together with continents where they occur) moves as a series of units, or plates. Older ideas of 'continental drift' (p. 15) suggested that the continents moved across the oceanic crust. The vertical scale is at least five times the horizontal in this diagram.

ocean-floor rocks are nowhere more than 200 million years old. The ocean-floor–continent division is thus the most significant breakdown of major relief features on the Earth (Figure 1.5).

This is the essence of the idea which provides a new understanding of the distributions of a variety of geological phenomena at the Earth's surface, and which will be developed in the next two chapters. It demonstrates that the Earth's surface undergoes continual change, with continents manoeuvring as the ocean basins change in size and shape whilst their materials are circulated above and below.

Continental drift

The idea that continents have moved around on the Earth's surface is not new. There is some dispute about the earliest mention of the idea, but it seems that this must be attributed to Antonio Snider-Pellegrini in 1858. Francis Bacon, who has also been associated with the idea, merely noted that the outlines of South America and western Africa were complementary in shape: he was writing in 1620. Snider-Pellegrini's suggestion was that the catastrophic events taking place during Noah's flood caused a rupture and pulling apart of the continents from an original single landmass. The idea then became associated with those who suggested that world relief features had originated largely in such catastrophic events, and lost popularity with the decline in support for that view. Yet it was kept alive by speculations following the notion that the Moon was formed by material removed from the Pacific Ocean, and by studies of climatological data.

In 1910 F.B. Taylor, an American scientist, began from a map of the world's highest mountain ranges and from the idea that these had been formed by lateral compression. He suggested that there was 'a mighty creeping movement' from the north, leading to a piling up of rocks against the ancient massifs of Africa and the Indian peninsula; the New Guinea fold belts showed that Australia had moved to the north-east; and South America and Africa had moved apart (Figure 1). Taylor did not suggest any mechanism to account for the movements.

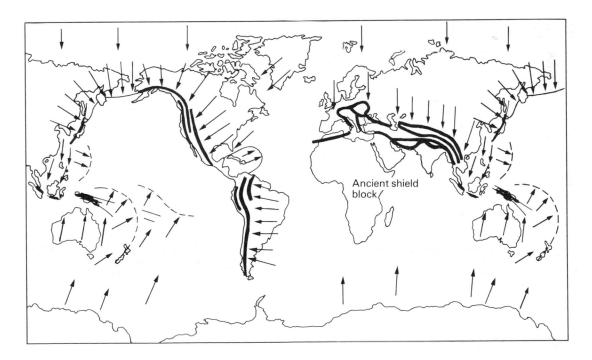

Figure 1 Taylor's map to show the directions of Tertiary crustal movement. He said little about the mechanism causing such movements, but the arrows seem to indicate movement from the poles. (After Taylor, 1910, in Hallam, 1973)

The concept of continental drift is mainly associated with Alfred Wegener, a German meteorologist, whose ideas, worked out at about the same time as Taylor's, were put over in an age when vertical movements due to contraction of a cooling Earth were in favour as the mode of mountain range formation. The Austrian geologist Suess had previously explained the coincidence of fossil types in Africa and India by collapse of sections of a former continent (called Gondwana) covering the area. Wegener was struck by the congruence of coastlines around the Atlantic and the intellectual problems raised by the 'land bridge' explanations of links between continents seemed insuperable. Such land bridges were imagined island chains, or long-since foundered continents, which facilitated the migration of animals and plants in the past, and could thus explain the occurrence of similar creatures in widely-separated continents. He began by

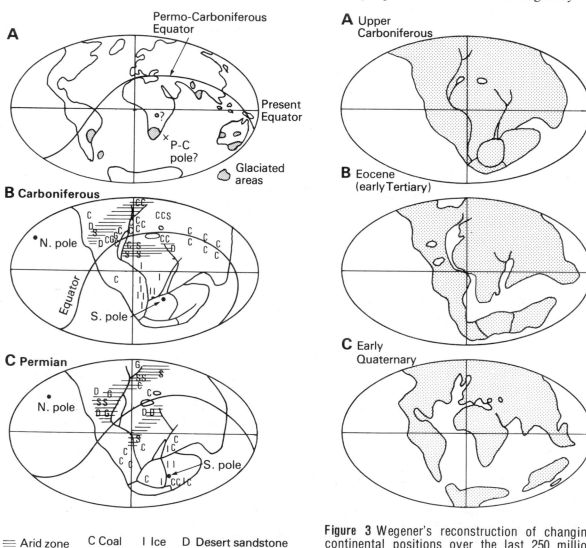

=== Arid zone   C Coal   I Ice   D Desert sandstone
                  S Salt   G Gypsm

Figure 2 Climatic belts of the Permian and Carboniferous.
(A) The glaciated areas in the southern hemisphere.
(B) Carboniferous climatic belts in rearranged continents.
(C) Permian climatic belts in rearranged continents.
(After Wegener, 1929, in Hallam, 1973)

Figure 3 Wegener's reconstruction of changing continental positions over the last 250 million years. Compare these with those shown in Figure 3.24. (After Wegener, 1929, in Hallam, 1973)

putting forward continental displacement as a working hypothesis in lectures in 1912, and then added increasing quantities of evidence as the published account of his ideas went through several editions (the 4th edition of 1929 is widely available). He preferred to explain the formation of fold mountain belts by continental movement rather than by contraction, and envisaged the continents moving around on the basaltic crust. Wegener put forward a wide range of evidence to show that continents could have been close together in the past — evidence of similarities in fossils, rocks and ancient climates on opposite shores of oceans (Figure 2) — and suggested a progression of events for the break-up of a master continent during the last 250 million years (Figure 3).

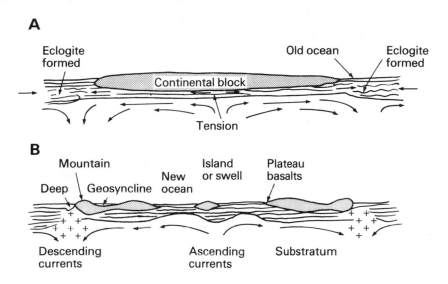

Figure 4 Holmes's interpretation of continental drift. Compare this with the explanation suggested by plate tectonics. (A) An early stage in convection. (B) Vigorous currents drag the continents apart; mountain-building takes place where currents descend and new ocean-floor is formed in the pulled-apart section where the currents ascend. (After Holmes, 1929, in Hallam, 1973)

These ideas met with varied responses at the time, but most were hostile, and many scientists merely ridiculed his concept. Every point made by Wegener, including the Atlantic 'fit' was questioned, particular play being made with the supposedly inadequate geophysical mechanism envisaged for the movements. Jeffreys, an eminent Cambridge geophysicist, led this side of the campaign against continental drift, beginning his publication of arguments in 1924 and continuing through into the 1960s. His point was that the forces tentatively suggested by Wegener to account for the movements (a rather vague tidal force combined with movements towards the Equator due to rotation) were insufficient, and that in any case the rocks at the continental-crust/ocean-crust division were rigid and would not permit the drifting of continents over the ocean floors.

Support for Wegener came from a South African geologist, Du Toit, who amassed more and more evidence to show links between continents in the past, and from the British geologist Arthur Holmes. Holmes was involved with applying his studies of radioactive elements to internal energy sources, and suggested a model which directed attention to a zone along which movement might take place somewhat lower in the Earth than the continental-crust/ocean-crust boundary. His first suggestions in this area were published in 1929 (Figure 4) and his text-book '*Principles of physical geology*' also contained these ideas. But most scientists rejected the idea of continental drift — largely on geophysical grounds — until the 1950s and later, when, strangely enough, it was the geophysical evidence from palaeomagnetic (chapter 4) and ocean floor (chapter 2) studies which led to the synthesis of ideas in plate tectonics. It can now be seen that Wegener and Holmes had gained early insights to the pattern of events, but that the fuller story had to await further developments in scientific investigation — just as Darwin's theory of natural selection had to await developments in genetics nearly fifty years after the public presentation of the idea.

Plate 1 The trace of the San Andreas Fault and its surface expression just south of San Francisco. Compare the map and the radar side-scan image. (USGS)

Plate 2 Elkhorn Scarp, between Elkhorn Plains (left) and Carrizon Plains. How have movements along the San Andreas Fault affected the relief and drainage pattern? (USGS)

Plate 3 The 1906 earthquake offset this fence 2.5 m in Marin County, California. (USGS)

Plate 4 Aerial view to the south-east along the San Andreas Fault and San Andreas Lake (locate on Plate 1). How would movements along the fault endanger the communities which have sprung up around it? (USGS)

Plate 5 The summit of Kilauea, Hawaii, with the Halemaumau pit. This has become the most closely studied of all volcanoes. In 1919 and 1921 lava rose and overflowed across the floor of this pit, but in 1924 it sank and the central pit was enlarged by collapse of the margins. Lava rose and spilled out again in the 1950s, but the major eruptions have occurred along fissures to the east. (USGS) See also colour section.

Plate 6 Shishaldin volcano in the Aleutian Islands. This has the typical form of an andesitic cone: compare it with the Hawaiian cones. (USGS)

Plate 8 Novarupta, Katmai National Park, Alaska. The central dome of rhyolite is 60 m high and 250 m across, and grows from the inside; it is in a cone of pumice. (USGS)

Plate 7 Mount Erebus, Antarctica. The summit is 3794 m, and this volcano is still active: the modern cone fills an older crater. McMurdo Sound and the Ross Sea ice are in the background. (USGS)

2

Ocean basins

Over the past 200 years geologists have studied the features of the continental areas with increasing rewards in terms of both a greater understanding of the forces at work and of the search for economically useful minerals. A detailed knowledge of many, seemingly unconnected, aspects of the geology of the continents had been built up by the 1950s, but their relationships and origins were often the subject of contradictory and highly speculative theories. This is scarcely surprising for only a limited understanding of Earth history could be gained from the study of the continents, which make up less than one-third of the Earth's surface. This deficiency is now being made up by the exploration of a vast realm of new evidence in the ocean basins, the study of which has flourished since the International Geophysical Year (IGY) in 1960 – 62 (although scientific investigation of the ocean-floor goes back to the voyage of HMS Challenger in 1872–76). As a result new and more soundly-based theories have been formulated, and it is not overstating the case to say that oceanographic studies have provided the key which has unlocked many of the secrets held by the rocks of the continents.

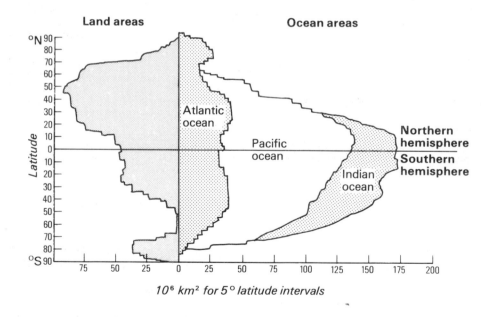

Figure 2.1 Continent and ocean. Contrast the areas occupied in different zones of the world. (After Gross, 1972)

The ocean basins are bounded by the slopes rising to the continents, and occupy two-thirds of the Earth surface. They are especially dominant in the Southern Hemisphere (Figure 2.1). At present the continents are distributed in three complementary north–south groups: North and South America; Europe and Africa; and Asia and Australia. The study of continental distribution has fascinated men since the first world maps were drawn in the late sixteenth century, but the questions which have been asked could not be answered until it was possible to study the ocean basin features.

Ocean basin relief features: general review

The major relief of the ocean basins, mapped in outline largely in the 1960s (Figure 2.2), shows features which contrast markedly with those on the land. The Atlantic Ocean has the simplest pattern, and has been studied most intensively. Its main features are as follows.

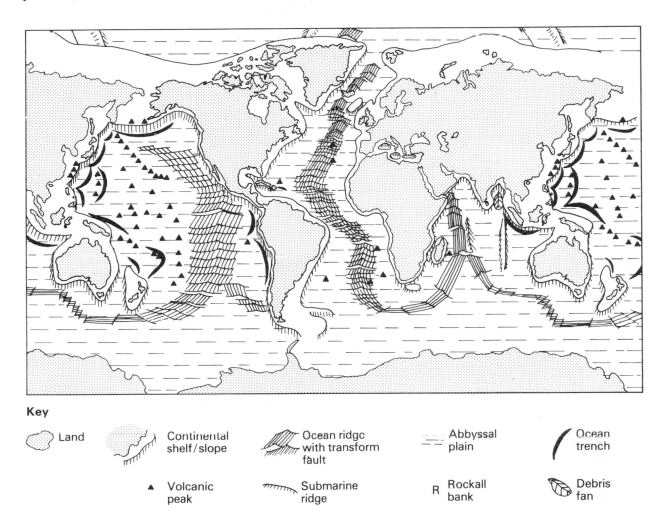

Key

- Land
- Continental shelf/slope
- Ocean ridge with transform fault
- Abbyssal plain
- Ocean trench
- ▲ Volcanic peak
- Submarine ridge
- R Rockall bank
- Debris fan

Figure 2.2 Major features of the ocean floors. This is a generalised impression (e.g. only a representative group of volcanic peaks are shown) without names. The North Atlantic is the best known section, and some regions are still very poorly explored.

1) A central ridge (the Mid-Atlantic Ridge) occupies up to one-third of the ocean area, and follows the S-shape of the ocean margins from north to south. It is formed of a series of parallel ridges, declining in height east and west from the centre, but along the central axis there is a deep rift valley. In the North Atlantic the ridge rises on average 1500–3000 m above the ocean-floor, but may still be 800–1500 m below the ocean surface; it rises above sea-level to form Iceland. The ridge is crossed by transform faults, offsetting the opposite sides by hundreds of kilometres to the east or west.

2) Marginal flat plains at depths of 3500–6000 metres.

3) Isolated peaks rising from the ridge or plains, and sometimes breaking the ocean surface to form islands.

4) Deep, narrow trenches descending to over 8000 metres. The main one in the Atlantic Ocean is north of Puerto Rico in the West Indies.

5) There are also broader areas at shallower depths, including the Rockall Rise in the North Atlantic, which is only 200 m below the surface over much of its area and has a tip above sea-level.

Comparison of the Atlantic floor relief with the other oceans shows that the same features are found in all the oceans, but that there are differences of emphasis (Figure 2.3): the Pacific Ocean, for instance, does not have a central ridge. The widespread occurrence of these ocean-floor features, together with the variations, and their relationships with continental features, hints at an underlying pattern which is being revealed by detailed studies.

Ocean	Continental shelf and slope (%)	Continental rise (%)	Ocean basin (%)	Volcanoes, volcanic ridges (%)	Ocean ridge (%)	Trenches (%)
PACIFIC	13.1	2.7	43.0	2.5	35.9	2.9
ATLANTIC	19.4	8.5	38.0	2.1	31.2	0.7
INDIAN	9.1	5.7	49.2	5.4	30.2	0.3
WORLD	15.3	5.3	41.8	3.1	32.7	1.7
EARTH SURFACE	10.8	3.7	29.5	2.2	23.1	1.2

Figure 2.3 The areas occupied by major ocean basin relief features. The Antarctic Ocean area is divided between the three main oceans; the Arctic Ocean is included with the Atlantic Ocean. Using this information and the map of Figure 2.2, contrast (a) the positions, widths and heights of the ridges in the three main oceans; (b) the number of deep trenches in the oceans; (c) the numbers and distribution of isolated volcanic peaks. (After Menard and Smith, 1966, in Gross, 1972)

The ocean-floor plains

These occur at depths of between 3500 and 6000 metres, and are known as **abyssal plains**. They are extremely flat and monotonous in relief, extending out from the low gradient margins of the continental slope and continental rise (see page 44) towards the ocean ridge. The flatness is due to the accumulation of fine sediment up to thicknesses of 1 kilometre, burying most relief features. The sediments include a variety of materials (Figure 2.4). A good proportion of the detritus is derived from the erosion of continents and finds its way to the ocean-floors via canyon-like gashes in the continental margins: clay, silt and sand are moved in this way, flowing out over the abyssal plains, or piling up at the foot of the continental slope to form a continental rise. In the areas away from this inflow of sediment deposits form due to the almost continuous rain of skeletal material from surface-living creatures, plus fine dust from the atmosphere. Deposits of this type are known as **oozes**, and their composition is zoned with depth, since the different substances involved in the skeletons are taken into solution at different depths: calcium carbonate is restricted to higher levels and silica becomes important at medium depths where the carbonate is dissolved (Figure 2.5). Red Clay predominates in the deepest areas, being formed of insoluble, wind-blown dust from deserts, volcanoes and meteorites; it may also include large boulders dropped from icebergs and manganese – iron nodules, resulting from the chemical reactions which take place in such deep water.

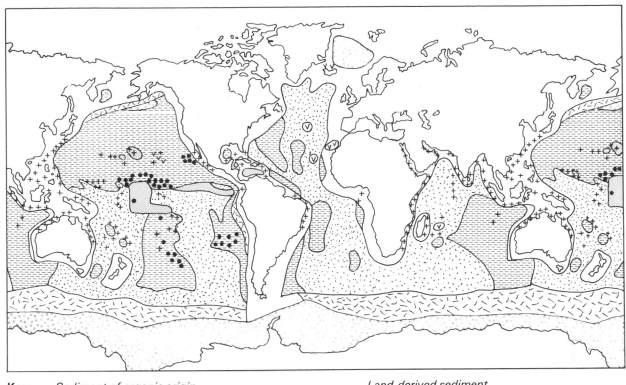

Key   *Sediment of organic origin*   *Land-derived sediment*

	Oozes, largely foraminifers: calcareous
	Oozes, largely diatoms
	Oozes, largely radiolaria
	Coral reefs

Siliceous (bracketing diatoms and radiolaria)

	Deep sea muds (red clay)
	Volcanic muds
	Glacial/marine sediments
	Abundant iron/manganese nodules

Figure 2.4 The distribution of ocean-floor sediments. Compare the distributions of the siliceous sediments. Is there any correlation between sediment type and relief features (Figure 2.2)? (After Gross, 1972)

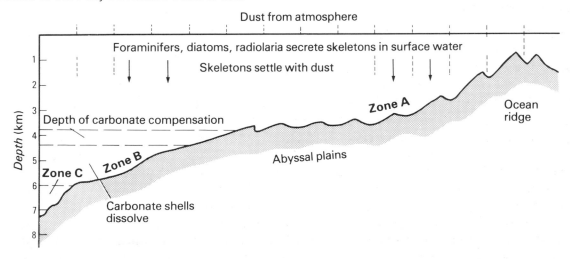

Figure 2.5 The depth distribution of ocean-floor sediments. The area at each depth above the ocean floor is approximately equal to its frequency of occurrence in all the world oceans. Zone A: mainly calcareous oozes, accumulating at 2–40 mm/1000 yr. Zone B: mainly siliceous oozes, with silica and deep sea mud increasing in proportion with depth as carbonate dissolves. Zone C: deep sea muds (red clay) dominant as silica dissolves; accumulation rates less than 1 mm/1000 yr.

The deposits of sediments so far investigated on the ocean-floors are normally less than 500 metres thick (although some exceed 1 kilometre), and become thinner towards the ocean ridges. Given the slowest rates of deposition (i.e. 1 mm per 1000 years, or 1 m per million years), the history of ocean sedimentation could be traced back 500 million years. This is only one-eighth of the geological history recorded in the continental rocks. In fact rates of deposition have been more rapid, and the skeletal remains of the oozes show that no known ocean-floor sediment can be traced back beyond the early Jurassic (i.e. less than 200 million years ago). This sort of calculation is backed up by recent drillings in the sediments, which have been dated by radiometric methods (chapter 4). In the Atlantic Ocean the oldest sediments (Upper Jurassic age) are found only along the margins of the ocean, farthest from the constructive boundary (Figure 2.6).

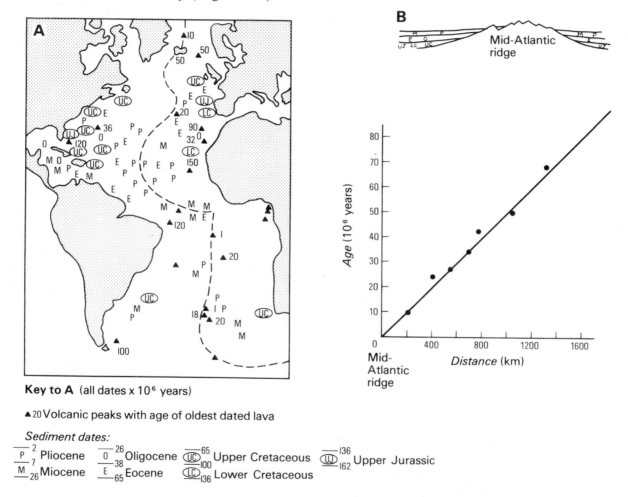

Key to A (all dates x 10⁶ years)

▲20 Volcanic peaks with age of oldest dated lava

Sediment dates:

Pliocene, Oligocene, Upper Cretaceous, Upper Jurassic, Miocene, Eocene, Lower Cretaceous

Figure 2.6 Evidence for the ages of the ocean-floor rocks in the Atlantic Ocean.
(A) A map of sediments and volcanic islands with their ages.

(B) Graph to show the oldest sediment in relation to distance from the Mid-Atlantic ridge. (After Maxwell, in Clark, 1971)

Ocean ridges

These are the main features rising above the abyssal plains. The Mid-Atlantic Ridge was known for some time before the IGY investigations revealed a system of similar ridges connecting through all the oceans (Figure 2.7) to form a worldwide feature nearly 60 000 km long. Each ocean ridge has its own distinctive features (Figure 2.2), but all are composed solely of the basaltic lavas typical of ocean-floors: there is no mixture of sedimentary, igneous and metamorphic rocks characteristic of the continental areas.

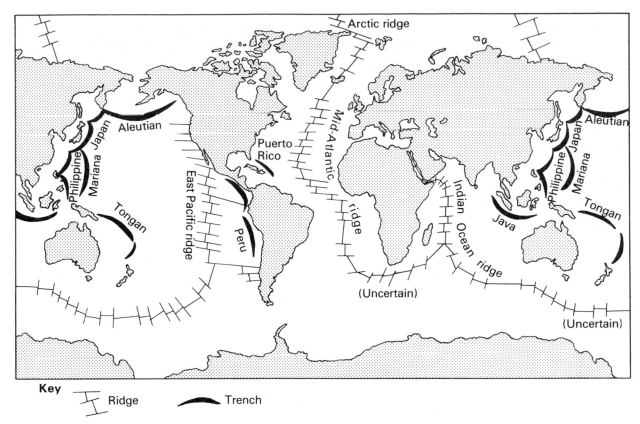

Figure 2.7 Ocean ridges and trenches: world distribution.

The closer study of the relief features of the ocean ridges has cleared up a mystery which puzzled oceanographers in the 1950s, and has led to a deeper understanding of the origins of ocean-floor rocks. The mystery concerned patterns of magnetic readings (Figure 2.8) with stripes of alternating positive and negative anomalies (i.e. differences from an average value). The clue to the answer came when mirror-images of such patterns were recorded across the

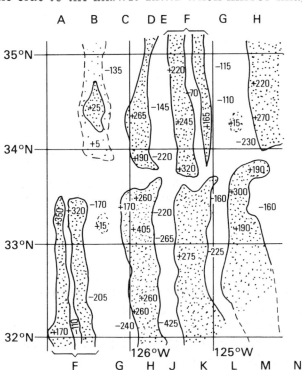

Figure 2.8 A puzzling pattern of magnetic field readings made off California in the 1950s. Sudden changes from positive to negative values of magnetic intensity were mapped, resulting in a series of 'stripes' on the ocean floor. Each stripe had a characteristic wideth and shape, so that offsets along east–west fractures could be measured (e.g. stripe F). One degree of latitude is approximately 110 km. (After Vine, in Gass, Smith and Wilson, 1972)

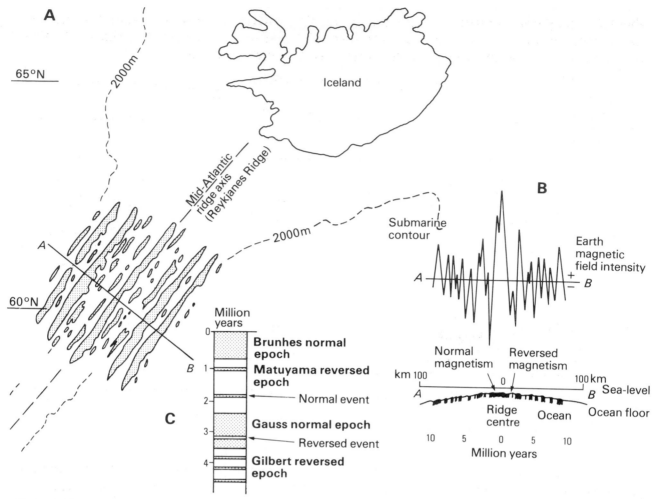

Figure 2.9 Magnetic stripes across the Reykjanes Ridge, south-west of Iceland. (A) Map showing the typical stripe pattern (positive anomaly values shaded). Notice the mirror-image effect. (B) The results of an aeromagnetic survey A–B across the ridge (top), related to a scheme of dating established in the Pacific Ocean region (lower). (C) A time-scale of magnetic polarity reversals for the past 4.5 million years. (After Vine, 1966, in Cox, 1973)

Mid-Atlantic Ridge south-west of Iceland (Figure 2.9), and it was discovered that a comparison of radiometric ages and magnetic directions of lavas erupted on the land indicated that the Earth's magnetic field reversed its polarity (i.e. the north magnetic pole became a south magnetic pole) approximately once every million years for the last four million years. Although the period involved in magnetic reversals may be quite long, it must be short compared with the intervals between reversals. This, combined with the slow rate of movements of the solidified rocks, leads to a pattern of apparently sharp reversals. The process has clearly been going on for a longer period than the measured four million years, and these continuing reversals have been correlated with the magnetic stripes mapped from the ocean-floors (Figure 2.10), giving a sequence for the history of the ocean ridges. The most recent lavas occur in the centre, and they become progressively older away from this line. The records of magnetic data had accumulated at oceanographic institutes and once the idea of magnetic reversals had been accepted they could be processed. The relationship of magnetic stripes to ocean ridge development was established on a worldwide scale. Such evidence showed that new lava was erupted along the axis of the ocean ridge, solidified, and then was carried outwards as a further mass of fresh lava was erupted in the centre. The distance apart of the magnetic stripes enabled the rates of ocean-floor spreading to be calculated. This showed that rates of 5–9 cm per year in the Pacific Ocean contrasted with 1–2 cm per year in the Atlantic Ocean. Drillings into the ocean-floor rocks have shown that this rate of movement continues to the margins. As the ages of the sediments suggested, so the basalt lavas confirm, that the ocean-floors are no older than the early Jurassic.

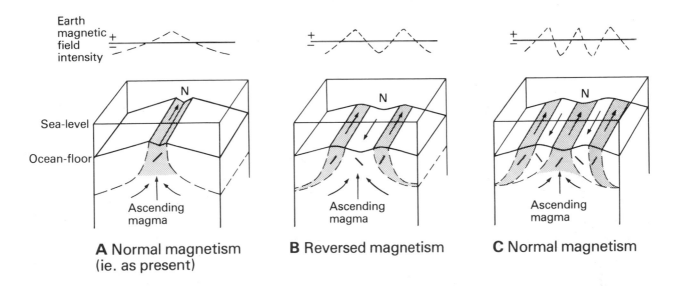

A Normal magnetism (ie. as present)   **B** Reversed magnetism   **C** Normal magnetism

Figure 2.10 The formation of ocean-floor stripes. Molten magma rises at the ridge centre (A), and on cooling it takes on the magnetic orientations imposed by the Earth's field at that time. When the magnetic field is reversed (B) the magnetic orientations forming in new rocks changes accordingly. Compare (C) with Figure 2.9B.

Another point to record here is that the ocean ridges come into contact with the continental rocks in some areas, even disappearing beneath the continents. The two most conspicuous are in California, USA, and along the line of the Red Sea. The East Pacific Ridge disappears into the Gulf of California, and the San Andreas fault system is related to transform faults crossing the ridge (see p. 10). The Indian Ocean Ridge continues into the line of the Red Sea at its northern end, and many consider that the Red Sea is the beginning of a new ocean which will develop as Africa and Arabia move apart.

Volcanic peaks

The isolated peaks marked on the map of ocean-floor features are all formed of volcanic rocks, like the ocean-floors themselves, and mark localised centres of lava extrusion. Some reach the surface to form islands, whilst others remain submerged peaks and are known as seamounts or guyots. Some of these peaks have level summits, suggesting erosion by the sea of the uppermost portion (chapter 10).

Ocean trenches

The deepest parts of the oceans are elongated troughs descending to depths of over 10 000 metres. They are hundreds of kilometres across and thousands of kilometres long. Sediments accumulating on the trench floors are relatively thin today. Nearly all the trenches occur around the margins of the Pacific Ocean, and an arc of volcanic islands is commonly present on the continental side of the trenches. The trenches are interpreted as the lines along which the ocean-floor, manufactured at the ocean ridges and spreading out towards the continents, plunges beneath lower density continental or ocean-floor rocks at the subduction zone to a depth where the rock materials are fused, resulting in volcanic and earthquake activity.

Volcanic and earthquake activity in the ocean basins

If the distribution of earthquake and volcanic activity is related closely to the plate margins (cf. Figure 1.1 and Figure 1.3), the closer study of particular features should confirm the details of plate tectonic theory. Apart from a thin veneer of sediment on the ocean-floors, all the rocks of the ocean basins are igneous in origin. This means that rock material has become molten beneath the surface, and has migrated up towards the surface, perhaps reaching it, before solidifying. Such rocks are formed of interlocking crystals, the sizes of which are determined by whether they cooled rapidly at the surface in contact with the atmosphere or ocean waters (when they will be small), or slowly beneath the surface (large crystals). The composition of igneous rocks, and thus the name given to particular specimens, is controlled by the point of origin of the molten rock material, or magma, and the thickness of overlying rocks through which it has to pass. The igneous rocks of the ocean basins appear to have formed from the melting of the upper mantle rocks in local sections subjected to unusual heating, and they contain high proportions of the dark iron and magnesium silicate minerals (Figure 2.11).

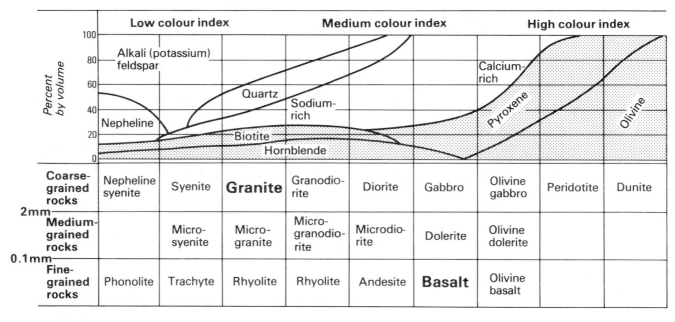

Figure 2.11 A classification of igneous rocks related to mineral composition and grain-size. The colour index is the proportion of dark minerals (those shaded in the chart) in the rock. Granites and basalts make up approximately 80 per cent of all igneous rocks. (After Woolley and Bishop, 1974)

Ocean ridge activity

The ocean ridges are formed by fissure eruptions with a fairly uniform rate of molten rock emission. In combination with fracturing along the central rift valley sides this leads to the formation of the parallel ridge system, and the lavas produced are known as tholeiitic basalts, probably originating from a partial melting of the upper mantle.

Iceland is a particularly interesting place in this context, since it is built astride the Mid-Atlantic Ridge system (Figure 2.12). It is formed mostly of lava flows, and volcanic activity, including hot springs, is a constant feature of the island's environment. Great fissures cross the island and basaltic material wells up along these at temperatures of 1200° C; it cools and blocks the fissure, forming wall-like intrusions, or dykes. At times volcanic cones are formed after explosive activity: Hekla was the best-known of these until the eruption of Helgafell in 1973. The intrusion of sheets of basaltic material gradually leads to widening of the island. Calculations show that it has widened by approximately 400 km since the beginning of the Tertiary period (65 million years ago), or at approximately 0.6 cm per year.

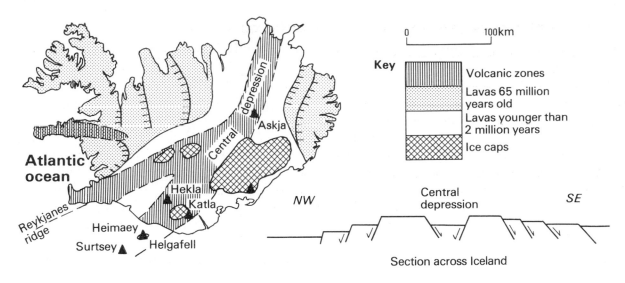

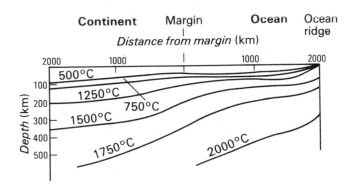

Figure 2.12 The volcanic features of Iceland. Compare this diagram with Figure 2.9 and suggest how the geology of Iceland can be related to ideas of sea-floor spreading.

Other features of ocean ridges indicate that there is a flow of energy along their length. Shallow earthquakes are located commonly beneath the centres of ridges, with their foci (Figure 1.4) at 25–35 km depth. Since earthquakes take place where stresses are relieved inside the Earth, and are known to occur where movements of molten rock take place beneath a volcano prior to eruption, it can be inferred that similar movements are taking place beneath the ridges. This is confirmed by the subsurface temperature pattern (Figure 2.13), which suggests a strong heat flow concentrated at the ridge and giving rise to the melting of rock material beneath the crust.

Figure 2.13 Subsurface temperatures beneath continent and ocean. Notice how the rate of temperature rise decreases after a rapid initial rise beneath the ocean margin, and how the ocean ridge is a point of maximum heat flow.

Activity and the volcanic peaks

Volcanic islands often occur along the ridges, especially in the Atlantic Ocean, but are also found scattered over the ocean-floors (e.g. central Pacific Ocean) and in volcanic arcs on the continental sides of ocean trenches. All these oceanic volcanic islands appear above sea-level as the tip of a vast mountain, often well over 3000 m above the ocean-floor when it reaches sea-level and perhaps rising a further 1500–3000 m above sea-level. As the volcanic peak is built up to this level only after thousands of years of eruption, the later molten material has to pass through an increasing thickness of older lavas. The magma cools as it rises through such an edifice, whilst the less dense materials separate and rise above the more dense magmatic matter. The result of this process is an increasingly explosive form of volcanic activity, together with the eruption of lavas with more low density and light-coloured minerals containing higher proportions of sodium, potassium and calcium relative to silicon: these are known as alkali basalts. The Azores group, near the centre of the Mid-Atlantic Ridge, are formed of a mixture of basalts and pumice ashes — the latter having been erupted in a phase of explosive activity.

The volcanic islands of the Atlantic Ocean are mostly associated with the ridge, and the active ones are generally those nearest the ridge. As the ocean-floor spreads from the ridge, the volcanic peaks are carried outwards too, away from the source of magma. They often subside beneath the waves due to the weight they impose on the underlying crust together with the drying up of the magma supply, and because the strength of the force uplifting the ridge (i.e. rising magma) decreases as one goes away from the ocean ridge crest. A comparison of the ages of the oldest lavas on the islands (Figure 2.6) shows that the oldest lavas on the Azores, astride the ridge, are 4 million years in age, whilst those of the Cape Verde Islands, close to the African coast, are 120 million years.

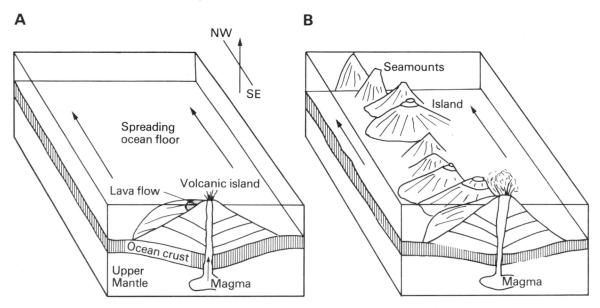

Figure 2.14 Ocean-floor spreading and the relationship of a 'hot spot', or 'plume', to the formation of volcanic islands and seamounts. The plume is a magma source, which supplies molten rock material from a particular point over a long period. This process can be seen to account for the Hawaiian Islands–Midway Islands–Emperor Seamounts chain of volcanic peaks on the floor of the northwestern Pacific Ocean.

(A) Approximately 70 million years ago: the first volvanic island is formed.
(B) Today: ocean-floor spreading (at 9 cm/yr in the Pacific Ocean) has carried a series of volcanic peaks away from the plume. Subsidence of the oceanic crust beneath the weight of each volcanic pile has caused many former islands to sink so that they now exist as seamounts.

The Hawaiian Islands are a group which is far from the East Pacific Ridge, but it is only the south-eastern extremity of a line of volcanic peaks extending north-west through the Midway Islands and the Emperor Seamounts to the Kamchatka Peninsula of Siberia (Figure 2.14). The most recent rocks of the line occur in the Hawaiian group, which contain the only active volcanoes, whilst the most ancient are in the sunken north-west portion of the ridge. It seems that beneath the Hawaiian Islands lies a fixed zone of active lava production which has poured out molten rock for at least 70 million years. As the ocean-floor has moved north-westwards, away from the East Pacific Ridge system at 9 cm per year, it has picked up volcanic peaks produced above this 'plume' and as it carried them away from this centre of activity they have become dormant.

Activity in the trenches and island arcs

Island arcs of volcanic peaks and their adjacent trenches are formed where the ocean-floor material is subducted beneath another plate along what is known as a 'Benioff Zone'. The friction resulting from this subduction gives rise to earthquake shocks of increasing depth down to 700 km, and generates heat which leads to the fusion and melting of upper mantle material combined with the ocean-floor lavas and sediments. The volcanoes in the island arcs are built of sodium-rich basalts, formed after the molten basalts reacted with the seawater into which they were erupted, and these are often surmounted by piles of andesite lavas. Andesite is less dense, and has a higher proportion of silicon, than the basalts: the difference is caused partly by the distance the magma has to rise through overlying rocks. During the course of this ascent the magma separates into fractions according to density, and mixes with the rocks through which it passes. The composition of the andesite varies as the depth of the subduction zone and distance from the trench increases (Figure 2.15). This is a further confirmation of the connection between plate activity and igneous rock type.

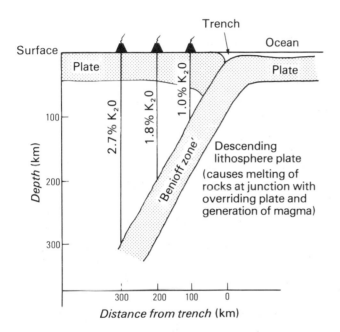

Figure 2.15 Changes in the composition of andesite lavas on the continental side of an ocean trench. The 'Benioff Zone' is the name given to the descending plate, after the man who first located such activity in 1954. The K_2O forms a small proportion of andesite lavas (average 1.5–2.0% by weight), so the variation shown represents a major change.

Many volcanic eruptions associated with island arcs are exceptionally violent, as has been shown by Krakatoa, in the Indonesian arc, Mount Taal, in the Philippines arc, and Mount Katmai, in the Aleutian Islands arc, all three having experienced huge explosions in the last 100 years which resulted in the complete removal of the upper part of the cone and the formation of a caldera.

Ocean basins and plate tectonics

The combination of discoveries which included the creation of new ocean crust at the ridges, and its destruction at trenches, with movement in rigid sections between, led to the idea of ocean-floor spreading in the early 1960s. Consideration of the full implications of this concept suggested that the Earth's surface layers could be categorised as plates, and this led to the hypothesis of plate tectonics (cf. Figures 2.16 and 1.3).

It has been estimated that 0.5 km³ volcanic rock is added to Iceland (350 km long) each year. This is probably an unusually active section of the ocean ridge system, and it has been suggested that 0.02 km³ per year is an average for similar strips of 350 km along the total length of 60 000 km. This gives a total of 3.4 km³ per year, which is approximately equal to the rock material which should be added to the oceanic crust and destroyed in the subduction zones to account for the plate movements.

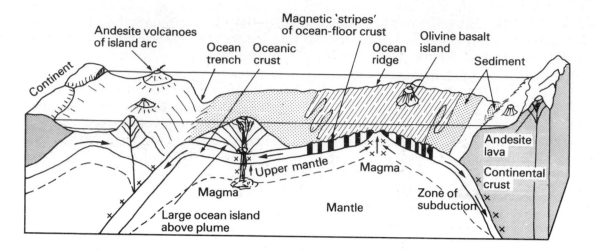

Figure 2.16 A summary of ocean-floor features, related to the ocean-floor spreading theory of their origin.

The ocean basins vary considerably in size, shape and distribution of relief features. The features of other areas, like the Red Sea and the Mediterranean Sea, also contain some resemblance to the large oceans: it has been suggested that the Red Sea represents an early stage in ocean development, opening along a line of rift valleys, whilst the Mediterranean is seen as the closing remnant of a much larger ocean which once existed between Eurasia on the north and Africa–Arabia–India on the south. Certainly the movements which are involved in plate tectonics lead to the redistribution of continents and ocean basins. A tentative history of ocean basin development can be devised on the basis of forms present on the Earth today (Figure 2.17), emphasising this evolutionary pattern of events. The opening and closing of ocean basins provides a foundation for the understanding of the evolution of continental relief features at least as far back as 200 million years ago, and possibly involving more ancient events as well.

Stage	Process	Sediments	Typical age (million years)	Examples
Embryonic	Uplift and rifting	None	10	East African rift valleys
Young	Rifting with slight lateral movement	Little	25	Gulf of Aden; Red Sea; Gulf of California
Mature	Extensive lateral movement	Moderate: thickest at margins	150	Atlantic Ocean
Declining	Extensive trenches; over-riding of ridges by crustal blocks	Extensive at margins	over 200	Pacific Ocean
Closing	Isolated basins with deformation of sediment and crustal rocks	Thick, deformed	over 200	Black Sea; Caspian Sea; Mediterranean Sea

Figure 2.17 A suggested series of events in the evolution of ocean basins, based on the evidence of features present in the world today. (After Gross, 1972)

Water in the ocean basins

So far the ocean basins have been discussed in terms of their shape, relief and origin, but their most obvious characteristic is that they form the most important place for the storage of Earth surface water: 97 per cent of the world water is contained in the oceans, a volume of 1350 million km³.

The liquid nature of oceanic water means that it will adjust its level throughout the world as the shape of any one ocean basin varies, or as the rate of return from the continents varies. Thus the deepening of an ocean trench, or the widening of an ocean, may lead to a general lowering of the sea-level, whilst sedimentation on the sea-floor, or the eruption on to it of volcanic material on a large scale, may cause a general rise. If an ice sheet accumulates on a continent more water will be abstracted from the ocean than is returned to it, and its level will be lowered; melting ice sheets result conversely in rising sea-levels. Such changes in sea-level are termed eustatic, and are of worldwide occurrence. It is often difficult to determine the full significance of such effects, apart from those associated with the changes which took place during the Ice Age of the last 2 million years, since there has been so much regional tectonic activity (i.e. earth movements which raise or lower portions of the land) in different places at different times — and especially around the Pacific Ocean margins — that any worldwide effects have been obscured. It is important to realise, however, that the ocean levels do vary over time, and this has extremely important implications for the study of surface processes in PART II of this book.

Ocean-floor minerals

The increasing needs by man for mineral resources have prompted him to look away from the continental resources, which are being used at ever-increasing rates. Studies of the oceans have led to a greater knowledge of the possibilities existing there, and to a technology which could make mining of the ocean-floors a practical proposition. This has resulted in business men, like Howard Hughes, chartering vessels to survey the prospects, and in governments regarding an agreement on 'who owns what' on the sea-floor as important.

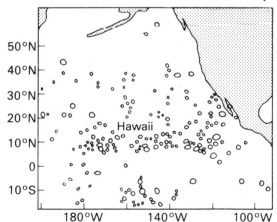

Figure 1 The distribution of manganese nodules in the North Pacific Ocean. The largest symbols represent over 30 per cent manganese by weight; the smallest under 10 per cent. (After Horn, 1972, in Press and Siever, 1974)

The most valuable ore which can be dredged from the ocean-floors at present occurs as concentrations of manganese nodules on the abyssal plains and ridges at depths of 2000–5000 m (Figure 1). The nodules are rounded aggregates of manganese, iron and other metal oxides (e.g. copper, nickel, cobalt) weighing from less than 1 g to several hundred kilograms: most are a few centimetres in diameter. On average they contain 20 per cent manganese, 6 per cent iron and 1 per cent copper and nickel. Mining of these will begin when the technology is refined, and the USA and Japan have been carrying out trials since 1970.

Changes in the future, including the exhaustion of continental ores, the availability of cheap power, and the increasing importance of preserving land resources, may lead to the extraction of minerals from the seawater itself, or from the red clay deposits at depth.

3

Continents and mountains

The continent–ocean boundary is generally defined as the present shoreline. On this basis 71 per cent of the Earth's surface is ocean and 29 per cent is land. Sea-level, however, varies with time. It has been calculated that the ocean trenches hold enough water to raise the sea-level 200 metres around the world if they became filled with sediment, and we know that during the recent Ice Age the level fell over 100 m below the present. The shoreline is also an unsatisfactory boundary on the basis of geological structure: the continental terraces and continental rises (Figure 3.15) are formed of similar rocks to the continental areas, and should therefore be included with them.

If the continental terrace and rise areas are added to the continents the world distribution is changed to approximately 65 per cent ocean basin and 35 per cent continent. The continental masses and ocean basins can then be seen to be quite distinct in terms of relief features, rock-types, and the range of processes operating (Figure 3.1). Whilst the ocean basin relief features result largely from the activity of internal Earth energy sources, those on the continents are due to the interaction of the forms produced by internal energy with the surface processes powered by external energy from the Sun (Figure 3.2).

	Continents	Ocean basins
Processes		
Internal energy source	Marginal volcanic activity. Deep burial of rocks: metamorphism, granitisation.	Creation of ocean-floor Subduction of ocean-floor Plate movement
External energy source	Subaerial weathering, erosion, transport, deposition by running water, ice, wind and along coasts.	Waves Tides
Relief features		
Positive	Fold mountains. Uplifted plateaux. Volcanoes.	Ocean ridges Volcanoes
Plains	Platform areas, lowlying shield. Continental shelves, coast plain.	Abyssal plains
Negative	Rift valleys. Eroded valleys: by streams, glaciers. Deflation hollows.	Ocean trenches Submarine canyons
Rock-types		
Sedimentary	A wide range of limestones, sandstones, shales, conglomerates.	Calcareous and siliceous oozes Red clay Flysch association
Igneous	A wide range of basalt and andesite lavas, ashes, and intrusive granites.	A more limited range, largely basalts
Metamorphic	A wide range: gneiss, schist, slate, marble.	None

Figure 3.1 The major differences between the continents and ocean basins. Relate these to the positions of the continents and ocean basins with respect to plate margins.

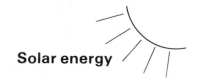

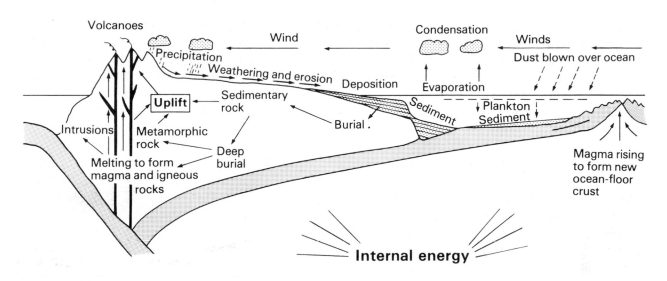

Figure 3.2 The rock cycle. Compare (a) the roles of the two main sources of energy in the origin of the main rock groups, and (b) the processes of rock formation on continent and in ocean basin.

The continental relief features contrast markedly with those of the ocean basins. They can be divided into five major groups. The distribution of these can be studied on a world-scale (Figure 3.3), or in more detail in relation to a smaller region (Figure 3.4).

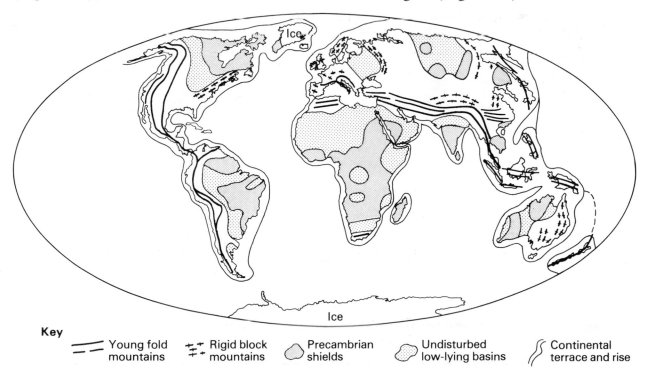

Key

—— Young fold mountains

+–+ Rigid block mountains

Precambrian shields

Undisturbed low-lying basins

Continental terrace and rise

Figure 3.3 The major structural regions of the continents. Compare these features with those of the ocean basins, and relate their distribution to plate margins.

Young fold mountains

Young fold mountains, like the Himalayas, Alps, Rockies and Andes, are the most impressive features of the continental relief. They mostly exceed 3000 m above sea-level, and include the highest point on Earth (Mount Everest 8848 m in the Himalayas). Most of the continents possess such a series of long, relatively narrow ranges or cordilleras. They are termed 'young' because they have mostly been uplifted during the last 50 million years and consist of rocks formed in the Mesozoic and Cainozoic eras (the last 225 million years). One of the two main groups of young fold mountains occurs along the west coast of South America and extends through Central America into the western part of North America, and thence into the island arc festoons around the north and west of the Pacific Ocean. The other consists of the Atlas – Alps – Himalayas ranges between the Eurasian continent to the north, and Africa,

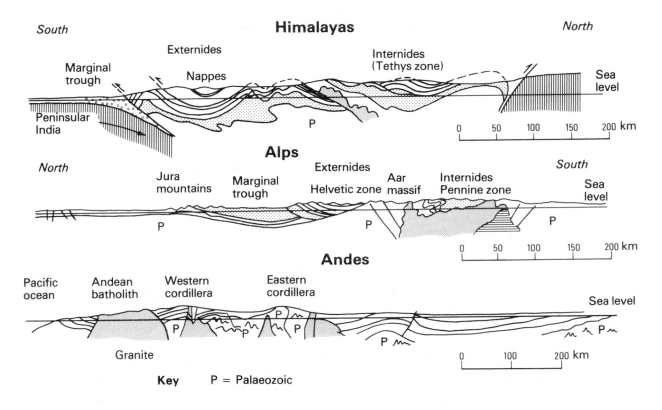

Figure 3.4 Sections across three major fold mountain systems. Relate the variations to the concept of internide and externide zones. How different are the Andes from the other two? Can this be explained?

Arabia and peninsular India to the south. These can be followed into Burma, Malaya, Sumatra and Java. Only in Africa and Australia are the highest points not formed by young fold mountains: thus in Africa Mount Kilimanjaro (5895 m) is a volcanic cone perched on a plateau, and in Australia Mount Koscuisko reaches only 2230 metres. The high relief of fold mountains encourages vigorous dissection by streams and glaciers; scree, landslides and rock collapse are common on the steep slopes. In many areas the rate of uplift may keep pace with the rate of downwearing. On the other hand the breakdown of rock has been assisted by the fact that the earth movements giving rise to the fold mountains often crushed sections of rock into small fragments ready for removal.

The young fold mountain regions are characterised by a variety of structural situations and rock-types (Figure 3.4), but several features are common to all the areas.

1) They all have a great thickness of sedimentary rocks compared with those of the same age in adjacent regions. Thus the maximum thickness of Mesozoic and Cainozoic rocks is over

10 km in the Alps, but scarcely 2 km in Europe to the north of the young fold mountain zone.
2) Massive granite intrusions also occur, up to several hundred kilometres in length and with the longest direction along the trend of the mountain ranges.
3) Structural features are related to conditions of intense pressure, and include large-scale recumbent folds, and thrusts which cause great wedges of rock to travel over the top of younger rocks. Nappes are sheets of rock which have become detached from their zone of origin (Figure 3.5). It is now thought that the uplift of the folded rocks led to large masses breaking away and slumping downwards, over the top of younger groups of rocks: this

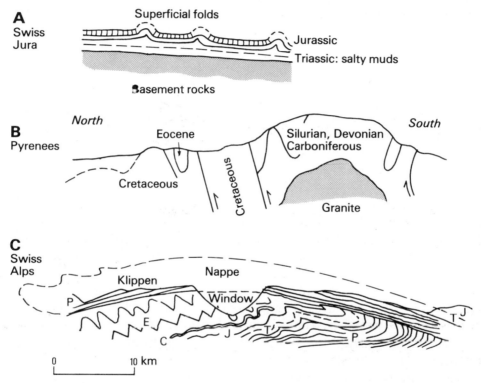

Figure 3.5 Structural features associated with young fold mountains. (A) The Swiss Jura: folds affect only the surface rocks, which have slid over lubricating salty muds under compression. (B) The Pyrenees: uplift has been caused by the injection of a mass of basement rock, and this has resulted in folding of the marginal rocks. (C) The Alps: nappes are huge slices of rock which have travelled over the top of younger rocks and have become detached from the group of rocks with which they were originally formed; this process is thought to occur after uplift, when these masses break away from the highest areas and slump under gravity. (P = Permian; T = Triassic; J = Jurassic; C = Cretaceous; E = Eocene)

process is known as gravity sliding. In areas of nappe formation the valleys are carved into an area which was once capped with almost horizontal layers of rock.

Within the fold mountain belts there is a major division between the 'internides' and 'externides' (Figure 3.4). The internides is the zone where earth movements have been most intense: thrusting and folding are complex, and much metamorphism and granite intrusion has taken place. The sedimentary rocks include mixtures of sand–silt sediments with graded bedding (Figure 3.6), known as greywackes, together with alternating layers of fine, clay-grade shales. This association is known as flysch in the Alps. Such sediments are thought to have been formed in deep water by turbidity currents — sediment-charged water masses flowing down steep slopes and spreading out to deposit the sediment. Other sediments found with these greywackes and shales include thin, dark limestones and silica-rich cherts, formed

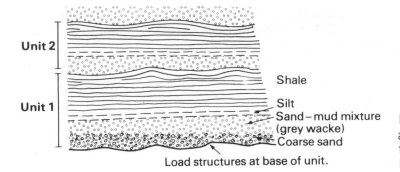

Shale
Silt
Sand–mud mixture (grey wacke)
Coarse sand

Load structures at base of unit.

Figure 3.6 A flysch association of greywacke and shale. Within each unit the sediment gets finer upwards: this is known as graded bedding.

largely of the skeletal remains of minute planktonic organisms and resembling hardened forms of modern oceanic oozes. Lava flows, erupted on the sea-floor, are commonly found with such sediments and have the form of a pile of rounded masses known as pillow lava.

The externides zone is formed of less altered and deformed sedimentary rocks. These include massive limestones, sandstones and thinner-bedded shales, and all may contain fossils of bottom-living and swimming animals, suggesting that the rocks were formed in shallow water, coastal or land environments. There are few volcanic, or intrusive igneous, rocks.

Rigid blocks

In Britain areas like central and northern Wales, the Southern Uplands and the Scottish Highlands, are rigid blocks (Figure 3.7). The Central Massif of France, the Rhine Highlands and the Appalachian Mountains are also of this type. All of these rise to less than 3000 m above sea-level, and are plateaus cut deeply by valleys. They are composed of rocks deposited largely during the Palaeozoic era (570–225 million years ago), and many of the areas contain structural features which resemble those in the young fold mountains. There are the eroded remains of complex fold structures (Figure 3.8), unusual thicknesses of sedimentary rocks including the varieties formed by rapid accumulation (alternations of greywacke and shale very much like the flysch of the Alps), volcanic rocks, large numbers of granitic intrusions and the metamorphism of extensive areas. In short, these rigid blocks are just older fold mountains, where the rocks are indurated by age and where subsequent earth movements have led to their being broken into faulted blocks, parts of which were uplifted whilst parts subsided beneath more recent sediments.

One former range of mountains, which included Wales, the Lake District, Scotland and Norway, was uplifted in the earlier part of the Palaeozoic era, probably reaching its greatest heights as mountain ranges 400 million years ago, with the ranges trending north-east to south-west. The name Caledonian Mountains has been given to these long-destroyed ranges. During the later part of the Palaeozoic era further sediments were formed from the erosion of the Caledonian Mountains, and these eventually gave rise to another series of mountain ranges, known as the Hercynian (or Variscan) Mountains. These have general east–west trend across central Europe from Southern Ireland, Brittany (Armorica) and Cornwall, across France (Central Massif, Vosges), Belgium (Ardennes), Germany (Rhine Highlands, Harz) and Czechoslovakia (Bohemian Massif). These, too, have long since been eroded. Somewhat similar events were responsible for the Appalachian Mountains in eastern USA, but the two phases are contained within the one major range.

The later, Hercynian, mountains in Europe were worn down by erosion during the Mesozoic and Cainozoic eras, providing the materials for another sequence of new rocks deposited on their margins and in local basins. When the young fold mountains like the Alps were uplifted these older rock areas often reacted to the stresses imposed on them by fracturing along major faults, and some of these ancient resistant rocks were uplifted as

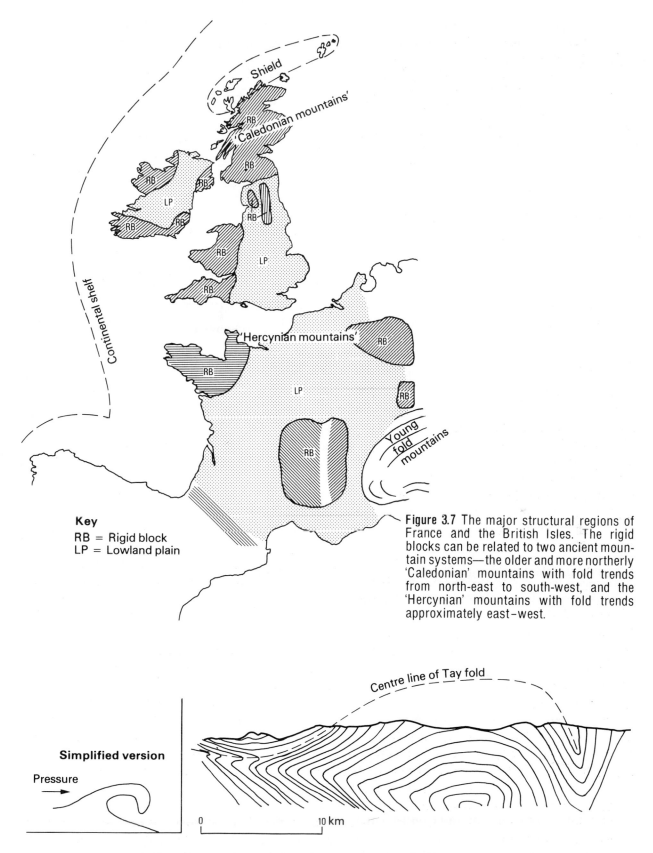

Figure 3.7 The major structural regions of France and the British Isles. The rigid blocks can be related to two ancient mountain systems—the older and more northerly 'Caledonian' mountains with fold trends from north-east to south-west, and the 'Hercynian' mountains with fold trends approximately east–west.

Key

RB = Rigid block
LP = Lowland plain

Figure 3.8 The Tay 'nappe' in the Grampian Highlands of northern Scotland. This structure was produced by fold mountain formation over 400 million years ago. Erosion has removed the upper part.

blocks, separated by downfaulted areas, including rift valleys (Figure 3.9). The previously worn-down surfaces thus became plateaus with accordant summit levels, cut into once again by streams and glaciers.

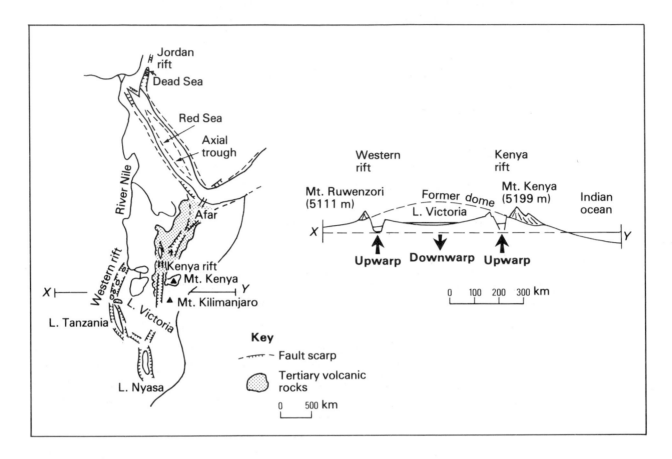

Figure 3.9 The East African rift valley system. This extends from south of Lake Nyasa to the Jordan rift. It is widest along the Red Sea (250–500 km), but averages 30–65 km in Africa. The Red Sea axial trough contains intrusive igneous rocks on its bed, together with brines and metallic deposits. At its southern end the Afar triangle is an area of volcanic activity and subsidence. The East African rifts (section X–Y) are related to a 'slipped keystone' structure, following warping of the ancient shield rocks. (After Brunsden and Doornkamp, 1974)

The study of the rocks in these ancient and worn-down mountain ranges led to the concept of the geosyncline although this is now less favoured in comparison with the implications of plate tectonics theory (page 51). A geosyncline is thought of as an area of crustal subsidence in which an unusual thickness of sediment accumulates (i.e. up to 10 km, compared with 2 km elsewhere in the same time). After a period of folding and crushing the rocks are uplifted to form fold mountains. Geosynclinal zones are several thousand kilometres long and a few hundred kilometres across. They are divided into two types, according to the rock-types which have accumulated in them (Figure 3.10).

Other rigid blocks, which form dissected plateaus, like the Pennines of northern England, are blocks of little-disturbed continental shelf sediments of Palaeozoic age. These have become hardened by age, and have been uplifted more recently in a similar fashion to the ancient, worn-down and block-faulted mountain ranges.

	Geosyncline	
	Miogeosyncline or Miogeocline ('lesser')	Eugeosyncline or Eugeocline ('true')
Rocks	Limestones, including reefs. Well-sorted sandstones. Shales. Fossils, including bottom-living and swimming forms. (No volcanic rocks)	Thin, dark limestones; cherts. Greywacke-shale association ('flysch'). Few fossils: mostly planktonic forms — e.g. radiolaria in cherts; graptolites. Volcanic rocks interbedded.
Structures	Less intense folding and faulting. (No metamorphism or granites)	Nappes and recumbent folds. Metamorphism and granite intrusions.

Figure 3.10 Typical features of geosynclinal zones, as found in ancient rigid blocks. How are the miogeocline and eugeocline related to the internides and externides of the young fold mountains?

Precambrian shield areas

The shield areas have a surface outcrop over 25 per cent of the world's land areas. They are formed of rocks 570–4000 million years old. These areas vary considerably in height, from the plateau surfaces of eastern Africa rising above 1500 m, to the submerged areas beneath the Baltic Sea. The East African rift valley system occurs where the rigid shield rocks have been subjected to tensional (i.e. pulling apart) stresses (Figure 3.9).

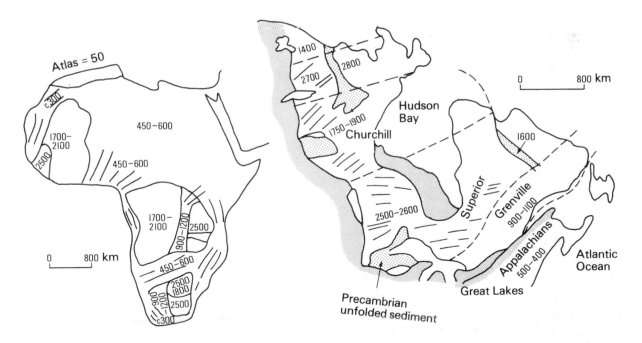

Figure 3.11 Ancient fold mountains in Africa and Canada. The figures are radiometric dates in millions of years for the rocks of the respective zones.

Notice the frequency of mountain-building movements and any correlations between these continents.

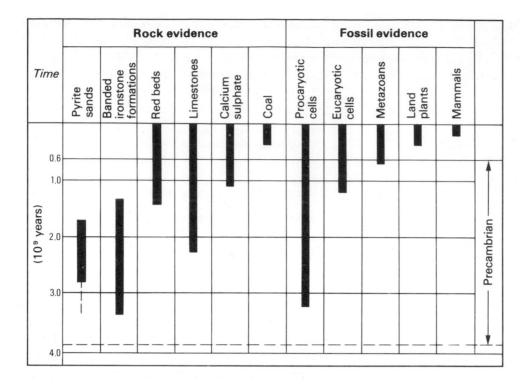

Figure 3.12 Precambrian and later rocks. What are the differences in rock types and fossil evidence? N.B. Procaryotic cells do not have nuclei which split on reproduction; eucaryotic cells do; metazoans are creatures with many cells.

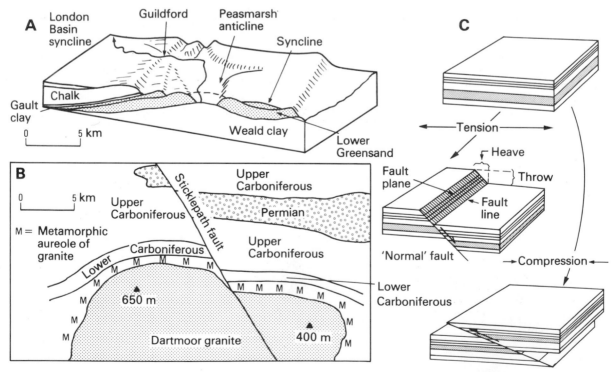

Figure 3.13 Simple folds and faults, commonly found in areas of lowland basin type, or as fractures in older areas.
(A) Open anticlines and synclines. The Chalk and Lower Greensand both hold water: where would it be best to sink a well for local water supply? How do the anticlines and synclines relate to the relief features?

(B) A tear fault, in which there is horizontal movement—to the right in this case as one looks from one side of the fault across to the other side. How has this fault affected the rocks' outcrops?
(C) Simple faults. These may be associated with throws of a few metres, but throws may also be over a thousand metres.

Close study of the often highly metamorphosed rocks in these areas shows that they are also arranged in a series of elongated structures similar to those in the young fold mountains and rigid blocks (Figure 3.11). Each area contains the record of several mountain-building periods. There are, however, important differences between the rock-types occurring in the shields and those in the more recently formed mountain zones. The Precambrian rocks include extensive mineral ore deposits, due partly to enrichment of metal content during metamorphism, but also to the original processes of sedimentation which cannot be paralleled today. Thus the vast deposits of iron being tapped in the shield areas of north-eastern Canada, west Africa and north-western Australia were formed as distinctive banded deposits in environments with very low oxygen contents (Figure 3.12).

The study of these shield regions, which form the geological foundations of the continents, shows that the processes of rock and mountain formation have been effective through the period of geological history recorded in the surface continental rocks. The accumulation of sediments, followed by folding, metamorphism, uplift, the intrusion of granites and finally erosion, has occurred again and again in the past.

Lowland plains and basins

The rocks beneath the lower-lying parts of the continents are often similar in age to, but in much thinner sequences than, the rocks which make up the young fold mountains. In addition they are disturbed only slightly by folding and faulting (Figure 3.13). They are composed of types of rock (quartz sandstones, fossiliferous shales and limestones) which occur only in the externides of young fold mountains. South-eastern England, the plains of northern France, the Low Countries, Germany and Poland, central North America and much of Africa south of the Atlas Mountains, have such rocks at the surface. The relief of these areas is affected by

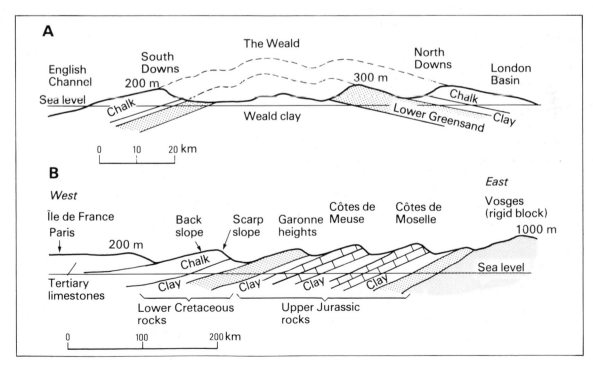

Figure 3.14 Cuesta topography is commonly found in areas of lowland basin. Notice which rocks form the cuestas (escarpments), and which form the vales between.
(A) The Weald of south-eastern England: a dome type of structure with smaller anticlines producing a 'corrugated' effect. The rocks are of Cretaceous age.
(B) The eastern part of the Paris Basin in northern France: a synclinal basin structure. The rocks are of Tertiary, Cretaceous and Jurassic age.

broad folding movements, and the wearing away of the softer sediments in the sequence of rocks results in landforms dominated by cuestas (escarpments) and vales (Figure 3.14).

A variety of events are responsible for these situations. In many cases the downwarping of an area of ancient shield rocks has led to the invasion of the sea and the deposition of a few hundred metres of sediments. These sediments include those formed by continental surface processes (rivers, ice, wind), and salt layers precipitated from evaporating lakes or isolated arms of the oceans. The rigidity of the underlying 'basement' rocks has prevented the accumulation of excessive thicknesses of sedimentary rocks and the transmission of intense stresses which might otherwise have given rise to complex rock deformation. In other areas, like the Deccan Plateau of south India, outpourings of basalt lava have covered the basement rocks. The most recent sediments in such areas are the river alluvium, which becomes thicker beneath deltas (e.g. Mississippi delta sediments are over 3000 m thick), and the covering of glacial till over the northern parts of Europe and North America.

Continental margins

The continental shelf, continental slope (together forming the continental terrace) and the continental rise are all below sea-level, and grade downwards into the abyssal plains of the ocean-floors. Study of these zones is less advanced than of the regions lying above sea-level, but the increasing use of boreholes in the search for oil, and of geophysical methods of surveying buried structures, has led to recent improvements in knowledge. Off the eastern coast of the USA the continental terrace deposits are a continuation and thickening of the rock layers found beneath the coastal plain (Figure 3.15). On the surface a variety of sediments are accumulating today (Figure 3.16). The continental terrace forms a wedge up to 5 km thick, composed of sediments laid down in the last 150 million years. These are shallow water, coastal and continental types of sandstone, limestone and shale. Similar deposits form most continental shelves (Figure 3.17).

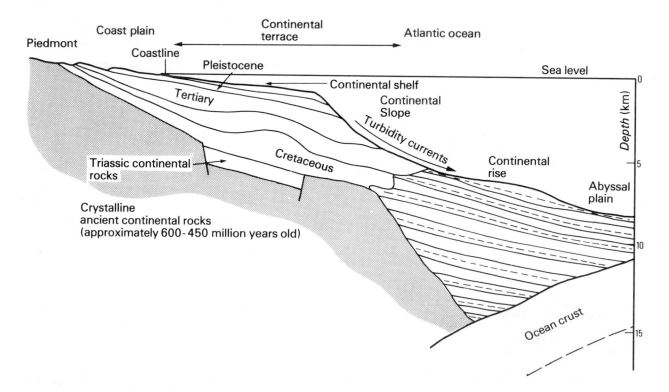

Figure 3.15 The continental margin of eastern USA: the construction of the continental terrace and continental rise. The continental terrace includes the shelf and slope. (After Dietz, in Wilson, 1973)

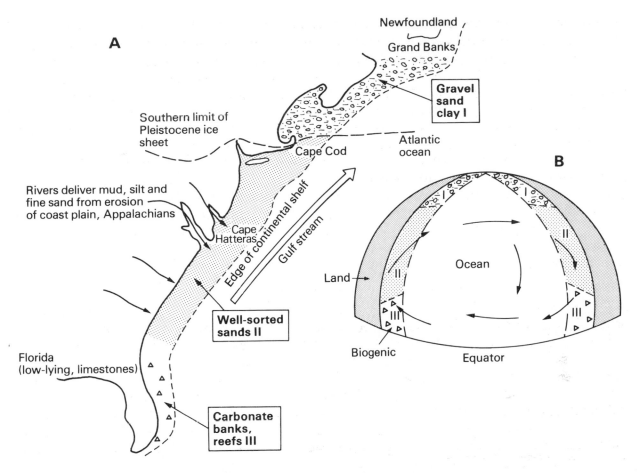

Figure 3.16 Continental shelf sediments. How are the sediments off the east coast of North America (A) related to the geology of the nearby land, the recent geological history of these areas, and lati-tude? (B) is a generalised model of the North Atlantic situation: how can it be applied there and elsewhere? (After Emery, in Moore, 1971)

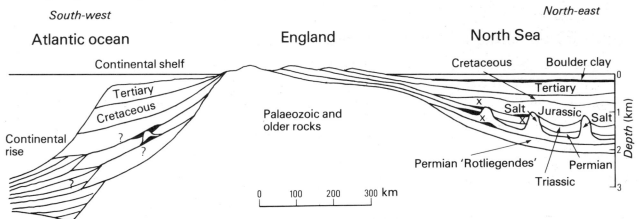

Figure 3.17 Continental margin sediments around the British Isles. Compare the thickness and the completeness of the sedimentary successions of Mesozoic and Cainozoic age on the land and off-shore. The North Sea sequence is better known than that off the western coasts due to the search for oil and gas (found at x). Compare these succes-sions with that of the east coast of the USA (Figure 3.15).

The continental terrace is inclined towards the deeper ocean-floor and at the foot of the continental slope (the steeper part of the continental terrace) is the continental rise. This is the upper surface of another wedge of sediment, but in this case it is up to 10 km thick and the thickest part is near the continent. The sediments of which it is composed are formed of coalescing fans of graded sand–silt mixtures interbedded with fine clays. These are of land-derived detritus which has slumped from the edges of the continental shelf and down the continental slope (maximum gradient 1 in 40) in masses of sediment-charged water known as turbidity currents. On reaching the low gradient at the slope foot the coarser sand and silt are deposited rapidly, followed eventually by the finest clay particles. This process gives rise to the alternate layers of coarse and fine sediment, thinning away from the slope foot. Few animal remains become trapped in these sediments, although a thin layer of ocean-floor calcareous or siliceous ooze may form between the turbidity flows, since the latter are formed by occasional events, rather than a continuous process.

The continental terrace and continental rise sediments resemble closely those of the miogeocline and eugeocline (Figure 3.10). At the present time there is some debate over whether the geosyncline concept is worth retaining. Some scientists suggest that the continental terrace and continental rise off eastern North America are deposits which will eventually be formed into mountain chains when the Atlantic Ocean closes again. There is no major sag, or trough, in the Earth's crust at this point, which would have been expected from some definitions of geosynclinal activity. Other scientists look to the ocean trenches as possible sites for geosynclines, but these have few sediments, although they do have the 'correct' shape. This type of discussion is typical of a fairly common occurrence in geological studies: an elegant model (like the geosyncline) is constructed from the evidence available in the rocks, but it may be difficult to recognise present day examples of such features unless one is content to modify the model.

It is important to realise that the continental shelves vary in form and composition. Their width varies from almost nothing up to almost 700 km from the coastline, and the edge ('shelf-break') averages 130 m in depth at the present time. Only 15 000 years ago, during the Devensian glaciation (page 207), sea-level was lower than this, so that the surfaces of continental shelves such as the floor of the North Sea and the shelf off the eastern coast of the USA were exposed to subaerial denudation. Some continental shelves have igneous and metamorphic rocks at the surface, indicating that they are erosional, rather than depositional, forms, but such outcrops are normally small in extent.

Volcanic and earthquake activity on the continents

The occurrence of volcanoes and earthquakes in the ocean basins is related closely to plate margins. In general the continents are affected by such activity only when they are close to such plate margins, and hence volcanoes and earthquakes occur largely on their margins (Figure 3.18).

Activity in young fold mountains

Most volcanic and earthquake activity in the continental areas is concentrated along the young fold mountain ranges, which may form a continuation of an island arc and run parallel to an ocean trench, as in the western Pacific Ocean, or, as on the eastern side of that Ocean, may be flanked by continental mountains such as ranges of the Andes. Some of the world's major earthquakes, like those experienced in Peru, Central America and Alaska, as well as a whole string of historically active volcanic peaks, occur through the length of these mountains. The other major young fold mountain chains of the world, the Alpine–Himalayan system, are less noted for their volcanic eruptions than for their earthquakes, although there are local centres of volcanic activity, as in southern Italy. Turkey and Iran are frequently affected by earthquakes at the present time, although historical studies have shown that other sectors of the Middle East have been more greatly affected in the last few thousand years.

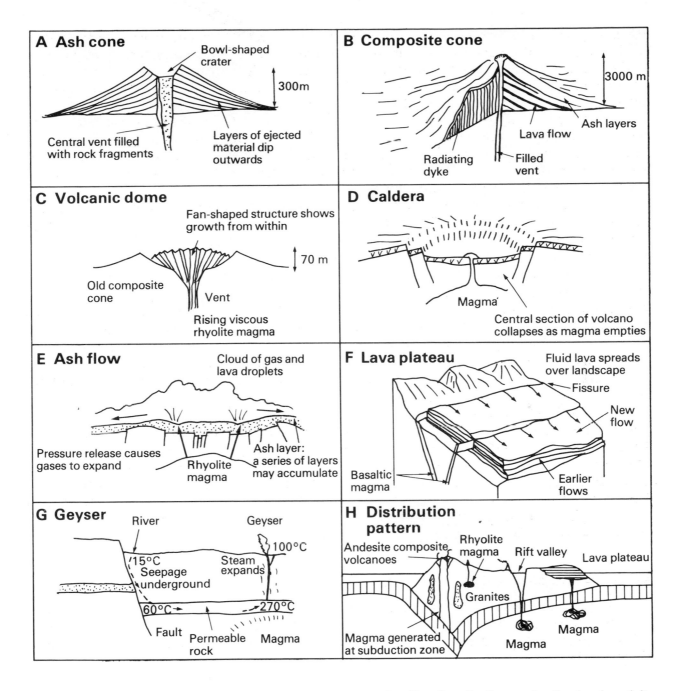

Figure 3.18 The variety of continental volcanic activity. Compare this variety with the volcanic features produced in the ocean basins. Apart from the rift valley situation nearly all volcanic activity is marginal to the continents. (Some of the data from Schmidt and Shaw, 1968)

The forms produced by continental volcanic activity are more varied than those found in the ocean basins. This is because there is a greater variety of lava and also of surface processes which can greatly modify the volcano outline between eruptions. Many of the volcanoes produce tall cones, with concave slopes. These resemble the volcanic cones of island arcs, with andesite, rather than basalt, lavas interlayered with ashes. Eruptions often take the form of an explosive ejection of ash from the top of the cone, with lava flowing out through vents in the cone side. The ash and lava together build up what is often known as a composite cone. At times the ash cone may be formed without the emission of lava, but such features tend to be eroded rapidly due to their unconsolidated nature. Volcanic domes result when the lava

9

10

Plate 9 Arenal Volcano, Costa Rica. This is one of the most active and explosive volcanoes in central America: a small flow is seen descending the steep cone. (USGS)

Plate 10 A crater formed by a volcanic 'bomb' erupted from Arenal. The mantle of ash has killed local vegetation, but has already been channelled by streams. (USGS)

Plates 11/12 Cerro Negro volcano in Nicaragua, showing the 40-day eruption of December 1968 by oblique and vertical views. Notice how the ash and gas issue from the vent, whilst lavas erupt from the base of the ash cone. (USGS)

11

12

13

Plate 13 The natural geyser area in Sonora County, California. This is in the Coast Range bordering the Pacific Ocean, and the steam is used for generating electricity. Intensive efforts are now being made to use this relatively cheap source of energy more fully. (USDASCS)

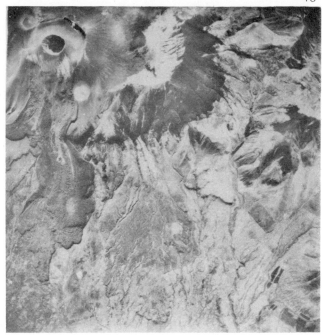

14

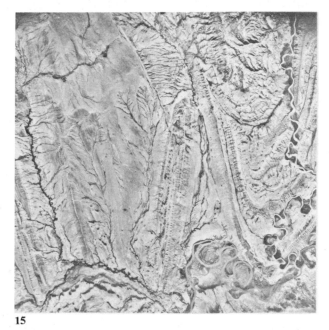

15

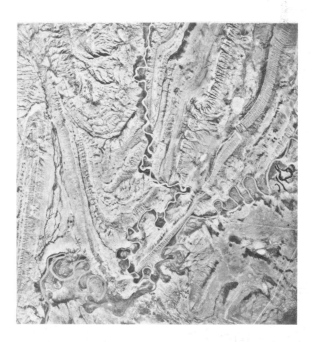

Plate 14 A stereo group of Asama-yama, one of Japan's most active volcanoes, on Honshu Island. The most recent crater is in an older caldera. Much of the ground is covered by ash. The most recent lava flow occurred in 1783, and is of andesite with a rough (blocky) surface: it partly fills a trench carved by a hot mass of molten lava droplets and gas known as a 'nuée ardente' (it is thought that this type of activity could have been responsible for forming rilles on the Moon).

Plate 15 Stereo group of folded rocks in the Rocky Mountains of Wyoming: note the types of fold present and the relationship between stream patterns and the structure. (All USGS)
N.B. A number of stereo groups occur in this book, and they should be viewed through special glasses. These are quite cheap and the name of a supplier is given at the end of the Bibliography. Pick a distinctive feature which occurs twice and place the glasses over the two images. Look through the glasses, concentrating on one of these and adjusting the glasses if there is a double image close together. Once these coincide the relief should stand out in three dimensions. The large-scale photographs used in this book give an exaggerated view of the relief, the vertical scale of which is enlarged by about six times.

reaches the surface in an almost solid state: the remaining liquid solidifies at once and further growth is internal, leading to the expansion and cracking of the outer layers. Domes are associated with lavas containing high proportions of silica and low proportions of iron and magnesium silicates (Figure 2.11). Such rocks are known as rhyolites.

The volcanoes of young fold mountain ranges are not so permanently active as, say, the oceanic Hawaiian volcanoes. After the eruption of a continental volcano the vent will become filled, and when molten magma rises once again it will build up pressure beneath the blocked vent until there is sufficient to overcome the resistance of the solid rock. An explosive eruption follows with an intensity varying according to the forces released. It may merely clear the vent, but it may also blast the entire volcanic cone into fragments, leaving a vast hole, known as a caldera. Crater Lake, Oregon, USA, is one such feature. When this type of explosion takes place quantities of molten lava are broken up by the expanding gases (rather as lemonade froths out of a bottle or can when opened) and the mixture of hot gases and lava droplets flows downhill to form layers of ash as they cool. These are known as ash flows. Some areas, such as North Island, New Zealand, Iceland, and parts of California, are also marked by hot springs and geysers, due to the heating by hot magmatic bodies of groundwater.

The magma which gives rise to this variety of volcanic landforms may be generated beneath the crustal rocks, where the ocean crust is carried down in the subduction zone. Such magmas are andesitic in composition, like those of island arcs. As the magma rises towards the surface it may have to force its way through extra thicknesses of continental crust (up to 30 km) and will become mixed with them. The deep burial of continental rocks in the fold mountain zones also leads to melting and the generation of magma which is very different in composition and characteristics from that which gives rise to the ocean basin volcanoes and the andesitic volcanoes at the subduction zones. Continental rocks contain high proportions of low density materials, such as silica and aluminium oxides (respectively the main constituents of sands and clays, the main sedimentary rocks). When these are melted a magma rich in such compounds is formed, and since it is of low density it rises in the crust, helping to raise the fold mountains and intruding at depth large masses of crystalline rock composed of feldspar and quartz — i.e. granite. Granite batholiths exposed by surface processes are a feature of fold mountains, ancient and modern, and commonly cover areas of thousands of square kilometres. When such magmas reach the surface they may form rhyolite domes, or are involved in particularly explosive eruptions.

Volcanoes and man

Volcanic activity can be both a destructive hazard and an important source of resources for man.

Volcanic eruptions have destroyed towns and villages. St. Pierre, Martinique, in the West Indies, was a town of 30 000 people which was overwhelmed by a cloud of hot gases and molten lava droplets on 8 May 1902; settlements around Vesuvius and Etna in southern Italy have been destroyed by ash falls and lava flows at various times; Tristan da Cunha had to be evacuated in the 1960s, and the 1973 eruption of Helgafell in the Westermann Islands, off the south-west coast of Iceland, led to the temporary loss of that country's main fishing port.

On the other hand volcanic soils are rich in plant nutrients and weather in a few years, especially in warm, moist climates, to provide fertile soils. That is why people continue to live in dense concentrations around volcanic areas in Java, Italy and central America. Geothermal power tapped in volcanic areas has become important in Iceland, New Zealand, and in Sonora County, California. The chemicals produced in volcanic activity, like sulphur and boron, and the building materials, like crushed pumice, also provide important resources. Volcanoes also present the local inhabitants with a source of tourist revenue, especially in areas accessible to population centres like Japan, Italy, the western USA — and even Hawaii.

Activity resulting in lava plateaus

Huge plateaus of basalt are built up from many successive lava flows, each up to 15 m thick. In places the total may exceed 1000 m. The rocks are sometimes known as 'flood basalts', since they have been erupted in a very liquid state and have buried much of the former landscape. Some of the largest outpourings are those of the Deccan Plateau of India, south-west of Bombay, and of Brazil inland from Rio de Janeiro. The flood basalts of the Columbia River Plateau in north-western USA were formed between 10 and 18 million years ago, covering 350 000 km² to depths of up to 600 m. Similar, but smaller, quantities form much of the islands of Mull and Skye off the west coast of Scotland. In each of these cases the outpouring of basalt lava can be associated with the breaking apart of continents (i.e. South America splitting from Africa; India moving north from Australia–Africa; Europe sundered from North America), or with magma rising up from a subduction zone or ocean ridge (north-west USA). It would seem that oceanic basalt was erupted on the adjacent continental margin during plate movement.

Activity in rift valleys

The margins of the East African rift are lined by volcanoes including the huge Mounts Kilimanjaro and Kenya. The characteristic of these areas is a great variety of volcanic rocks, ranging from lavas extremely low in silicon to those which have very high proportions. In addition there are unusual varieties containing high proportions of carbonate minerals, not normally found in magmas. This results in a great variety of landforms. These rifts and their nearness to the Red Sea area are clearly associated with the early stages of ocean formation (Figure 2.17), and the variety of rock-types can be explained by the fact that the rifts give access to deep-seated rock material sources as well as encouraging the melting of material nearer the surface.

In each case, therefore, volcanic or earthquake activity which affects continental areas can be traced to an association with plate margins which are essentially oceanic features! The concept of continents being passively carried on the conveyor-belt oceanic plates shows that these masses of relatively low density rocks are the quiet zones of the world, geologically speaking. The eruption of certain types of volcanic product adds to the material, which becomes part of the continental mass and is not returned to the upper mantle source like most of the oceanic crust. Whilst granites are merely remobilised continental crust rocks, the flood basalts and the andesites originate beneath the crust. It has been estimated that over 4 million km³ plateau basalt have been erupted during the last 180 million years (although this is small when compared with the fact that the granites of the British Columbia Coast Range batholiths occupy 3 million km³ alone), and that the andesitic volcanoes of fold mountain ranges have added between 70 and 150 million km³ in the same period (i.e. a rate of 0.3–0.7 km³/year). Whilst these give lower figures than those of material added to the oceanic crust along the ocean ridges, the ocean-floor material is being recycled to the upper mantle in large measure, whilst the continental volcanic rock is recycled by erosion and deposition largely within the continental areas.

Fold mountains and plate tectonics

Young fold mountain ranges are thus zones of volcanic and earthquake activity, and these in turn are associated with plate boundaries, a realisation which has led to a revision of the former geosynclinal theory of mountain formation. The young fold mountain ranges are associated exclusively with convergent plate mountains: the collision of the plates leads to the building up of stresses within the plates and any continental rocks on top, resulting in folding, faulting and igneous activity. Mountain chains may be formed in a number of situations: the process is known as orogenesis and at least four types can be recognised.

The island arc situation

Island arcs are often related closely to fold mountain ranges. They are found today mostly around the western margin of the Pacific Ocean and in the north-east of the Indian Ocean. They form where a section of the ocean-floor is subducted in the ocean basin away from a continent — i.e. where ocean-floor crust is on either side of the convergent plate boundary.

Japan is the largest area of land formed in this way, and is a mountainous country. Mount Fuji reaches nearly 4000 m, and several other peaks on Honshu, the main island of the group, top 3000 m. They are all volcanic in origin, and in this sense the island arcs cannot be regarded as fold mountains like the Alps and Himalayas. They do exhibit a number of features akin to fold mountains, however, and their overall characteristics must be considered in this context. The island of Honshu (Figure 3.19) is largely a pile of basalt and andesite lying between the Japan Trench and the Sea of Japan. The Pacific Ocean floor is subducted at the trench, a hypothesis which is supported by the greatest volume of volcanic rocks being erupted over

Plate tectonics and mineral resources

The concept of plate tectonics has been discovered to be not just a basis for theoretical discussion, but a means of pointing out important relationships reflecting the distribution of mineral ore deposits around the world (Figure 1).

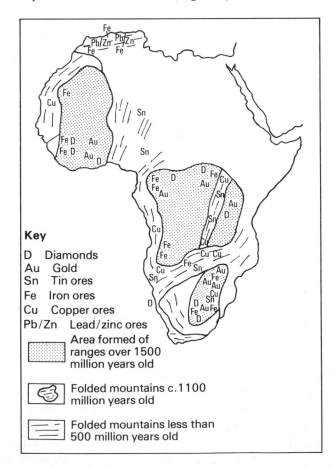

Key

D   Diamonds
Au   Gold
Sn   Tin ores
Fe   Iron ores
Cu   Copper ores
Pb/Zn   Lead/zinc ores

  Area formed of ranges over 1500 million years old

  Folded mountains c.1100 million years old

  Folded mountains less than 500 million years old

Figure 1 Mineral deposits and the structural units of Africa. Which minerals occur in the oldest rocks, and which in the younger groups? Account for the anomalies in the general pattern—e.g. diamonds on the coast of south-west Africa. (After Clifford in Gass, Smith and Wilson, 1972)

In addition it is now realised that many of the world's petroleum reserves are associated with continental shelf areas, formed as major sedimentary basins at the trailing edges of continental masses (Figure 2).

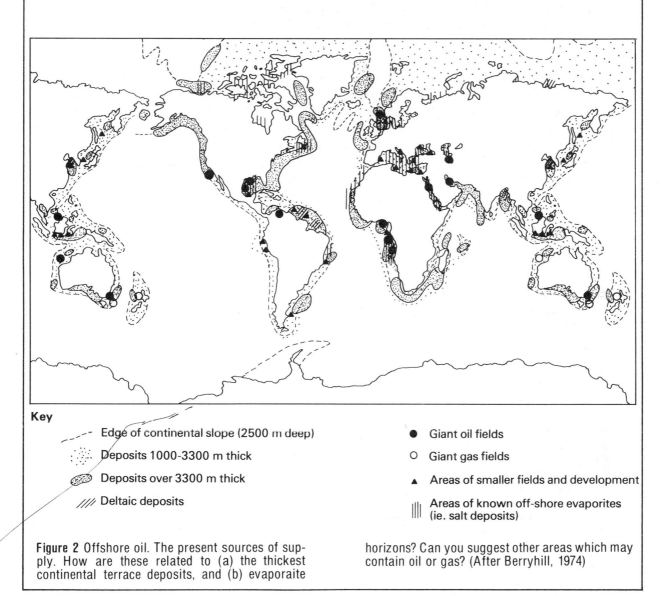

Key

- – – Edge of continental slope (2500 m deep)
- Deposits 1000–3300 m thick
- Deposits over 3300 m thick
- //// Deltaic deposits

- ● Giant oil fields
- ○ Giant gas fields
- ▲ Areas of smaller fields and development
- ||| Areas of known off-shore evaporites (ie. salt deposits)

Figure 2 Offshore oil. The present sources of supply. How are these related to (a) the thickest continental terrace deposits, and (b) evaporaite horizons? Can you suggest other areas which may contain oil or gas? (After Berryhill, 1974)

western Japan during the last 2 million years. It seems that the subducted ocean-floor becomes mobilised as magma when it reaches 120 km depth, and this is why the east of the island experiences little volcanic activity compared with the west. The evidence of the double metamorphic belts (Figure 3.19) suggests that Japan has been formed by the driving together of two island arc masses at different times, and just to the south of Japan today are the two separated arcs of the Philippines and Marianas. The Sea of Japan now forms an effective sediment trap for material eroded from Asia as well as from western Japan. On the Pacific Ocean side a wedge of flysch-type sediment is accumulating. Thus the volcanic rocks form the first land, but the atmospheric processes act on these rocks and produce sedimentary rocks.

It would seem that the trench develops as plate descent begins. Ocean-floor crust together with any covering sediment is thrust beneath more ocean-floor crust, giving rise to rocks metamorphosed under high pressure conditions on the continental side of the trench. When the descending plate reaches over 100 km in depth partial melting takes place, magma rises to form a pile of volcanic rocks and the island arc begins to rise. The heat generated in this way results in further metamorphism as a result of combined heat and pressure. The greater the depth reached by the plates, the more intense the volcanic activity and the farther the magma has to ascend. This affects the type of lava produced: basalts are poured out first, followed by andesites which become increasingly alkaline and are erupted farther from the trench.

The cordillera situation

Ocean-floor crust may be subducted near the margin of a continent, where sediments of the type accumulating in the continental terrace and the continental rise of eastern North America today (Figure 3.16) become involved in compressional movements. The oceanic crust is thrust beneath the continental margin, as is happening on the west coast of South America, forming the Andes. The process begins with events very like those taking place in island arcs (Figure 3.20), but with volcanic rocks and mobilised continental crust rising through the

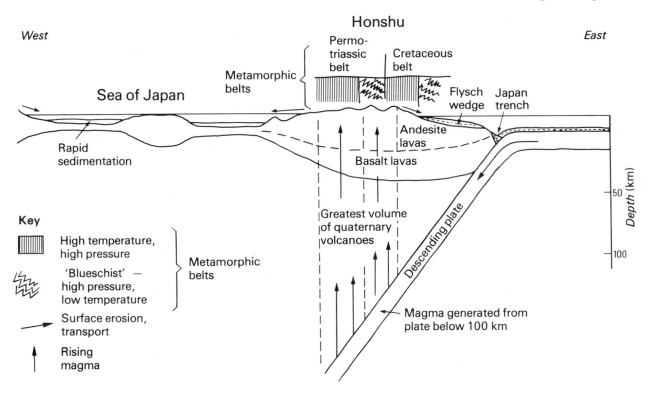

Figure 3.19 Island arc orogenesis, Japan. Note the association of maximum recent volanic activity and depth of plate subduction. 'Blueschist' metamorphism is related to compression at the trench; high temperature and high pressure metamorphism is associated with rising magma. The twinned metamorphic belts (Cretaceous and Permo-Triassic) suggest that Japan has been formed by two island arcs being pushed together. The Philippine and Mariana arcs diverge to the south, and may formerly have been separate in the Japan area. (After Dewey and Bird, in Cox, 1973)

shelf/rise sediments and leading to compression of the adjacent continental area. Oceanic crust underthrusting is the early mechanism affecting the surface relief, but later the growth of the 'orogenic welt' (i.e. the updoming vaused by the increasing zone of mobilised rock material) leads to overthrusting of the rocks towards the continent and to nappe formation.

This model of the development of a cordilleran system is sufficiently complex to explain most of the observed evidence. It also suggests that there can be a gradation from the island arc situation to the cordilleran, depending on the relationship of the point of subduction to the continental margin. This model can be seen to fit the concept of geosynclinal zones as found in the ancient rigid blocks (Figure 3.10), with a division between miogeocline and eugeocline occurring approximately at the boundary between shelf and rise sediment.

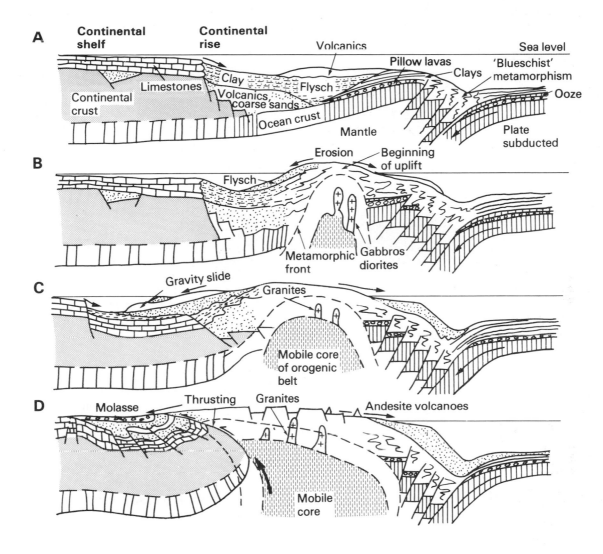

Figure 3.20 The development of a cordillera-type fold mountain system, based on the evolution of the Appalachians (Ordovician stage) and the USA western cordillera (Cretaceous).
(A) An oceanic plate descends beneath the continental rise to over 100 km, leading to the eruption of submarine volcanic rocks and high pressure/low temperature deformation at the plate margins.
(B) Rising magma generates heat and an expanding dome leads to an early zone of uplift with gabbro/diorite intrusions.
(C) High temperature/high pressure deformation becomes extensive as the mobile core of the orogenic belt grows; this begins to affect the continental rocks; the continental shelf subsides and gravity slides occur as uplift continues.
(D) The mobile core drives metamorphosed thrust sheets towards the continent. Granites are emplaced at high levels in the mobile core zone.

The New Guinea situation

New Guinea was formed by the collision of a continent and an island arc approximately 20 million years ago (Figure 3.21). The southern part of the island is really part of Australia, whilst the north is the old island arc. More recently the whole mass has been underthrust from the north — a change of subduction direction. This situation produces a different pattern of relief features and rock-types in terms of their succession across the belt affected.

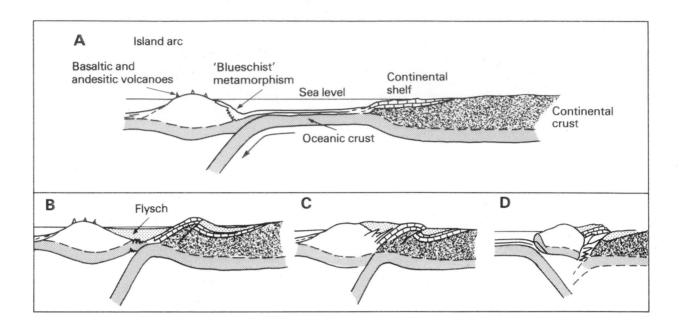

Figure 3.21 The 'New Guinea' type of collision between an island arc and a continental margin.
(A) A volcanic island approaches a continent by consumption of the intervening ocean floor.
(B) Flysch fills the reduced ocean just before collision

(C) The continental crustal rocks do not underthrust the island arcs due to their low density, but are deformed by thrust slices.
(D) A new trench may develop on the far side of the arc. (After Dewey and Bird, in Cox, 1973)

The Himalayan situation

A further alternative occurs where two plates carrying continental masses collide. The relative buoyancy of the continents restricts the amount of continental material which is underthrust with the oceanic crust of the plate, so that the uppermost rocks are crushed together, leading to the formation of thrust slices and thickening of the crustal rocks. Varying types of continental margin may be involved in such collisions, and this provides a number of different final situations, as can be seen today along the Alpine–Himalayan ranges. The Himalayas are taken as the general, or type, situation, since they exhibit a fairly simple occurrence with wedges of ocean-floor sediment and crust being thrust upwards between the two continental margins (Figure 3.22). Flysch, produced by erosion of the rising mass, and molasse, accumulating in marginal troughs, overlie these. The Himalayas were formed when the ancient continental mass now forming southern India was driven northwards and beneath the Asian continent. In Iran the Zagros Mountains have been underthrust by the Arabian massif, but the salt layers at the base of the sediments acted as a lubricant and simple fold structures resulted at the surface (cf. Figure 3.6A). Farther west the Alpine ranges of southern Europe and north-west Africa were formed by the collision of a smaller mass of continental material (approximately along the line of the present Adriatic Sea) with a continental margin, leading to a complex pattern of ranges. At present the African plate is moving northwards and is being subducted along a line south of the Aegean arc (Figure 3.23). The Mediterranean Sea is thus a closing ocean (Figure 2.17).

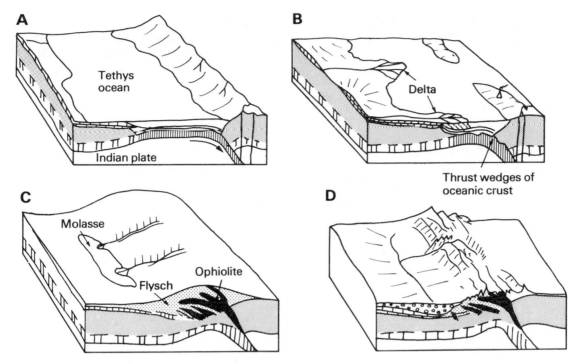

Figure 3.22 The 'Himalayan' type of orogenesis: two continents collide. Notice the time period over which the events take place.
(A) Approximately 65–70 million years ago. The Tethys ocean is contracting as the Indian plate brings its continent towards Asia.
(B) Approximately 30–60 million years ago. The two continents close and the oceanic crust begins to fracture into thrust wedges.
(C) Approximately 2–30 million years ago. The Himalayas emerge as mountain ranges and early erosion leads to the deposition of molasse in a basin to the south.
(D) Today. Erosion has already worn down much of the range. (Partly after Dewey and Bird, in Cox, 1973)

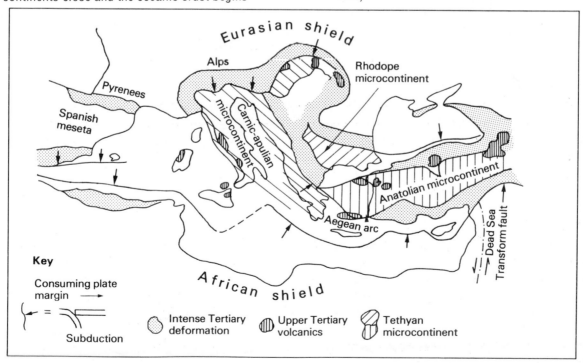

Figure 3.23 The evolution of the young fold mountains around the Mediterranean Sea can be related to a complex series of events during which the African shield has driven towards the Eurasian, causing small continental masses and island arcs to collide: this explains the arcuate loop of the Alps. This complicated picture contrasts with the simpler Himalayan type region. (After Dewey and Bird, in Cox, 1973)

Summary

The four situations can be summarised in two groups, based on the driving force involved. The island arc and cordilleran systems are driven by thermal energy generated by the underthrusting of one section of oceanic crust beneath another. They have extensive suites of rocks derived from oceanic crust materials: these are dark igneous rocks including lavas and intrusions and known collectively as ophiolites. The collision types (continent/continent or island arc/continent) are dominated by the mechanical energy of collision and have smaller proportions of high temperature metamorphism and of ophiolites.

When these various models are applied to the mountain systems of the world today, however, complexity is the order, rather than the seeming simplicity of a twofold division. Some belts have been built by a combination of situations: the Appalachian Mountains experienced an island arc and cordilleran situation in the Ordovician Period, followed by continental collision in the Devonian. Others are built over a long time: the Alpine–Himalayan ranges have been developing for the last 200 million years by a multiple series of collisions in which smaller continental blocks and island arcs have become involved. The main point about the four models of orogenesis, therefore, is that they seem to include the main separate ways in which mountains can originate, even if several are combined over a period of time to give rise to a major system of fold mountains.

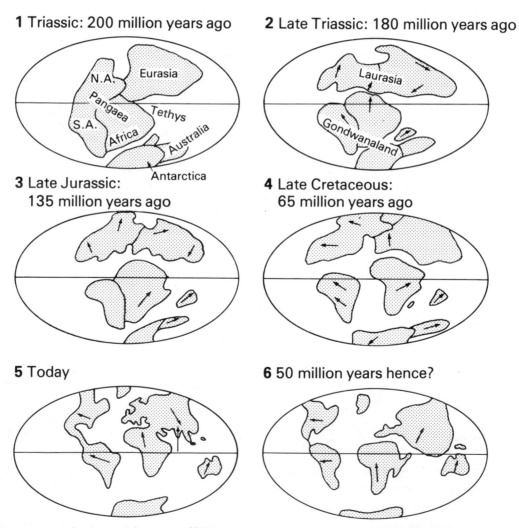

Figure 3.24 The evolution of the continents by plate tectonics since the Triassic, and the possible future pattern of events. Describe the break-up of the continents, and the features which suggest what could happen in the future. (After Dietz and Holden, in Wilson, 1973)

Mountains, plates and the history of the continents

The evidence studied in relation to the major features of the ocean basins and continents makes it clear that neither are permanently fixed. The ocean basins have been seen as opening and closing: the Mediterranean Sea is the remnant of a larger ocean, and the Pacific Ocean is gradually beginning to close as its ridge — the constructive margin — is overrun and consumed by the American plate. At the other end of the series the Atlantic has been opening for some 200 million years, and the Red Sea has been interpreted as the beginnings of an opening ocean basin. As the ocean basins open or close, the continental masses are also carried around, being brought nearer to one another, or moved farther away.

The ocean-floor evidence relates only to the last 200 million years of Earth history, but the evidence from the continents makes it possible to suggest a series of events during that period, and then to project these back into the more distant past. Before 200 million years ago there was a vast, single continent (known as Pangaea) which first broke into a northern (Laurasia) and southern (Gondwanaland) pair, and later into the broken and widely distributed pattern found today (Figure 3.24). At first the Pacific covered over half of the world, and an east–west ocean (Tethys) opened between the northern and southern continent, but the Atlantic became established following a rift valley stage, and the Americas moved westwards with the gradual consumption of the Pacific, whilst the Tethys closed again. This pattern can be used to make a prediction of future patterns.

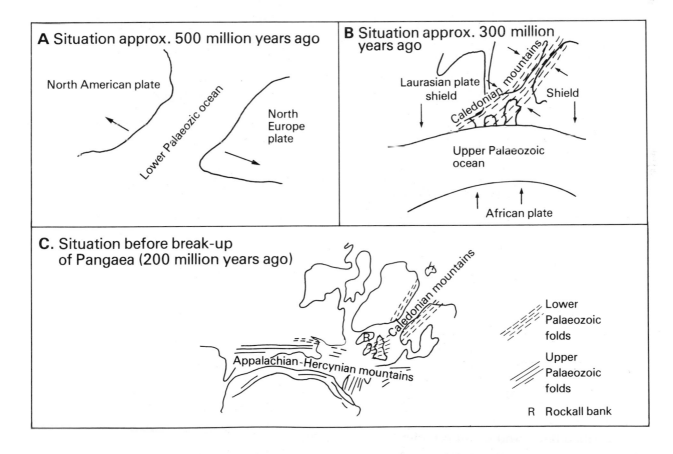

Figure 3.25 Palaeozoic plate movements and mountain formation. A generalised attempt to apply the principles of plate tectonics to the period before the break-up of Pangaea (Figure 3.24). This would suggest that continents were in existence before the present phase of plate movements began, and came together at the end of the Palaeozoic to form the single mass of Pangaea.

Less evidence is available for the reconstruction of events before 200 million years ago, and all the evidence is confined to the continental rocks. The facts that fold mountains are formed along the convergent margins of plates, and that wedges of continental shelf/rise sediments occur on the 'trailing' edges of continents, allow some degree of reconstruction to be made. The North Atlantic margins have been studied most thoroughly, and it seems that the Lower Palaeozoic (570–400 million years ago) was a time during which that ocean opened (Figures 3.25 and 3.26). A zone of subduction then developed along the north-western margin and resulted in the formation of cordilleran ranges some 400 million years ago. These are known as the Caledonian Mountains. The ocean then closed with collision-type folds forming the Hercynian ranges of central Europe approximately 300 million years ago. A general world pattern can be suggested for the last 700 million years (Figure 3.27).

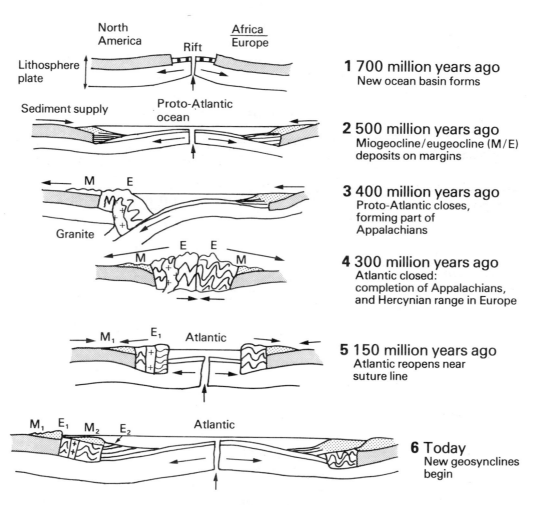

Figure 3.26 The history of the Atlantic Ocean over the last 700 million years. Relate this to the maps of Figure 3.25. (After Dietz, in Wilson, 1973)

The Precambrian shield areas record the oldest events in the evolution of the Earth's surface and contain evidence of fold mountain belts, although these have been worn down by long ages of erosion. These areas also have different types of rock associations (Figure 3.12) from those formed later, and it seems from this evidence that conditions were somewhat different in terms of atmospheric composition (i.e. a lower proportion of oxygen) and the extent of the continental masses. The oldest known rocks are highly metamorphosed, but originated from others which had been formed after exposure to the atmosphere. It is thought that the Earth once resembled the Moon, with a crust purely of the ocean-floor type. Continents began by

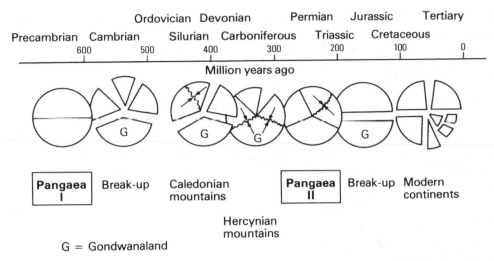

Figure 3.27 Suggested continental movements over the last 700 million years. It is possible that there has been more than one period of break-up and re-assembly of the continents. (After Valentine and Moores, 1970, in Hallam, 1973)

the formation of island arcs and the erosion of their volcanic materials exposed above sea-level to form sedimentary rocks. There is no record that this was the case, but the Precambrian continents were probably smaller than those of today (Figure 3.28). Addition to continental area takes place by the formation of island arcs offshore, or by the collision of an island arc with a continent. The cordilleran or Himalayan type of fold mountains add nothing to continental area, and these latter types must have become more important as the volume of sediment produced by continental erosion has increased.

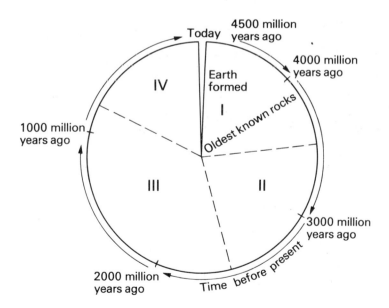

Figure 3.28 Changing conditions of Earth surface evolution since the formation of the oldest known rocks.
Phase I. The oldest Earth rocks: these are always highly altered granitic materials, enclosing some intrusions of dark, heavy mantle rocks. These rocks suggest that they are the remnant of an early granitic crust.
Phase II. The rocks are also highly altered, but include recognisable former basalt pillow lavas (i.e. formed in water), and piles of lava which could have been volcanic islands. The rocks of this phase also contain cherts with the first signs of fossil evidence for life on the Earth.
Phase III. Small areas of rigid continental crust. This phase may have seen the beginnings of plate movements, involving smaller sections of the crust than at present. The oldest kimberlite (rock containing diamonds) was formed during this phase, and indicates a deep origin for magmatic materials. The smaller time also supports this suggestion.
Phase IV. Fewer, larger plates acting as modern plates appear to. Extensive zones of high pressure/low temperature alteration are related to the higher stresses imposed by the clash of large plates. (After Sutton)

4

Internal Earth energy

The major relief features of the Earth's surface — ocean basins, continents, ocean ridges and fold mountains — owe their origins to processes which originate within the planet. It is impossible to observe directly the general internal structure, the sources of energy or the movements taking place there. Observations made at the surface of the Earth enable geophysicists to make an assessment of the physical properties of the Earth as a whole and of some aspects of the Earth's interior. Such observations include the shape of the Earth; the emission of radioactive energy; earthquake shock wave records; the Earth's magnetic field; and the Earth's gravitational field. Tentative conclusions concerning the internal structure of the Earth, the internal processes, the origin of the planet and its early development can be drawn from a study of these physical properties. Other assistance can be gained from astronomical studies of the Solar System and of the behaviour of other stars throughout the universe.

The shape of the Earth

The Earth is spherical, as has been demonstrated convincingly by the Apollo moonshot photographs. Practical disciplines like surveying and navigation have, however, worked successfully on this assumption since the early circumnavigations of the globe.

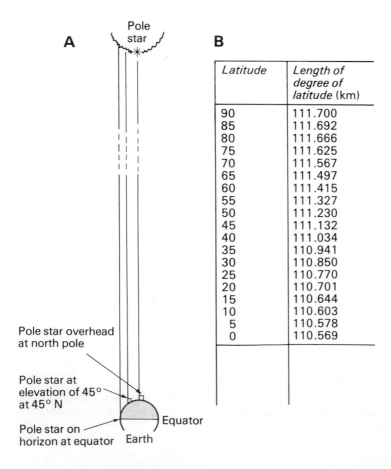

Latitude	Length of degree of latitude (km)
90	111.700
85	111.692
80	111.666
75	111.625
70	111.567
65	111.497
60	111.415
55	111.327
50	111.230
45	111.132
40	111.034
35	110.941
30	110.850
25	110.770
20	110.701
15	110.644
10	110.603
5	110.578
0	110.569

Figure 4.1 Degrees of latitude are calculated by reference to the Pole Star in the northern hemisphere (A). When distances between them are measured (B), they increase towards the poles, demonstrating the polar flattening.

The Earth is not a perfect sphere. Measurements of the lengths of degrees of latitude by reference to the Pole Star confirm this (Figure 4.1). There is a flattening at the poles, and a slight bulge at the Equator, of the order of approximately 1 in 300 (polar diameter 12 714 km; (equatorial diameter 12 757 km). This was caused by the Earth's rotation on its axis. If the Earth had a uniform interior it has been calculated that this deformation would be only 1 in 230, but the smaller amount measured suggests that the interior is composed of different materials from those at the surface.

The flattened sphere is known as an oblate ellipsoid, and its precise shape is now calculated from satellites orbiting the Earth. This method is an improvement on earlier surface-based surveys (Figure 4.2). Surveyors, mapmakers and astronomers refer their calculations to another figure, the geoid. This is the imaginative extension of sea-level through the continents, and is calculated by comparing gravity observations. The geoid and the ellipsoid show a number of differences (Figure 4.3).

Surveyor	Date	Area	Semi-major axis (metres)	Semi-minor axis (metres)	Flattening	Fraction
Everest	1830	India	6 377 276	6 356 075	0 003 324	1/301
Bessel	1841	East Asia	6 377 397	6 356 079	0 003 343	1/299
Clarke	1866	North America	6 378 206	6 356 584	0 003 390	1/295
Clarke	1880	Central and South America	6 378 301	6 356 584	0 003 408	1/293
International	1909	Oceanic areas	6 378 388	6 356 912	0 003 367	1/297
Fischer	1960	Satellite data	6 378 160	6 356 778	0 003 352	1/298

Figure 4.2 The Earth ellipsoid. Varied measurements made with increasingly advanced instruments, and in different areas. The maximum variation in the non-satellite surveys was 1112 m on the semi-major axis and 837 m on the semi-minor axis.

Radioactive energy and the Earth's age

It was not until the early twentieth century that the nature of radioactive elements was discovered. Some elements were found to have the property of decaying to 'daughter' elements, and of releasing nuclear energy in the process. The rate of decay is constant for each element: it is calculated as the half-life of the element — that is, the time needed for half of the parent atoms to decay. This has important implications for two aspects of Earth evolution: the source of internal heat energy, and the time-scale of the events which have taken place.

Internal Earth energy sources

The Earth's internal heat is the only source of energy sufficient to bring about the processes involved in mountain formation. Volcanic activity demonstrates that the Earth's interior is hot, and mine-shaft evidence suggests an increase of 10–30°C/km as one descends from the surface. This geothermal gradient suggests that there is a flow of heat by conduction from the Earth's interior to the surface, although the amount of heat reaching the surface in most places is too small to contribute significantly to the heating of the atmosphere. It does not melt the

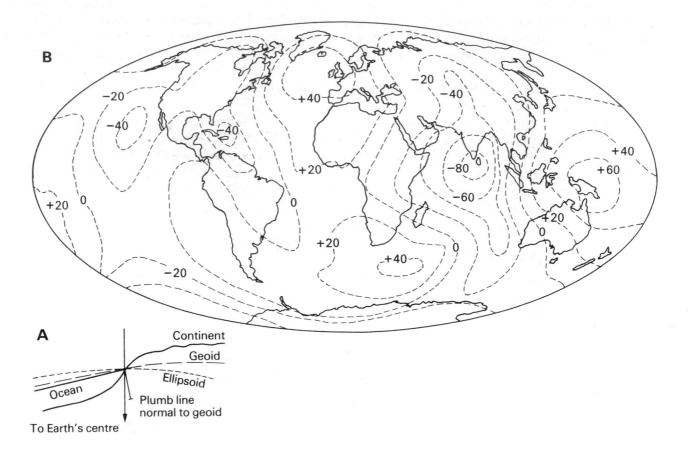

Figure 4.3 The geoid. (A) The relative positions of ellipsoid, geoid and sea-level. (B) A map of the geoid: elevations are shown in metres above the ellipsoid, based on a polar flattening of 1:298.25.

polar ice, for example, although it may contribute to changes within the ice (chapter 8). The heat flow is greatest in areas of recent volcanic activity, and along the ocean ridges (cf. Figure 2.13). It is not possible to measure the temperatures deep in the Earth. Current suggestions are that there is a steep initial temperature rise with depth just below the surface, but that the rate of temperature increase then falls so that the temperature at the centre is about 6000° C. The source of this internal heat is also something which is not fully understood. Some of it is thought to have been left over from the gravitational work done at the time of the Earth's formation, but a source about which there is greater certainty is the energy released by the decay of radioactive elements in the rocks: the radiation emitted is converted to heat. Present sources of such energy would be those elements which have decayed extremely slowly — i.e. with half-lives of 10^9–10^{10} years like the isotopes U^{235}, U^{238}, Th^{232} and K^{40} (U uranium; Th thorium; K potassium). Other original sources have decayed to tiny proportions which do not now provide significant quantities of heat. It is calculated that radioactive energy can produce a large part of the Earth's heat energy, but it is also clear that the elements with long half-lives are now concentrated near the Earth's surface, since most occur in crustal rocks like granite. The radioactive heat production from granite averages 360×10^{-7} joules/g yr, but less is produced by basalt (53×10^{-7} joules/g yr) and little in the rocks of the types which are thought to exist beneath the Crust. The rocks of the interior thus have little energy supplied to them in this way today, and their high temperature must be a survival from intense heating at an earlier stage of the Earth's history.

Dating the rocks

The rates of decay of these same radioactive isotopes enable geologists to assign dates to the rocks containing them, Radioactive decay rates are not affected by heat, pressure changes or erosion, and have been proceeding since the origin of the universe. The method of calculating dates from radioactive isotopes involves a comparison of the proportions of parent and daughter atoms. Since the half-life is known, the length of time over which decay has taken place can be worked out. This period of time takes the investigator back to the point at which the daughter atoms ceased to be lost to the mineral in which the parent atoms occur. Crystallisation from a molten magma will often cause the daughter atoms to be trapped within the crystal lattice, and igneous rocks — especially those cooled at depth — are the most useful for this method of dating rocks. Dates given to igneous rocks, reflecting the crystallisation of the minerals at a relatively sharp point in time, can be used to work out the dates of major stratigraphical events (Figure 4.4). Metamorphic rocks give dates which are related to the act of metamorphism and re-crystallisation in new mineral forms, which is often associated with rock burial and stress-application during mountain-building. Geological events can thus be set within a definite time-scale (Figure 4.5). There are difficulties which attend this method: the proportions of radioactive isotopes are so tiny that the analytical chemist finds it difficult to isolate them, and some of the daughter atoms may be lost from the minerals (thus giving rise to inaccurate results).

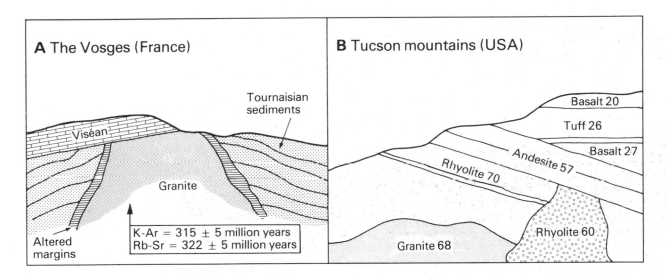

Figure 4.4 The applications of radiometric dating. (A) The granite was intruded into Lower Carboniferous (Tournaisian) sediments. Erosion took place before the Viséan sediments (also Lower Carboniferous) were formed. Radioactive elements in the granite enable the boundary between the two groups to be placed at approximately 320 million years ago.
(B) An area of unfossiliferous, largely igneous rocks, in which a detailed series of dates has been obtained. Figures in millions of years.

Earthquake shock wave records

Earthquakes occur beneath the surface of the Earth, where the rocks yield suddenly after a prolonged build-up of stresses (Figure 4.6). They are often associated with fault-lines, which provide a zone of fracture and easy yielding. Those associated with the ocean ridges are fairly shallow in origin (up to 60 km deep), but those around the convergent plate margins may be extremely deep (up to 700 km, with the deepest recorded at 720 km). The Pacific Ocean area is particularly important: 80 per cent of the more important shallow shocks, and virtually all the deep shocks, occur beneath this ocean and its continental margins. The shock waves passing through the Earth are the result of its being an elastic body — i.e. when a stress is applied it

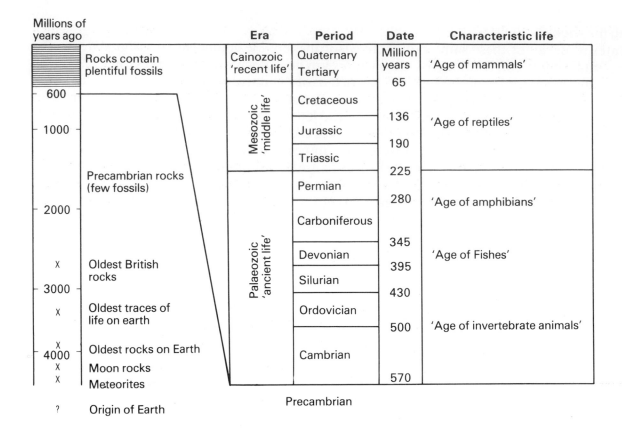

Millions of years ago		Era	Period	Date	Characteristic life
	Rocks contain plentiful fossils	Cainozoic 'recent life'	Quaternary	Million years	'Age of mammals'
			Tertiary	65	
600		Mesozoic 'middle life'	Cretaceous	136	'Age of reptiles'
1000			Jurassic	190	
			Triassic	225	
	Precambrian rocks (few fossils)	Palaeozoic 'ancient life'	Permian	280	'Age of amphibians'
2000			Carboniferous	345	
X	Oldest British rocks		Devonian	395	'Age of Fishes'
3000			Silurian	430	
X	Oldest traces of life on earth		Ordovician	500	'Age of invertebrate animals'
4000	Oldest rocks on Earth		Cambrian	570	
X	Moon rocks				
X	Meteorites				
?	Origin of Earth	Precambrian			

Figure 4.5 A geological time-scale

becomes deformed more and more, but when the stress is removed it returns to its original shape. If the stress applied is too great the elastic body will yield permanently. All elastic bodies can be subjected to two types of deformation — compression and shear — and earthquake shock waves are related to these. The P, or 'push' waves move through the materials of the Earth's interior by a series of alternate expansions and compressions in the direction of travel (cf. sound waves move in a similar way). The S, or 'shake' waves move like the oscillations in a piece of string vibrated sideways from one end. Thus P waves are compression–expansion stresses in the direction of travel, whilst S waves are shearing stresses at right angles to the direction of travel. Both wave types are affected by the medium through which they travel, and speeds vary accordingly (faster in denser materials), although in general the P waves travel at 1.7 times the speed of the S waves. The S waves do not pass through liquids because these cannot sustain a shearing stress.

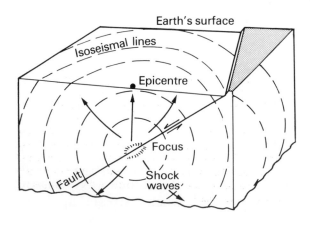

Figure 4.6 Earthquake focus and epicentre. Shock waves radiate in all directions from the focus. Isoseismal lines join points of equal arrival time of the shock wave front.

Earthquake shock waves are recorded on seismographs (seismos — an earthquake), which are made of delicate instruments (seismometers) anchored in rock together with apparatus for providing a continuous record (the seismogram) of the movements taking place (Figure 4.7). Usually at least three seismometers are used to give components of movement measured in three dimensions. Both P and S waves may be refracted and reflected at the boundaries of different materials as they pass through the Earth (Figure 4.8): if they pass into material of gradually increasing density at greater depths they will be refracted continuously, following a curved path.

Similar distribution patterns of records are obtained around the world wherever the earthquake focus is located, so leading to the conclusion that the Earth's interior is composed of a series of concentric shells of differing materials (Figure 4.9). There is a marked 'shadow

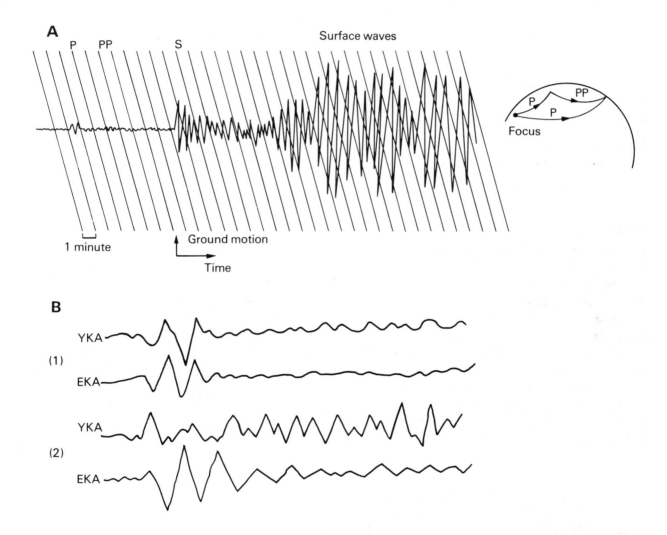

Figure 4.7 Shock wave records.
(A) A seismogram, obtained at a station in Queensland, Australia, from a magnitude 5.9 earthquake off northern Sumatra, on 21 August 1967 (distance 6100 km). Nearly 8 minutes elapsed between the arrival of the P and S waves. The PP waves are P waves reflected from the surface, as shown by the inset diagram: P and S waves can also be transformed into each other by such reflection.
(B) Nuclear explosion (1) and earthquake (2) shock wave records. The nuclear explosion took place on 29 October 1965, and the earthquake on 18 February 1965: both were centred at Amchitka Island in the Aleutians (North Pacific Ocean). The records were obtained at Yellowknife, Canada (YKA, 4000 km away), and at Eskdalemuir, Scotland (EKA, 8100 km away). Explosives generate only P waves (and some surface waves), whilst earthquakes give a more varied range of wave forms. (After Thirlaway, *Discovery,* 1966)

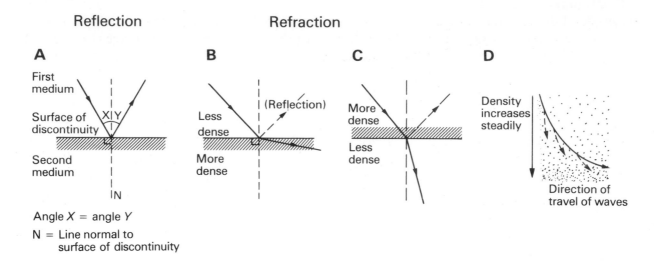

Reflection

Refraction

A
First medium
Surface of discontinuity
Second medium

X Y

N

Angle X = angle Y

N = Line normal to surface of discontinuity

B
Less dense
More dense
(Reflection)

C
More dense
Less dense

D
Density increases steadily
Direction of travel of waves

Figure 4.8 Reflection and refraction. Radiating waves are reflected back from a surface of discontinuity between media of differing density. They are also refracted, or bent, as they pass through an interface: notice the differences between the entry to more dense (B), less dense (C) and in the passage through a medium which increases gradually in density (D).

zone' where only faint reflections of P waves arrive in the hemisphere opposite the one in which the earthquake focus is located, and beyond this a zone where only P waves arrive. It is concluded that this situation is due to refraction at a depth of 2900 km and that there is possibly molten rock beneath this level, preventing the S waves from passing through. This boundary zone, the Weichert-Gutenberg Discontinuity, marks the division between the core and mantle, the two fundamental sections of the Earth's interior.

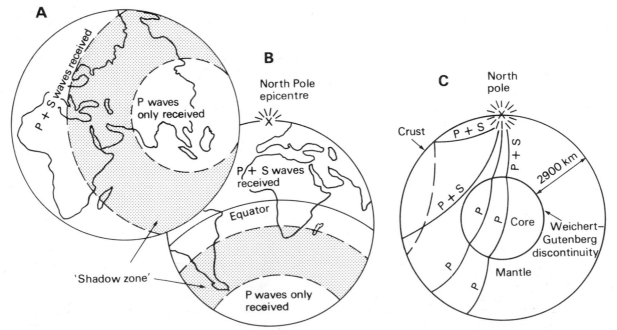

Figure 4.9 Earthquake shock wave passage through the Earth.
(A) The pattern of records following an earthquake centred near Peru.
(B) The similar pattern of a hypothetical earthquake beneath the North Pole (few occur there, but the diagram helps to locate the extent of the zones).

(C) A section through the Earth, showing the paths of the shock waves: notice how the waves are refracted through the mantle and at the boundary with the core. How does the presence of a 'shadow zone' help in calculating the thickness of the mantle?

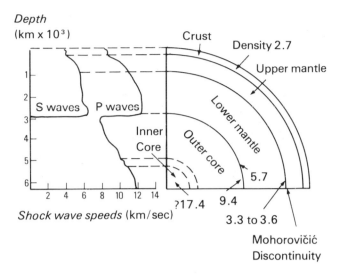

Figure 4.10 The shells of the Earth's interior related to earthquake shock wave speeds: the crustal thickness is exaggerated (it is only 5 - 10 km beneath the oceans, but up to 50 km beneath the continents). The densities suggested for the internal layers are related to the shock wave speeds at each level. Figures on the right hand side are in g/cm³.

The records permit amplification of this simple model. The P and S wave velocities vary with depth (Figure 4.10), and each change can be related to a change in materials. Each region of changing shock wave velocity is associated with a zone of discontinuity. Conclusions based on such evidence must be subject to the reservation that the extreme pressures in the Earth's interior have never been simulated in the laboratory. Pressures of only 100 000 times the normal atmospheric pressure have been obtained for laboratory experiments, but it is estimated that pressures reach 1 million atmospheres at 2200 km depth, and over 3 million atmospheres at the Earth's centre. At least two interpretations of the internal composition are possible.

1) The shells of increasing density towards the centre are formed of different materials. This conclusion would be based on the properties of materials as they occur under surface conditions of heat and pressure.

2) The shells of increasing density towards the centre are formed of similar materials, but with a more closely packed crystal structure. It is known, for instance, that carbon atoms can be bonded loosely together, as in pencil lead (graphite), or closely packed, as in diamond, and this affects profoundly the nature of the two materials. Laboratory experiments also show that some materials maintain their crystal structure until the pressure increases to a certain point, after which there is rapid change to a more compact structure. This process has not yet been confirmed for rock-forming minerals, but it is possible in theory.

Predicting and controlling earthquakes

Earthquakes are destructive natural hazards, along with volcanic eruptions, landslides and flooding. For long, man has been powerless to predict — let alone control — these events, but recent evidence suggests that some measure of success may be obtained in these areas. The USA is particularly concerned, and the US Geological Survey Hazard Reduction Program seeks to understand more concerning the basic principles of earthquake activity.

In the area of earthquake prediction it has been observed that large earthquakes have been preceded by local ground deformation, variations in well-water composition, or by changes in seismic records (Figure 1). These methods stem from an approach related to what has been movement patterns. These methods stem from an empirical approach related to what has been observed to happen before, and not from a fundamental understanding of earthquake causes and mechanisms. A recent suggestion that the San Andreas Fault will move again disastrously in 1982 relates fault movements to the triggering effect of extraterrestrial gravitational pull: at that date all the planets of the Solar System will be in line and pulling on the Earth. Once again, however, this suggestion cannot be tested before the event — and in any case very little could be done to avoid it.

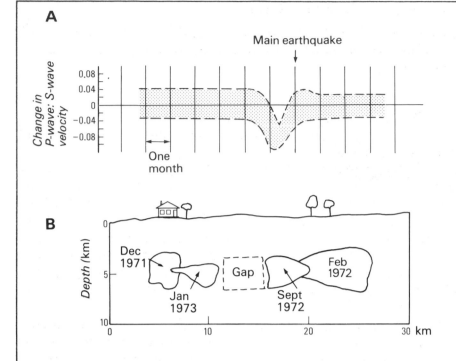

Figure 1 Predicting earthquakes.
(A) Small foreshocks are plotted as a ration of P:S wave velocities. This ratio drops some weeks before a major earthquake, and then rises sharply. This was discovered by USSR scientists. (After Press and Siever, 1974)
(B) Spotting the epicentre. Four moderate earthquakes along the San Andreas fault. The next may occur in the gap. (After Wallace, 1974)

An important monitoring experiment is in progress along the San Andreas Fault, analysing the occurrence of tiny earthquakes, strains and movements along that vulnerable zone, and developing methods of recording the changes taking place.

Controlling earthquakes has been related to the alteration of the frictional resistance within fault zones by changing the fluid pressures (Figure 2). These results give a simple relationship, but fault-zones are complex and the forces involved immense.

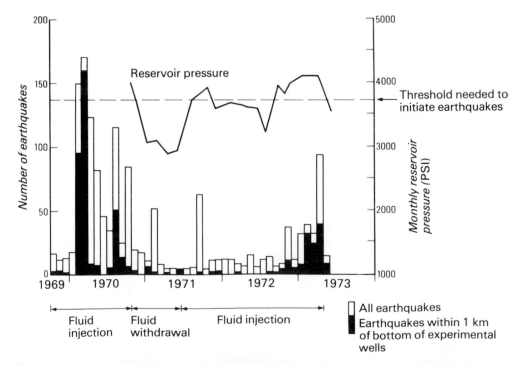

Figure 2 Controlling earthquakes. Earthquake frequency has been correlated with reservoir pressure at Rangely Oilfield, Colorado, USA. Pumping was controlled so that pressure was above or below the critical level. (After Wallace, 1974)

Magnetic and gravitational fields: the Earth's interior

The study of earthquake shock wave records has led to the establishment of a model of the interior structure. A variety of other physical observations can be used to determine further properties of the major interior zones.

The core

The deepest and most inaccessible part of the Earth is the least known, but it is clearly important to understand as much as possible about it in relation to internal processes and their effects. A number of facts are clear:

1) The density of the whole planet is approximately 5.5 g/cm³. This is greater than the average density of the Earth's surface rocks (2.8 g/cm³), and it follows that the core has a mean density of 10.6 g/cm³. It has been suggested that the material there could be a nickel – iron mixture, such as is common in meteorites, and the name NIFE has sometimes been given to this zone.

2) The pressure and temperature in the core are both very high. The high pressures (over 3 million atmospheres) will raise the melting points of the rocks, and the suggested temperatures of around 6000° C may not be high enough for the whole core to be molten, but this is difficult to confirm, since the S waves do not penetrate the outer core and can therefore provide no information about the inner core.

3) The fact that the S waves do not pass through the core indicates that at least the outer core may be molten. This is also supported by the way in which the rotating Earth wobbles on its axis like a spinning top: this is the sort of thing which happens to a body with a liquid mass inside.

4) The Earth's magnetic field can be plotted by comparing the directions of magnetic north and geographical north over the Earth's surface: the difference is known as the magnetic declination (Figure 4.11). Magnetic north can be found by using a magnetic compass.

This magnetic field is a relatively simple pattern, which can be simulated by the presence of a bar magnet at the Earth's centre (Figure 4.12). Studies of earthquake shock wave records have suggested a model of the Earth's interior composed of a series of concentric shells, rather

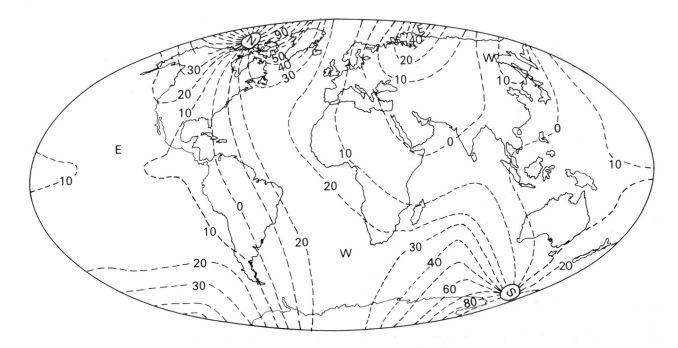

Figure 4.11 A world isogonic map showing lines joining places of equal magnetic declination (at 10 degree intervals).

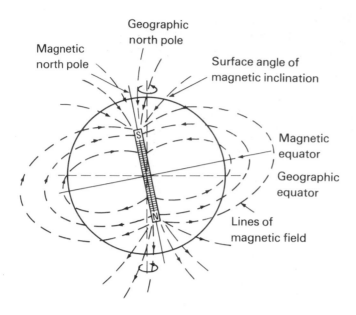

Figure 4.12 The geomagnetic dipole: an imaginary bar magnet at the Earth's centre relating to the lines of the magnetic field.

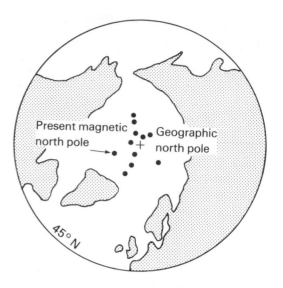

Figure 4.13 The movements of the magnetic North Pole. The dots represent positions over the last 7000 years, calculated from the residual magnetism in materials of up to that age.

than a bar-shape pattern. In addition iron loses its magnetism when heated to the temperatures found in the Earth's core. Another mechanism must be responsible for the Earth's magnetic field.

The magnetic field pattern is not static, and shifts gradually. The magnetic poles themselves move (Figure 4.13), although it seems that they do not depart widely from the positions of the geographical poles, and may approximate to this axial position over a long period of time. In addition there are also slight variations in the intensity of the magnetic field from place to place, and the centres of high intensity move westwards at 20 km per year, quite unrelated to any surface features or known subsurface movement (Figure 4.14). It is thought that this movement may be associated with the fact that the outer core is liquid and so will not rotate as rapidly as the solid mantle and crust. If this is the case the magnetic field would appear to be related to the core.

One modern theory suggests that convection currents set up within the outer core, together with the effects of rotation, could initiate a dynamo-type of situation in the core, and that this might give rise to the magnetic field. An enormous current (approximately 10^{10} amp) would have to be generated, but as the situation due to the circulation within the outer core is like that of a coil around the axis of Earth rotation (Figure 4.15), it seems that this is a distinctly possible answer. A weak magnetic field is needed to start such a system, and this could have been instigated in various ways during the Earth's formation: any suggestion made at this stage would be mere speculation. The system also has to be maintained, and one suggestion is that it may be fed by energy liberated as materials from the inner core change phase and become part of the less dense outer core. This whole scheme, however, is still a very tentative hypothesis, but it can help an understanding of the Earth. At the least it is fairly well-established that the core must have an important effect on the Earth's magnetic field, and in this way the centre of the Earth can have an effect on the external activity of the planet.

The mantle
The mantle has a mean density of 4.6 g/cm³, less than half that of the core. It may be formed largely of a silicate mineral rich in iron and magnesium, such as olivine. No actual samples of

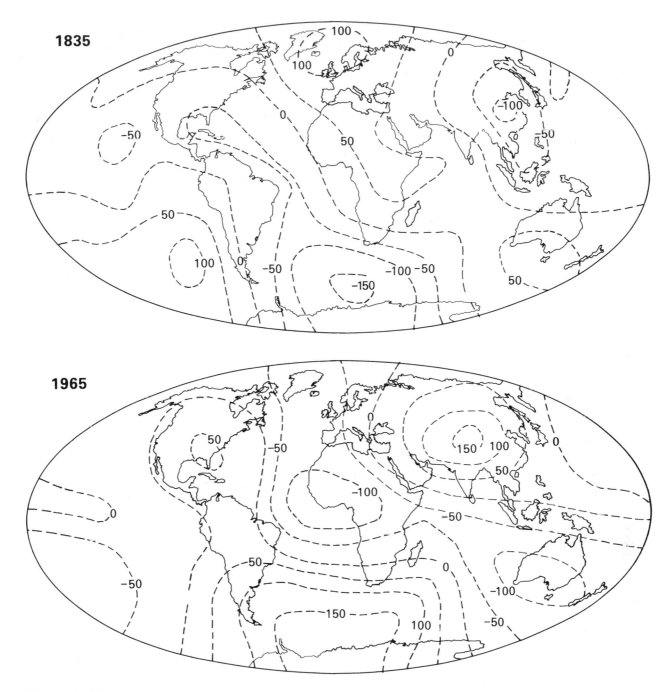

Figure 4.14 Maps of the Earth's magnetic fields in 1835 and 1965. Compare the movement of zones recording higher and lower values in that time.

the mantle materials have yet been obtained, although rocks erupted near the ocean ridges give strong clues that such conjectures are correct. It is hoped that scientists will be able to drill through the oceanic crustal rocks soon, since both USA and USSR teams are working on the project, and that they will obtain some of the mantle rocks. At the present evidence concerning the mantle comes from indirect sources.

The fact that the deepest earthquakes are no deeper than 700 km, and that most earthquakes occur at depths less than 100 km, suggests that the increasing pressure within the mantle results in changes of mineral properties with increasing depth. Closer atomic packing within a mineral like olivine may change it to spinel, which has a similar chemical composition

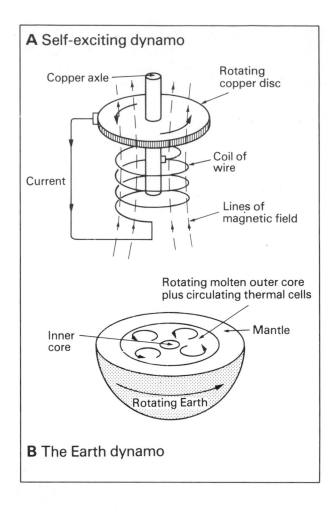

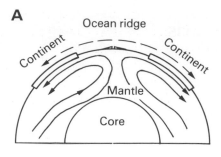

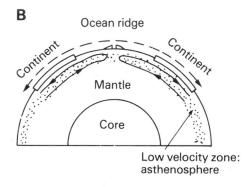

Figure 4.16 Convection current hypotheses to account for plate tectonics. (A) The whole mantle is involved. (B) Only the upper mantle with a base in the low velocity zone is involved.

Figure 4.15 Accounting for the Earth's magnetic field and reversals.
(A) Any minute magnetic field is sufficient to start a small electric current through the disc and coil. The field from this current will then produce a larger current, so the system builds up to a state of equilibrium in which the current could be quite large.
(B) In the Earth rotation of the core and in the thermal cells replaces that of the disc: the coil is replaced by currents in the conducting molten or semi-molten material itself.

but takes up 20 per cent less space. The progressive speeding up of earthquake shock waves as they pass through the deeper mantle also suggests this. In the upper mantle there is a slight slowing of the shock waves in a layer between 100 and 200 km deep. This layer has been shown up particularly by the monitoring of underground nuclear explosion shock waves with an increased network of seismograph stations. The S waves show signs of interference at this level, and it is possible that there is a plastic, lower viscosity layer where the rate of rise in temperature is momentarily greater than the rate of rise in pressure, and so the rocks come close to melting. If movement could take place along this zone under sustained pressure it could form the basis of convection cells which give rise to plate movements (Figure 4.16), and it may also be a level at which much magma is generated.

The crust

The crust covers the rocks of the interior thinly. At the most it is 50 km thick beneath the continental massifs, and it may be as little as 5 km thick beneath the oceans. This compares with a mantle thickness of 2900 km. Additional information about the crust has accumulated from the study of shallow focus earthquakes and artificial seismic explosions made by man in his search for mineral resources. A double set of P and S waves in the records of such shocks has led seismologists to conclude that there are two zones of crustal rocks beneath the continents, although only one occurs beneath the oceans (Figure 4.17). The zone between the crust and mantle is known as the Mohorovičić Discontinuity, after the Jugoslav who discovered it in 1919. This marks a fundamental change in the rock compositions, but the crustal rocks are here welded firmly to the underlying upper mantle: it is the low velocity zone between 100 and 200 km which forms the base of the lithospheric plates.

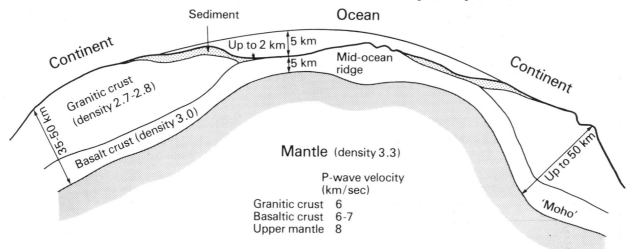

Figure 4.17 The Earth's crust. Relate the seismic velocities to rock density. Where does the division between continent and ocean lie? The vertical scale is exaggerated.

The crust-mantle relationship

The study of the Earth's gravitational field led to the realisation that the surface relief features are related to a state of balance existing at the base of the crust. As in the case of the magnetic field there are tiny local variations, which can be measured by a gravimeter. Where dense rocks, such as iron deposits, lie beneath the surface the gravitional attraction is greater and the instrument reads higher; by contrast, the attraction is less above a low density rock (Figure 4.18). Variations of this type are measured in units known as milligals, whereas

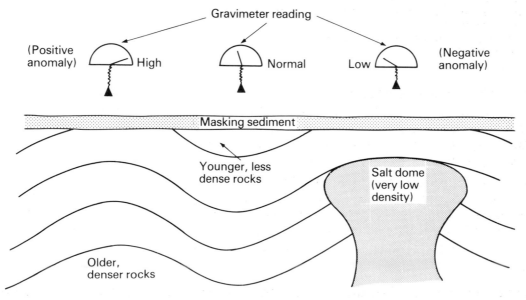

Figure 4.18 A gravimeter detects small, local variations in the Earth's gravitational field. These can be related to buried rock structures.

average sea-level gravity is 980 gals (1 gal= 1 cm/sec^2; it is a unit of gravitational acceleration, named after Galileo). The measured intensity, corrected to sea-level, is compared with the theoretical norm, and any difference is referred to as an anomaly: a positive anomaly is a value greater than the norm, and a negative anomaly is less than the norm. These can be plotted on a map, and, like the seismological and magnetic techniques of subsurface investigation, are used in the search for supplies of oil and metallic minerals. On the larger scale, it is found that continental areas have positive anomalies and oceans have negative anomalies: this confirms patterns of less dense continental rocks and denser ocean-floor rocks shown by seismic investigations. In particular the mountain ranges of the world have large negative anomalies, indicating that they are composed of an abnormal thickness of lighter crustal rocks extending down into the mantle.

It has been suggested that a state of balance, known as isostasy, exists between the lighter crust and the denser mantle. This would be of a similar nature to the state of balance existing between ice and water (Figure 4.19), although the adjustments to changes of balance in the solid rocks would be made more slowly. Confirmation of this effect is seen in the history of Scandinavia and of other parts of the world which experienced heavy loading by ice during the Quaternary Ice Age. Scandinavia was covered by ice 25 000 years ago. The weight of this ice caused sinking of the land, and when the ice melted, the immediate result was that the Baltic Sea area, which had been depressed beneath the ice centre, was drowned. The removal of the ice load by melting led to a slow recoil of the underlying solid rock: the Baltic area has witnessed rising land with rates of as much as 1 m/100 years in the Gulf of Bothnia and 30 cm/100 years near Stockholm in historic times. Beach deposits in the central parts are now 275 m above the present level of the sea, and it has been calculated that there are another 210 m to go in this region where uplift rates are now the greatest: much of the present Baltic Sea floor will emerge.

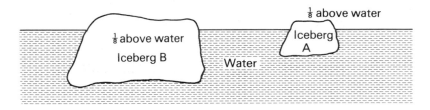

Figure 4.19 Isostasy (1) The state of balance existing between the ice and water. How is this related to their respective densities? What would happen if a portion of iceberg A was removed to iceberg B?

The principle of isostasy would suggest that the continents, ocean basins, mountain ranges and ice sheets must all be in, or tending towards, a state of equilibrium with the underlying mantle rocks. Thus the erosion of a mountain range and deposition of the transported material on the continental margin will result in compensating changes and the subterranean movement or rock material (Figure 4.20).

The Earth's origin and early development

The origin of the Earth is bound up with the Solar System as a whole, about which little can be known with certainty. All ideas concerning the origin of the Solar System (Figure 4.21) have their problems, and new discoveries often add different slants to the theories. Modern theories are those which envisage the planets originating in some way from a cloud of dust and gas.

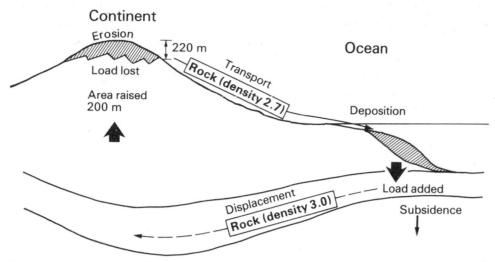

Figure 4.20 Isostasy (2) Movements of mass which take place as a mountain is eroded. The actual decrease of height in the mountains is only 20 m, although 220 m of rock have been removed. Relate the subterranean movement of material to the idea of plate tectonics.

Dust cloud to planets

In whatever way the cloud of dust and gas was acquired by the Sun it seems that it would have settled gradually into a thin, flat disc, rather like the one around Saturn, but with a radius equivalent to the orbit of Pluto or Neptune. Tensions would have occurred within this disc

Basis of theory	Theory title	Outline of theory	Comments
Planetary materials from sun or second star: molten masses cooled to form planets	Tidal theory	Another star neared Sun, leading to a massive tidal wave and a string of spray (0.01% Sun volume) pulled out and left in solar orbit, condensing to planets.	Tidal forces not sufficient to overcome solar gravity.
	Collision theory	Second star grazed Sun's surface, causing material to break away.	Orbits of planets not more than 1 solar radius.
	Double star theories	(a) Sun one of pair of stars, separated by radius of solar system; third star approached second, releasing material into Sun's orbit and drawing away second star. (b) Companion of Sun a double star, made unstable in cloud of gas and dust; broke up, with some caught in Sun's orbit. (c) Companion star denser than Sun: companion star exploded.	It is not known whether solar gases would hold together at high temperatures if released to space.
Planetary materials from cloud of dust and gas: planets originally cold	Sun passed through dust cloud	The Sun acquired planetary materials from a cloud of dust and gas through which it passed.	
	Sun originated from cloud as well	Mass of galactic cloud collapsed gravitationally, leading to rapid rotation and contraction; internal energy generated after nuclear fusion within sun. Planets then accreted.	This type of theory leads to problems of explaining the origins of double stars.

Figure 4.21 Some theories of the origin of the Solar System summarised.

and locally gravitational pressures within dust groups could overcome solar gravitation and lead to accretion. Numerous aggregates were formed, with larger ones growing more rapidly in size. The inner planets, nearer the Sun, could not retain any hydrogen atoms and so are formed largely of the relatively dense materials. The larger planets, like Jupiter and Saturn, were sufficiently large, far away from the Sun and cold in their outer parts, to retain the lighter gases. They have thus assumed larger sizes, and may have had a position where the lens-shaped dust–gas disc was thickest.

Many problems remain concerning the orbits and rotational times of the planets, and also the presence of satellites (Figure 4.22). All of these characteristics are the results of events in the distant past which cannot be investigated now, and many of the questions arising may never be answered. The origin of the Solar System continues to be a problem and even the

Planet	Mean distance from Sun (Astronomical Units: Sun to Earth = 1)	Approx. diameter (Earth = 1)	Mass (Earth = 1)	Average density (10^3 kg/m³)	Number of known satellites	Orbital period (Years)	Inclination of orbit to ecliptic (Degrees)	Inclination of equator to orbit (Degrees)	Rotation rate
MERCURY	0.38	0.4	0.05	5.1	0	0.24	7.0	?	59 days
VENUS	0.72	1.0	0.9	5.3	0	0.61	3.4	23	243 days
EARTH	1.00	1.0	1.0	5.52	1	1.00	0	23	1 day
MARS	1.52	0.5	0.11	3.94	2	1.88	1.9	24	24.5 hours
JUPITER	5.20	11.0	318.0	1.33	12	11.86	1.3	3	10 hours
SATURN	9.54	9.54	95.0	0.69	10	29.46	2.5	27	10 hours
URANUS	19.18	4.0	15.0	1.56	5	84.01	0.8	98	11 hours
NEPTUNE	30.06	4.0	17.0	2.27	2	164.79	1.8	29	16 hours
PLUTO	30.44	?0.5	?0.1	?	0	247.69	17.2	?	6 days

Figure 4.22 Planets of the Solar System: a comparison of selected characteristics. Compare the distances from the Sun, masses and rotation rates, in terms of the possible effects on climatic conditions on each of the planets.

most recent theories contain many points in need of verification. It is important, however, to note the change from theories involving very hot original planets shot out from the Sun, to those envisaging cooled material from other stars, or interstellar clouds, accumulating into small planets, which are solid near the Sun and more gaseous in the outer regions.

The Earth thus began as one of a series of accretions of dust and gas revolving around the Sun, which could have been a fully established star, or one which was only just beginning to warm up and radiate energy. The energy radiated by the Sun covers a wide range within the electromagnetic spectrum, including heat, light and gamma-rays, and it would have had the effect of sweeping away much of the gas still existing between the protoplanets.

The atmosphere

Planets of sufficient mass, like the Earth, were able to retain a significant envelope of gases because of gravitational attraction. This is in contrast to smaller satellites, like the Moon. At first this atmosphere would have been composed of gases like ammonia (NH_3), water vapour (H_2O) and methane (CH_4), resulting largely from volcanic activity and the original gases present in the dust–gas disc. Reaction with sunlight would cause these to break up, the lighter hydrogen atoms being subsequently lost to space. The condensation of water vapour to form the oceans took place as the atmospheric circulation was initiated. It is thought likely that the atmosphere did not become oxygenated until about 3000 million years ago, when photo-synthesising plants began to supply the oxygen, and the proportion then rose until it was possible for plants to survive on dry land some 500 million years ago. For many ages the atmosphere would have militated against the forms of life depending on oxygen, and even when they developed it was some time before they could stand the full glare of sunlight on land (Figure 3.12). The increasing quantities of oxygen enabled the atmosphere eventually to filter out the short-wave fractions of solar radiation, which are dangerous to living matter.

The solid Earth

The solid part of the Earth probably began as a contracting mass of cold dust. At a certain point the release of energy by radioactive minerals, combined with heat generated by the gravitational compaction, caused internal melting. The temperature may have risen to a few thousand degrees Celsius over a period of approximately 10^9 years. Melting occurred first at the centre, and the planet might have expanded slightly at this stage. The melting process would have been rapid at first, and if only 6 per cent of the total mass was affected it would cause a collapse, decreasing the Earth's radius by some 350 km. After this it is thought that stability would be established: the liquid core would increase gradually as the temperature rose, and the Earth might continue to contract slowly. The internal melting may also have allowed gravity settling of the densest material to form the core of the Earth, or there may have been internal changes in the phase of the materials under conditions of increasing pressure. The surface of the globe may never have been molten. After a time the release of energy began to slow down, in accordance with the process of radioactive energy release, leading to cooling and solidification of the mantle. Today only the radioactive elements with the longest half-lives are still releasing energy. The core probably remains, at least partly, molten, as witnessed by its effect on earthquake waves and by its possible connection with the Earth's magnetic field.

The evolution of the Earth's interior is thus in many ways as much a closed book as the origin of the Solar System. The statement of Eddington concerning the problems of astronomy can be applied to the Earth's interior: 'Practically the whole of the universe is tucked away and forever hidden from us behind impenetrable barriers.' Man's direct observations are thus confined to a narrow zone at the planet's surface. Even here the immense periods of geological time limit his conclusions concerning the changes which have taken place, but the picture of an evolving surface is becoming clearer.

Continental evolution

The earliest crustal rocks were probably like those which now compose the ocean-floors: dense basalt rocks erupted from the upper mantle. Perhaps there was a complete covering of ocean waters at an early stage. Two events were necessary before the continental masses could be formed. Rocks had to be brought into contact with the atmosphere, and then the reactive agents in the atmosphere had to work on them to decompose and erode those which appeared above the oceans. The first continents may have arisen from the accumulation of basaltic rocks around island arcs, which became subject to the atmospheric processes as soon as they rose above sea-level. There is little evidence for this, however, even in the oldest known continental rocks.

The development of the surface on the large scale, as well as on the smaller scale, has always depended on the complex interaction of internal forces, atmospheric processes, rocks, ocean waters and living creatures. Studies of Moon rocks from the restricted number of locations visited by man have suggested additional clues which tend to confirm such a view of early Earth development. The Moon relief is dominated by two types of features: the heavily cratered 'highlands' and the flatter 'mare'. One view of the Moon's history is that the highlands represent an old, original surface of volcanic peaks, greatly modified by meteoritic impacts, and that the mare are composed of lava poured out at a later stage (p. 324). Visits to both types of landscape by the Apollo astronauts have tended to confirm that this is the case: the rocks of the highlands have radiometric dates of over 4000 million years (older than any known rocks on the Earth), while those in the mare have dates as young as 2700 million years. Most Moon rocks resemble the Earth basalt lavas, though some of the constituents are in different proportions. The fact that the Moon has no atmosphere to react with these surface rocks, and that any volcanic activity has largely died out, means that the Moon now preserves a surface which may have been like an early stage in the development of the Earth.

Part II

Surface Processes

Introduction

The surface of the Earth is a zone where the rocks uplifted by internal Earth processes come into contact with the atmosphere and ocean waters. They then become subject to a range of processes powered by energy from the Sun. The greatest surface effects are caused by the movements of water through the hydrological cycles. The water causes chemical decay as it comes into contact with rocks, whilst surface streams and glacier ice carve into the underlying rocks. In addition there are changes of temperature and wind action, and it is becoming more widely recognised that organisms — particularly man — affect the rates at which these processes can act.

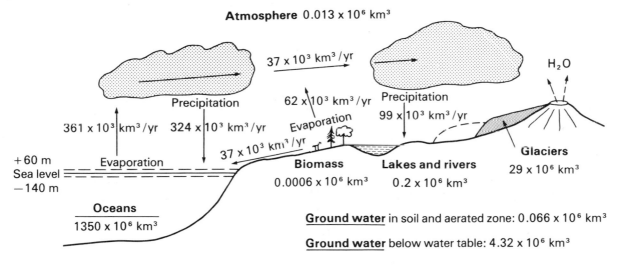

The hydrological cycle circulates water from the oceans to the land. A small proportion of the ocean store is circulated each year. Extension of glaciers has led to increasing storage on land and falling ocean levels; melting ice has the opposite effect. (After Bloom 1969)

The surface environment is very different from the one in which many rock-forming minerals originated. Those which solidified from the molten state at several hundred degrees Celsius in conditions of high confining pressure beneath the surface are exposed at the surface to low temperature and pressure conditions. Such a change in physical environment encourages changes to take place in the chemistry of the rock materials.

The surface processes act on the landforms to break up the rocks (weathering), to wear down the surface and carve out valley features (erosion), and to transport the debris produced to lower areas where deposition takes place. The whole landscape is lowered by the combination of these effects (denudation). Some of the processes, such as weathering and mass movement, affect the complete landscape, whilst others act only along a narrow zone at any one moment. Thus rivers and glaciers are confined to valley bottoms, whilst the wind and the sea act within a narrow vertical zone above the ground surface or at the tide level respectively.

Particular processes may be more important in different climatic realms, a fact which is related to the way in which water is transported to the continental areas from the oceans. In high latitudes, or at high altitudes, the water freezes and may be stored in ice masses on the surface. These may melt in warmer seasons, or when the ice moves down to lower altitudes, so that running water and ice may be important in the same area. In very dry regions there is so little rain that wind action becomes an important process, especially in moving the dry, broken rock material. Even in these areas, however, there is much evidence that running water is, or has been, important in moulding the landscape. It becomes clear from the study of cold and dry regions of the world that the relationship of process and climatic type is not simple. In other parts of the world precipitation is sufficient, and the temperatures high enough, to support permanent streams of running water. The theme which examines landforms in the light of present and past climatic zones is one which will be raised again in PART III.

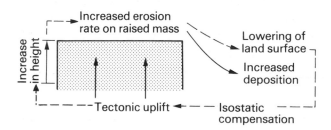

A negative feedback loop relating uplift, erosion, sedimentation and surface height of the land.

The immense variety of processes and landforms present at the Earth's surface can be seen within the context of a system in which rates of uplift and denudation are balanced against one another. This is a negative feedback system — i.e. where one process acting enables another to come into force, and this in turn acts to reduce the subsequent effect of the first.

The processes which affect the whole landscape directly — weathering and mass movement — are examined in chapters 5 and 6. These are followed by the study of processes which are confined largely to valley lines — running water and flowing ice — in chapters 7 and 8. It may be noted that ice sheets may cover and bury the entire landscape, but activity beneath them is often concentrated within narrow belts. Wind and the sea have an extensive horizontal, but limited vertical, zone of influence in eroding rocks and are the subjects of chapters 9 and 10. Finally there is the consideration, in chapter 11, of the effects of organisms and particularly of the increasing effect of man on the physical and chemical processes which have for long been considered the main processes affecting landforms. This may be the most important chapter in this part of the book.

5

Weathering and the regolith

Introduction

Weathering is the breakdown of rocks at the Earth's surface in contact with the atmosphere. It includes physical disintegration of rock masses into smaller particles, and chemical alteration so that particular minerals may be broken down into dissolved ions or clay minerals. It does not involve transport of the degraded rock materials, and the sand, clay and rock fragment particles accumulate on the spot to form a waste mantle, or regolith. If the regolith is not moved away rapidly by other processes, it will become the home for organisms and will develop into a true soil, unless the climate is too cold or too arid. The production of soil can be seen as a positive feedback process in the weathering system, since it is produced by weathering processes and organisms, and enhances the activity of weathering: rocks weather more rapidly beneath a true soil until it exceeds the thickness to which the agencies of alteration extend.

Weathering lowers the surface of the rock being attacked and produces broken or disintegrated rock material which can be transported more easily by running water, ice or wind. Some indication of the rate of weathering of bare rock faces can be gained from a study of unworked quarry walls: a pile of debris may accumulate after a series of frosty nights, or through one winter. Bare rock is also exposed to the atmosphere when used as a building stone, and it has been estimated that weathering has removed an average of 1 cm of rock from the surface of St. Paul's Cathedral in London since its construction 260 years ago. The zone of interaction between rock and atmosphere is thus one of considerable activity. This activity, however, is not easy to analyse, since a complex group of processes act on a variety of rock-types with different results in different parts of the world.

Weathering processes have been divided traditionally into those which cause the break-up of rocks by mechanical means, and those which involve chemical reactions (Figure 5.1 A,B). It is becoming increasingly clear, however, that plants, animals and bacteria are closely involved with many of the processes, and may be the controlling influences (Figure 5.1 C). In any case the individual processes cannot be separated in nature, since a complex variety may be involved in a particular case, depending on the local climate and rock composition. Chemical decay leads to mechanical disintegration; mechanical fracturing and organic activity break up rock masses and allow deeper penetration of chemical reactions; the accumulation of regolith and the growth of vegetation protect a rock from physical processes, but accentuate the chemical effects.

The zone affected by weathering below the surface also varies from place to place. The depth of regolith (i.e. the rotted rock, plus any residual or transported material which may overlie the solid rock) will depend on the nature of the rock (i.e. porous, permeable and chemically active materials will weather more deeply), on the intensity of atmospheric processes (tropical areas have deeper weathering profiles than temperate, and those of polar regions are shallower still), and on the rates of removal (areas of high relief lose weathering products rapidly and deep regolith cannot develop). Depths up to 100 m for the weathered zone have been recorded in parts of Nigeria and Czechoslovakia, whilst granite in New South Wales, Australia, is weathered to 300 m. Borings for a hydro-electricity project in Victoria, Australia, relate regolith depths to rock-type: mylonite 130 m; granodiorite 175 m; schist

A Physical or mechanical weathering

Process	Effect
Sheeting, unloading, spalling	Release of pressure by removal of overlying rocks. Spalling forms irregular shapes.
Crystal growth	Includes salt and frost weathering: expansion of crystals exerts pressure. Also volume changes during chemical alteration.
Insolation	Temperature changes at surface: cleavage, cracking.
Fire	Forest fires leading to cracking as above.
Moisture swelling	Changes in volume.
Wetting and drying	For instance, at water level on shore platforms.
Cavitation	Bubbles in turbulent water leading to collapse.
Abrasion	Friction of boulders or impact.
Mechanical collapse	Following undercutting.
Colloid plucking	Clay film dries out, plucking out grains.
Soil ripening	Evolution of fresh alluvium: soil colloids lose water and volume, becoming sticky and then solid.

B Chemical weathering

Process	Effect
Solution	The first stage of reaction of water with rocks: amount of change determined by solubility of material and amount of water passing. Solutions may also lead to precipitation when saturated. Solute concentration in stream reflects rocks in basin and rates of percolation.
Oxidation/reduction	Adding or removing oxygen. Common form of natural weathering via oxygen dissolved in water. Oxidation mostly in aerated zone above water table; also by bacterial action. Reduction in waterlogged sites, where red/yellow oxides change to green/grey forms.
Carbonation	Reaction of carbonate, bicarbonate ions with minerals: often a step in the weathering process, including breakdown of feldspars. Abundance of CO_2 in soils aids process.
Hydration	Addition of water to mineral: important in forming clay minerals; often with large volume change. Prepares mineral surfaces for carbonation/oxidation.
Chelation	Chelating agents extract metal ions from organic chemical structure: assist leaching from humus.
Hydrolysis	Chemical reaction between mineral and water, related to hydrogen ion concentration of water: increasing concentration makes silica dissolve.

C Biotic weathering

Process	Effect
Particles broken	Animal burrowing; growing roots exerting pressure.
Transfer, mixing	Animal movement: materials moved to areas of different process.
Simple chemical effects	Solution enhanced by respired carbon dioxide.
Complex chemical effects	For instance, chelation.
Soil moisture effects	Roots, humus holding moisture; shading by plants.
Ground temperature	Shade; fermentation increasing heat and activity rate.
pH effect	Respiration and absorption by plants affecting pH.
Erosional protection	Less exposure, lowered weathering rates.

Figure 5.1 Weathering processes: a summary of the main types.
(A) Physical or mechanical weathering,
(B) Chemical weathering
(C) Biotic weathering
Notice the importance of water in all groups.

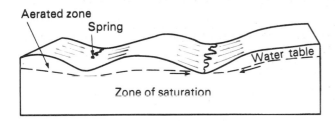

Figure 5.2 The water table is the upper limit of water saturation in a rock; below this level there will be slow movement of water between pore spaces, and the rock will become cemented as chemicals in solution are precipitated. The water table may fluctuate seasonally with the input of water, and above it there will be movement of water—downwards after rain, but often upwards in a dry spell.

200 m; copper lode 330 m; and gneiss 380 m. At the other end of the scale, weathering may affect only a few centimetres of compact, exposed rock. The processes involved in the weathering of rock are concentrated at the surface, but may be permitted to penetrate through porous or jointed rock, and will be carried out even beneath the level at which the rock becomes saturated with water (i.e. the water table, Figure 5.2). The existence of these localised lines of penetration often results in a highly irregular base to the weathered zone.

The geological variables

Earth materials subjected to weathering processes can be divided into solid rocks and unconsolidated sediments (i.e. river detritus, beach and dune sands, glacial debris, loess). Each of these has a particular character, but all are composed largely of a restricted variety of minerals with differing compositions and reactions to weathering (Figure 5.3). Minerals like quartz are broken down slowly in a temperate climate, whilst olivine, micas and feldspars are affected by solution and chemical alteration more rapidly. In short, those minerals which were formed at high temperatures (i.e. in contrast to those at the Earth's surface) are those which are affected most rapidly by chemical processes and altered to minerals which are more stable at the lower temperatures characteristic of the land surface.

Mineral	Occurence in rocks	Weathering characteristics
Quartz	Sandstones; granites; some metamorphics.	Last to form in cooling granite: stable at surface temperatures and resistant to chemical weathering. No cleavage or fracture and thus resistant to physical weathering, even attrition. Quartz is the standard against which others are measured.
Feldspar group Alumino- silicates	Igneous and metamorphic rocks; arkose sediments	Almost as hard as quartz, but well-cleaved, allowing rapid break-up and water penetration. Calcium forms weather more rapidly than sodium and potassium forms. Forms clay (often kaolin) and sand.
Mica group Aluminium silicates with sheet lattice	Abundant in granites and metamorphic rocks; some in sediments	Flaky cleavage: easily weathered to chlorite and clay minerals.
Olivine Magnesium-iron silicate	Basalt and basic igneous rocks	Uncleaved, but irregular fractures: weathers easily, being formed at high temperatures — alteration begins during the process of crystallisation. Forms clays.
Pyroxene e.g. augite	Basic igneous rocks	Similar to olivine in many ways, but with good cleavage to aid break-up. Clays and solutes formed.
Amphibole e.g. hornblende	Igneous and metamorphic rocks	Slightly slower weathering than pyroxene and olivine, but very susceptible with good cleavage.
Carbonates	Mainly sediments	The most soluble materials.

Figure 5.3 Rock-forming minerals and weathering. Minerals react in varying ways to weathering processes due to a combination of physical hardness, chemical composition and the presence or absence of internal lines along which water may enter.

In both rocks and unconsolidated sediments, however, other factors emerge which have an important bearing on the ways in which weathering occurs. The factors which influence the passage of water and air into and through these materials are particularly important. A rock which is porous (i.e. with a high proportion of empty space in comparison with that occupied by mineral matter) will also be permeable (i.e. will allow water to pass through it) if the pore spaces are large enough. Permeability arises in other ways, too, including the presence of fissures in the rock formed by bedding planes, joints or fault zones. Thus a combination of chemistry (as determined by the mineral composition) and the physical structure of the rock, will determine whether surface materials are going to be rapidly or deeply weathered (Figure 5.4). Rock 'hardness' thus depends on a variety of characteristics and must not be confused with mineral hardness, which is simply the resistance of the mineral surface to scratching. Rocks which are resistant to weathering contain high proportions of minerals which are chemically unreactive in the surface environment, and do not allow water to penetrate deeply. A quartzite (i.e. a rock formed of quartz sand grains cemented together by silica) is one of the most resistant rocks in this sense under any climate. Chemical weakness also occurs in coarse-grained rocks where one mineral is susceptible to chemical action: its removal or decay weakens the remaining rock structure, causing it to crumble. Thus, although granite contains some quartz, the feldspars are more important in terms of volume occupied and are susceptible to chemical reaction, making the rock as a whole liable to crumbling.

Rock-type	Composition	Forms and lines of weakness
Unconsolidated sediments	Loose accumulations of sand, gravel and silt in alluvium, dunes, scree, moraines.	Variable sorting and mineral composition, but with few definite internal structures.
Sedimentary rocks	Compacted and indurated sediments formed of clay minerals (shales), sands, calcium carbonate, pebbles and with mixtures. These materials have been through weathering processes once.	Bedding and jointing formed in deposition and emphasised during earth movements. Cross-bedding forms additional lines of weakness in some sedimentary rocks.
Igneous rocks	Formed of a range of minerals including those highly susceptible to weathering. Extrusive lavas and intrusive bodies.	Lavas and ashes layered. Lavas and intrusive sheets and massive bodies jointed — both cooling and unloading joints.
Metamorphic rocks	Altered forms of other groups — may contain minerals more or less susceptible to weathering	Often layered due to pressure cleavage in slates, or to heat and pressure foliation (schistosity).

Figure 5.4 Rock characteristics and weathering. The resistance of a rock to weathering processes is determined by its mineral composition and the lines of weakness which allow the penetration of water.

The rocks forming the highest land are not always those which are the most resistant. This may be the case in areas of scarpland topography, where clays and shales are weathered and eroded more rapidly than the sandstones and limestones which give rise to ridges (Figure 3.14). In general, however, the highest points are the result of tectonic processes, causing uplift due to internal Earth movements. Thus the highest parts of Britain occur in the north and west (Wales, Lake District, Scotland), but this is due less to the fact that ancient, highly-compacted and often resistant rocks occur there, than to the block-like uplift of these areas some 20-30 million years ago. Uplift and the fissuring which results from tectonic processes and the folding or rocks often produce more rapid rates of weathering and erosion. Thus weathering may take place more rapidly on resistant rocks in upland areas, than on softer rocks in the lowlands. Regolith accumulates more easily on the low-angle slopes of lowland regions. In mountainous areas the broken rock material is removed rapidly, exposing further rock to weathering.

At times the geological variables become part of the processes, rather than merely the passive material acted on by them. This is particularly true where a mass of rock is exposed at the surface following the erosion of overlying rocks: the release of pressure due to the removal of the weight of these rocks may lead to expansion of the exposed rock and the formation of joints parallel to the surface. The sheets of rock produced in this way are easily prised away from the main body of rock: this can be seen on Dartmoor, in the granite peaks of Arran, and around many hills of exposed rock in savanna areas.

The climatic variables

Whilst not all weathering processes are related to atmospheric events, the most important factors are the supply of water and the temperature régime. Both of these are determined largely by the state of the atmosphere, and interact with the geological variables.

The importance of water

Water is important in both mechanical and chemical weathering processes, and in supporting the living organisms which are bound up with these (Figure 5.1). Changes in volume caused by the addition or removal of water impose mechanical stresses on a rock and may cause it to split apart or crumble. Most of the chemical changes also require the presence of water: solution, carbonation, hydration and hydrolysis are obvious examples, but even oxidation occurs mainly by reason of the oxygen dissolved in water. Many of the reactions are controlled by the hydrogen ion (H^+) concentration in solutions. This is expressed as the pH value (i.e. the log of hydrogen ion concentration), which ranges from 1 ('acid') to 10 ('alkaline'). Typical values found in particular environments include 3-4 for mine water, 4-5 for peat, 5-6 for acid soils, 7 for river water and rainwater, 8-9 for seawater, and 9-10 for alkaline soils. The pH value affects the solubility of substances like iron, which becomes 100 000 times more soluble at pH 6 than at pH 8.5; iron in solution in river water will be precipitated when it reaches the sea.

Water on the land comes from precipitation, but it seems that even the small quantity available in deserts is effective in rock destruction, combining with the physical work of temperature changes. A seasonal distribution of rainfall is clearly important, since downward movement of dissolved material may alternate with upward movement in dry seasons, and alkaline soils will form if evaporation exceeds precipitation.

Temperatures

Temperature conditions are also important in all aspects of weathering processes. A considerable range of temperature (i.e. 40-50° C, diurnal or annual) subjects the surface layers of exposed rocks to expansion and contraction. Temperature changes at ground level are up to 50 per cent greater than those measured by standard meteorological instruments in a Stevenson screen (a shaded position 1.25 m above the ground). Perhaps the most important temperature changes are those which give rise to a continual series of freeze-thaw changes, leading to the freezing and expansion of water in rock or soil, and thus to mechanical destruction.

The rate of chemical reaction (and of biological activity) also tends to rise with increasing temperature. Organic matter is removed more and more rapidly as the temperature increases, and it seems that silica becomes more soluble. On the other hand carbon dioxide is more soluble at low temperatures, and a high concentration of carbonic acid aids carbonation — which becomes more intense in colder areas.

In hilly areas Sun-facing slopes may have significantly different conditions from slopes facing away from the Sun. In the northern hemisphere temperate south-facing slopes may become semi-arid because of high insolation levels and rapid run-off; in cooler areas the alternation of freezing and thawing may be more intense on slopes facing the Sun.

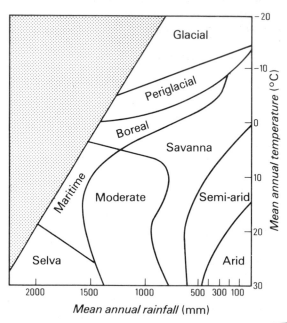

Figure 5.5 Climatic boundaries of regions giving rise to distinctive groups of landforms. Conditions of temperature and rainfall affect weathering processes: chemical weathering increases in importance from the top right to the bottom left of this diagram, whilst frost weathering increases towards the top centre. Places in the centre thus receive moderate chemical weathering and moderate frost action. (After Peltier, 1950, in Ollier, 1969)

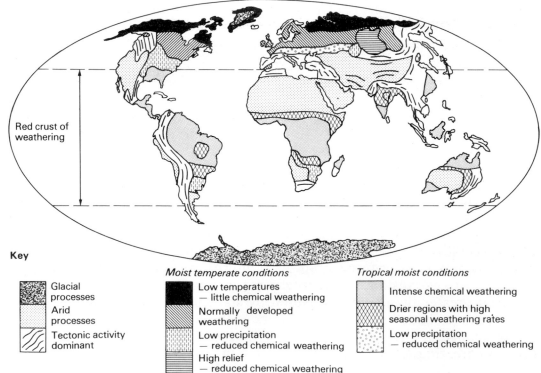

Figure 5.6 The world distribution of weathering types. Use this map as a basis for later comparisons between different groups of landforms associated with the different climatic regions. Arid, glacial and high mountain regions are distinctive, but is it easy to distinguish within the moist regions? (After Strakhov, 1967, in Olier, 1969)

The weathering of buildings

When stone is used in buildings it is exposed to atmospheric processes, and the interaction may cause it to crumble or be dissolved. This is particularly the case where acidic pollutants are more common, and it is important for the owners of the building to know how long the stone will last — as well as the possibility of it providing a pleasing appearance. Perhaps this is even more important today than formerly, since many buildings are now merely faced with stone and costs of replacement are so much higher.

Many building stones are limestones, since they can be quarried easily in massive, load-bearing blocks, but these have proved to be susceptible to weathering. The Houses of Parliament in London are built of Magnesian Limestone from Durham and suffer such effects: one memorable quotation from a letter to 'The Observer' by Professor A.B. Pite in 1925 stated that the building 'has been slowly and surely transformed into some sort of heap of Epsom Salts'. The Portland Stone of St. Paul's Cathedral in London is another limestone which has been reduced by an average of 1 cm in 250 years, although this has proved to be one of the most resistant stones in London's atmosphere. The limestone composing an old fortress in the Ukraine accumulated 30 cm of soil on it due to weathering — a depth greater than the soil in surrounding fields. The Great Pyramid of Giza, Egypt, is faced with limestone blocks, but many of these facings were removed 1000 years ago to expose underlying blocks of varying composition: hard and dense limestones have been little affected by the subsequent weathering, but softer limestones have pits 1–2 cm deep, a shaly limestone has niches and clefts up to 20 cm deep and a friable sandstone is reduced to rubble. An estimate of 50 000 m³ debris had accumulated around the base of the pyramid — i.e. a rate of 50 m³ per year, or a loss of 0.2 mm over the whole surface per year. Measurements on ancient tombstones formed of limestone show that a reduction of 1 cm was accomplished in 100–200 years in northern England.

Other rock-types are also susceptible, though less so unless they are poorly cemented. This is true of sandstones, which may show few signs of weathering if well-cemented by quartz, but which may crumble rapidly if not. Granite is another popular building stone and normally lasts longer than most. Thus granite facings at Giza are fresh on buildings 4000 years old and only slightly scored on those 1500 years older — although, significantly, there is a greater degree of alteration just below the ground surface where water is most commonly present.

All rocks are eventually subject to these changes, but some react more slowly than others. This type of information has been used to calculate weathering rates, but there is an important difference between weathering on rocks exposed at the surface and the changes proceeding beneath a cover of soil and vegetation: the latter often promotes more rapid disintegration of rocks.

Weathering environments

Because of the important roles of water and temperature changes in weathering processes it has been suggested that they can be related to climatic zones. Distributions of particular processes have been compared (Figure 5.5) and even mapped (Figure 5.6). A general account can be given in summary form (Figure 5.7), but it must be made clear that many of the statements still require quantitative verification. In addition, it is not always certain which of the features in an area are related to the present climate and which to a past phase of different climate. Some have attributed much of the formation of the tor features on Dartmoor to a past tropical climate, whilst others have looked to tundra conditions to explain them!

Weathering products and landforms

Weathering results in the accumulation of broken rock debris — the regolith (Figure 5.8). In general this is modified to a soil, or, transported away from the site of the weathering, often in solution but also in solid form. At times, however, a residual deposit may build up: this may be of boulders or scree, but may also include economically valuable mineral concentrations.

Weathering environment	Characteristics processes
Humid tropical	Chemical activity dominant in high temperature and high rainfall. Thick regolith and rapidly decaying vegetation releasing acids. Even silica may be dissolved under these conditions, leaving bauxite/laterite. Rivers carry more than ten times the solute load of temperate rivers.
Seasonal tropics with a marked dry season	The wet season has a marked leaching of salts brought near the surface in the dry season by high evaporation rates. Laterite forms commonly, impeding drainage. Some insolation weathering.
Arid	Capillary action leads to the surface concentration of salts forming desert varnish, calcrete. Water assists simple chemical processes, but little clay or solutes are produced.
Humid temperate	Water available for medium rates of chemical reactions. Low evaporation rates produce little capillary action. Freeze-thaw important in winter.
Periglacial areas	Freezing through long winter; freeze-thaw in spring and autumn; water available for chemical activity in summer.
Arctic	Snow-ice mantle. Nunataks frost-riven. Carbonation beneath ice when it melts.

Figure 5.7 Weathering environments. Some general statements concerning the emphases of weathering processes in different world climatic regions.

Weathering also results in landforms which are etched into the solid rock, and so may lead to features which show that the land surface has expanded and risen, instead of being worn away.

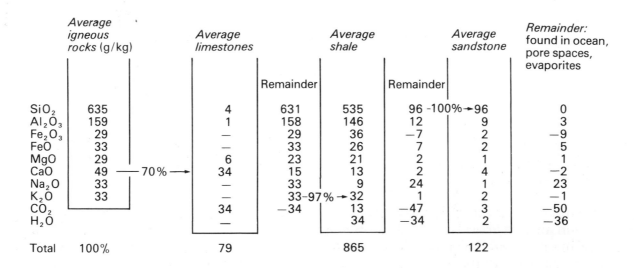

Figure 5.8 The conversion of igneous rock materials to the main varieties of sedimentary rocks. The main anomalies in the right-hand column involve Na_2O, CO_2 and H_2O. The last two come from seawater, CO_2 being required to convert MgO and CaO to carbonate minerals. HCl, is required to convert the Na_2O into NaCl which is so important a constituent of seawater. The deficiency of Fe_2O_3 is almost balanced by the FeO, reflecting the fact that the average sediment is more oxidised than the igneous rock. Apart from these cases, the igneous rocks can be seen to be the source rocks for the sediments. Summarise this information, and the processes involved, in a flow diagram. (After Garrels and Mackenzie, 1971)

Residual deposits

Residual deposits include the clay regolith of limestone areas, often known as 'terra rossa', and the disintegration sands overlying granites (Figure 5.9). At times large, angular boulders may be formed by frost action, and scree may accumulate at the foot of an actively weathered slope. Many such deposits may be due to former climatic conditions (e.g. scree may be the result of past tundra, or periglacial, conditions in areas which now have a temperate climate).

The weathering processes remove some minerals from the regolith, leaving it enriched in others which remain, so that the latter become an economically viable ore deposit. Bauxite, the main source of aluminium, is formed in tropical areas where heavy rains leach out other

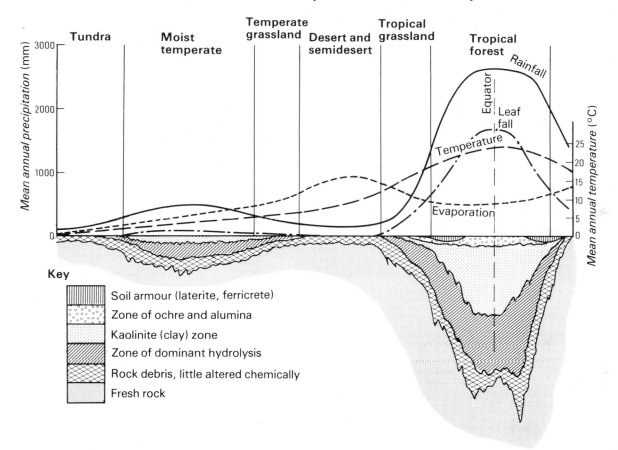

Figure 5.9 Zones of weathering and climate. This diagram emphasises the varying depths of the regolith in relation to climatic conditions. Compare the types of residual deposits in temperate and tropical areas. The original author gives no indication of the subsurface scale: the depths are thus a relative indication of differences. (After Strakhov, 1967, in Ollier, 1969)

minerals: igneous, metamorphic and sedimentary rocks including limestones and clays may be the parent rock in areas of low relief. Much of the world's bauxite seems to have been formed during the early Tertiary period (i.e. 65–30 million years ago). Half of the world's supply is associated with limestone rocks. Some iron deposits have been formed in a similar way, resulting in laterite concretions: the Cuban ores are found on serpentine plateaus, and other deposits of this type occur around the Caribbean Sea. Nickel, manganese and cobalt ores may also accumulate in this fashion.

Differential weathering

Weathering processes may etch the surface of the exposed rock unevenly due to differences in rock characteristic, local cementation, or the distribution of plant cover. Small weathering pits up to a few metres across and deep are produced by local flaking of the rock or breaching of a hardened surface. Other small-scale features such as ridges, trenches and cavernous weathering on steep slopes are produced in similar ways.

Exfoliation

Exfoliation is a term used for a variety of weathering effects at different scales. These have varied origins, and it would be better to adopt a more precise terminology for this group of features.

1) On the largest scale unloading of a section of the Earth's crust by erosion of the overlying rocks will lead to the formation of thick, curved slabs of rock up to several metres thick. This is common in granites, but occurs in other rock-types as well. The partings between slabs are often parallel to the ground surface and become closer-spaced towards the surface. They may be related closely to valley forms, indicating their very recent origin. Dome-shaped hills up to a few hundred metres high may result from the process as the rock sheets break up.

2) Surface flaking of rocks, particularly in deserts, has been attributed to expansion and contraction following day/night contrasts in heating by the Sun. Certainly boulders heated by, for instance, a forest fire, will crack and lose a surface flake, but laboratory experiments have shown that a small quantity of water and some degree of chemical activity are necessary

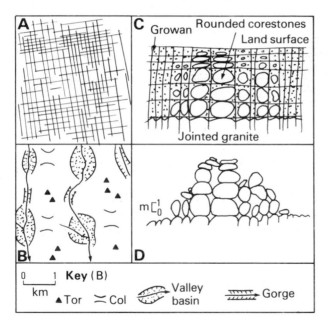

Figure 5.10 The formation of tors (1)
(A) A joint pattern map of a granite area. (B) Landforms of a granite area, related to (A). (C) Deep weathering of granite in a tropical climate produces growan and corestones; depth of weathering is controlled by the depth of the water table in this theory. (D) Subsequent removal of growan leaves the corestones as a tor feature. (After Linton, 1955, in Small, 1970)

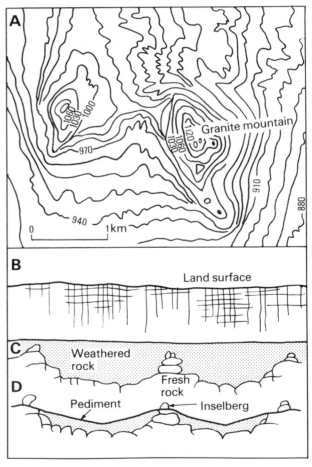

Figure 5.11 Inselbergs and pediments.
(A) A map of a pediment and inselberg landscape in Arizona. The contour interval is 10 m
(B) The suggested evolution of inselbergs and pediments in Uganda (B)–(D). Compare this sequence with the tor evolution as shown in Figure 5.10, bearing in mind differences of scale. The section in (B) shows subsurface jointing.
(C) Weathering of closely-jointed rock forms zones of broken debris.
(D) Removal of weathered rock material leaves pediment surfaces and inselberg remnants—the latter lowered by further weathering. ((B)–(D) after Ollier, 1960, in Ollier, 1969)

before rock flaking will occur. Flakes of a few millimetres to a few centimetres thick are produced by the chemical decay of minerals, particularly by hydration, but may also be formed by the growth of salt crystals in a layer beneath the surface.

3) Spheroidal weathering contrasts with the other forms of 'exfoliation' in that it does not involve expansion of the rock surface. Chemical migration of elements within the rock takes place, resulting in re-deposition in a series of rings marked by alternate chemical depletion and enrichment. The concentration of this effect at the corners of a boulder leads to the rounding of the layers within the block, and affects the way in which the boulder breaks up.

Tors

Many hills, slopes or spurs in Britain are topped by groups of boulders between 10 and 30 m high. These are particularly common and well-known on the granite of Dartmoor, but also occur on other granite areas (e.g. Arran, Cairngorms) and on other rocks which resist physical weathering and have wide-spaced jointing (e.g. quartzite in Shropshire; Millstone Grit of the Pennines; and altered dolerite at Cox Tor, a few hundred metres from the Dartmoor granite boundary). Similar piles of boulders occur in Africa (koppies).

One view of the origin of these features suggests that the Dartmoor area was subjected to deep weathering, possibly in a humid tropical environment, followed by the stripping away of the broken, weathered material to leave the pile of 'corestones' or other forms of upstanding rock mass (Figure 5.10). There are thus two stages in the formation of tors according to this view.

A second hypothesis relates tors to inselbergs (larger features rising to over 300 m and occurring in the tropics) in their modes of origin. Tropical areas with a seasonal rainfall, and particularly those in Africa, are commonly dominated by extensive plains which are weathered deeply, and from which rocky domes (inselbergs or bornhardts) rise (Figure 5.11). Deep chemical weathering is followed by the removal of the rock debris to form the plain. If the high level surfaces of Dartmoor are interpreted as having been formed in former tropical conditions, they may be of this type — known as pediplains — and the tors would be residual masses of slowly weathered rock.

A third view suggests a more local origin for the tors of Dartmoor, related to the nature of the granite itself and the most recent history of the area. The granite has often been found to be rotted, but this may be due to rising hot gases (pneumatolysis) rather than to deep chemical

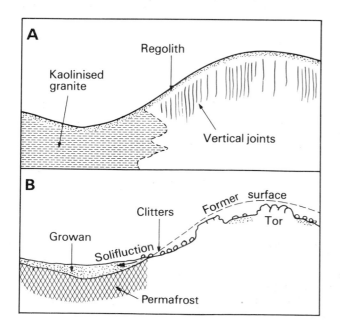

Figure 5.12 Formation of tors (2)
(A) Periglacial conditions lead to fragmentation of granite on the hills and breaking up of the granite rotted by hot gases at the time of formation.
(B) Periglacial solifluction removes regolith from slopes and exposes granite to frost action (especially where jointing is close), forming clitter blocks on slopes. (After Palmer and Neilson, 1962, in Small, 1970)

weathering. An examination of quarries where 'corestones' are still buried in rotted granite, or growan, shows that the granite is seamed with thin veins of tourmaline, a sure sign of pneumatolysis, and this is true of many tors which are characterised by angular blocks composing their superstructure and littering the surrounding slopes. The growan itself is sandy in texture, rather than the clay which tropical weathering would produce. The fact that tors are often surrounded by fields of angular blocks (clitters) strung out into lines down the slope suggests conditions of frost action and soil flowage which were prevalent in tundra environmental conditions, such as must have been experienced during the Quaternary Ice Age (Figure 5.12). Tor features have also been found to occur close to glaciers in Antarctica in sandstones and dolerites.

Tors, however, occur in so many varied circumstances that it is probably futile to expect a single theory to account for the origin of all of them. The nature of the rock — its composition and jointing pattern — is probably the dominant factor which determines the tor feature and its detailed morphology. Tors may develop in a variety of climates, involving a range of different weathering processes.

Coastal weathering

The coastal zone is often characterised by extensive exposures of bare rock. It has been common to treat the features of cliffed coasts as being due to marine erosion at high tide level, followed by the collapse of undercut masses of rock. In places this is supplemented by the action of short, steep streams running down the cliff face, and weathering processes are also important on the exposed rock. The continual wetting and drying of rock surfaces in the intertidal zone and above, where the wave spray reaches, gives rise to water layer weathering. This is now thought to be important in the formation of the shore platform. Wetting and drying leads to the lowering of rocks projecting above the platform and to the extension of rock pools. When the rock is reduced to a level where it is permanently wet, and possibly covered by seaweed, the activity ceases. Other facets of the shore zone are also affected by weathering (Figure 5.13), thus affecting the coastal landforms (chapter 10).

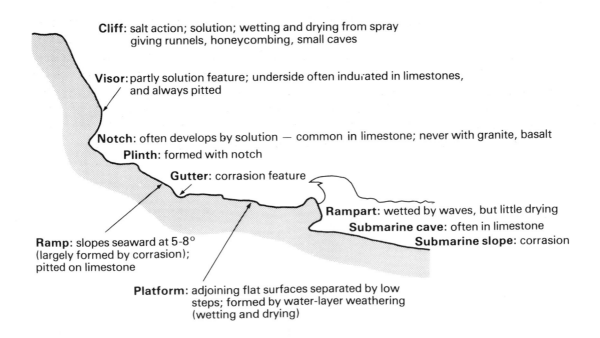

Cliff: salt action; solution; wetting and drying from spray giving runnels, honeycombing, small caves

Visor: partly solution feature; underside often indurated in limestones, and always pitted

Notch: often develops by solution — common in limestone; never with granite, basalt
Plinth: formed with notch
Gutter: corrasion feature
Rampart: wetted by waves, but little drying
Submarine cave: often in limestone
Submarine slope: corrasion
Ramp: slopes seaward at 5-8° (largely formed by corrasion); pitted on limestone
Platform: adjoining flat surfaces separated by low steps; formed by water-layer weathering (wetting and drying)

Figure 5.13 Some aspects of coastal weathering. (After Ollier, 1969)

Weathering in limestone areas

Limestone rocks are so affected by solution that a special name — karst — is given to the landforms which result. Since the weathering effects are closely connected with the effects of running water they are considered together (chapter 7).

Weathering and time

Weathering is the basic process by which atmospheric forces attack rock masses raised to the Earth's surface by internal activity. It is responsible for the break-up of rocks at rates depending on their exposure, composition and structure together with the presence of water and suitable temperature conditions for chemical reactions. Such rates can be calculated from the exposure of bare rock in buildings and gravestones (page 89), and from the way in which soils form on dated lava flows. Soils have begun to form on the ash deposits resulting from the 1883 volcanic explosion at Krakatoa, and measurements taken in 1928 after 45 years showed that a layer 35 mm deep had been affected by a surface loss in silica and enrichment in aluminium and iron oxides. A clay soil 2 m deep has developed in 4000 years on volcanic ash on the island of St. Vincent in the West Indies. It is clear also that weathering slows down as soon as the solid rock face is separated from the atmosphere by a deepening regolith and plant cover.

Rates of weathering are thus slow unless the rock is exposed or continually re-exposed at the surface, or unless it is covered by a soil with a high content of organic matter, but this must be seen in the context of the vast length of geological time. This introduces the complexities of changing climates over time and the fact that landforms produced by weathering in different climatic environments may occur in the same area. It is always possible that larger weathering features — such as tors, coastal rock platforms or underground caverns — in areas which have experienced a tundra climate during the Quaternary Ice Age, may have been formed largely under those conditions, since it is only 10 000 years since the last retreat of the ice — scarcely enough time for them to have been formed postglacially. There is also evidence that the British Isles have experienced different climates in the more distant past. Thus 50 million years ago laterite was formed on lava flows accumulating in Northern Ireland, and 250 million years ago in the Permo-Triassic period desert conditions obtained.

In extending interpretations based on weathering processes as observed today back into the geological past it must also be remembered that the atmosphere probably had a different composition in the early Precambrian (Figure 3.12), and that oxidising conditions became dominant later. Rates and types of weathering would also have been different before the spread of land plants in the Silurian–Devonian over 400 million years ago. These did not become widespread away from marshy lowlands until 150 million years ago, whilst turf grasses appeared only 20 million years ago. Man's own effect on the rates of weathering must also be taken into account when reasoning from the present to the past (chapter 11).

16

18

17

19

20

Plate 16 The avalanche on Nevado Huascaran, and the succeeding mudflow which buried Ranrahirca, Peru, in June 1970, following a major earthquake. The ice was partly melted by friction as it moved downslope, transforming the mass of rock and snow to a mudflow. (USGS)

Plate 17 A landslip in the cutting slopes of a freeway near Los Angeles. (USDASCS)

Plate 18 A landslide near Tepekoy, Turkey, in the heart of the area affected by the 1966 earthquake. (USGS)

Plate 19 A landslip in well-timbered country with steep slopes in Washington, USA. The ground had been saturated by moisture from snowmelt after the 1968–69 winter, and then disturbed by logging activity. Some 2000 m³ soil cascaded from the hillside to cover a small meadow at the foot. (USDASCS)

Plate 20 An earth flow on fallow land in Washington, USA: the slopes have been eroded by sheetflow and close-spaced rills following a heavy storm on frozen soil. (USDASCS)

A

B

Plate 21 Two movements in the same slide in Washington, USA.
The first (left) occurred in June 1970, caused probably by water
seepage high on the hillside, and had a total width of over 400 m. The
second followed in August 1971. The Yakima river flows at the base
of the slope. (USDASCS)

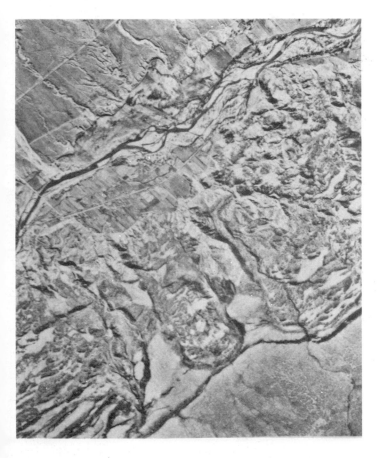

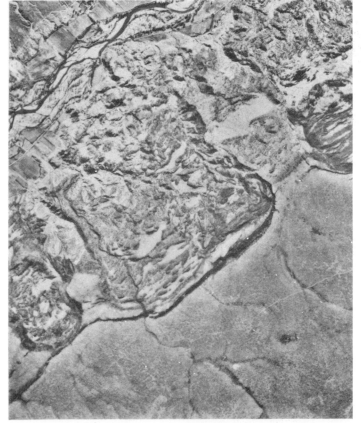

Plate 22 A landslide in New Mexico: stereo pair. Black Mesa is
formed of massive lava flows resting on poorly consolidated Tertiary sediments, and these are undercut by the Rio Grande at their
base. Large sections of broken lava caprock mantle the slides.
(USGS)

6

Mass movement and slope development

On 31 May 1970 a major earthquake rocked Peru, setting off an avalanche of rock and ice on the slopes of Nevados Huascaran, the highest Andean peak in the country. Between 25 and 50 million cubic metres of material moved down the slopes at high speed, largely burying the towns of Yungay and Ranrahirca. The ice was partly converted to water in the rapidly moving mass, resulting in a huge mudflow. More than 20 000 people died. Only eight years before an earlier avalanche had devastated the same area.

Such catastrophic events contrast markedly with the slow and relatively uniform rates at which weathering takes place. They occur especially in areas of high relief: landslides, sometimes of huge proportions, occur commonly in Switzerland. They have also occurred where men have produced an oversteepened or unstable slope, as at Aberfan in the south Wales coalfield, where a mass of coalpit waste, tipped for many years on a hillside over an unknown spring, became lubricated by the water and slipped downhill to kill 147 children and adults on 21 October 1966 (Figure 6.1). On a smaller scale the grading of a road cutting at an inappropriate angle will often lead to the collapse of a section of the bank (chapter 11).

Figure 6.1 The debris flow at Aberfan, south Wales. (After Strahler, 1973)

The rapid movement downslope of sections of regolith is the most noticeable aspect of a range of processes operating on slopes, and known collectively as mass movement, or mass wasting. Many of the processes work too slowly to be seen, but they act on every slope carrying weathered material away from its original site and downhill. Gravity is the main force acting on such material, so that the activity should be most rapid on the steepest slopes, but the movement is often assisted by the presence of water. An extensive cover of vegetation will slow such movement.

Mass movement processes

Mass movement involves the movement of regolith with varying amounts of water or ice. It occurs at speeds ranging from those which are so slow that the movement is imperceptible (creep) to rapid flow and catastrophic slumping and rockfalls (Figure 6.2). There is a continuum from one end of the scale to the other, but all types lead to the transportation down-slope of large quantities of rock material.

Nature and rate of movement		Glacial transport	← Increasing ice content	← Rock or soil →	Increasing water content →	Stream transport
Flow	Imperceptible		Solifluction	Creep (soil creep; rock creep)	Solifluction	
Flow	Slow to rapid		Debris avalanche		Earth flow Mudflow Debris avalanche	
Slide	Slow to rapid			Slump Debris slide Debris fall Rock slide Rock fall		

Figure 6.2 A classification of forms of mass movement. (After Sharp, 1938, in Bloom, 1968)

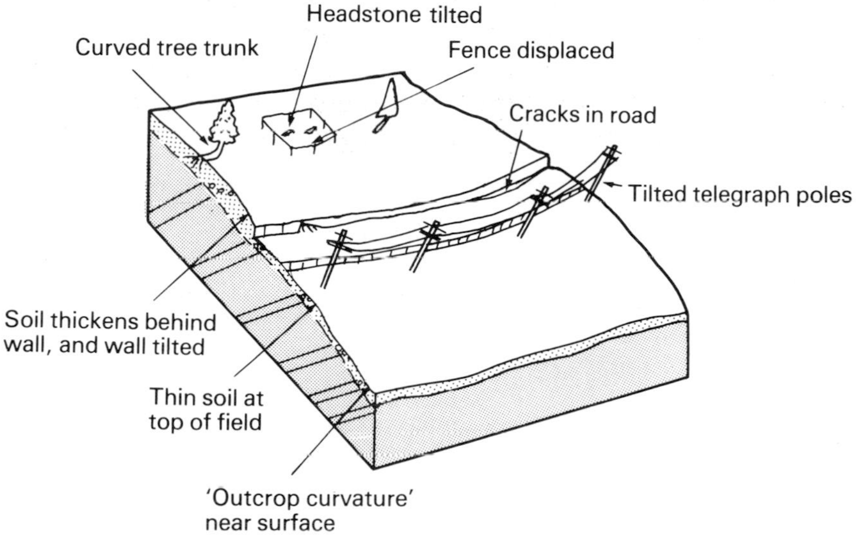

Figure 6.3 The effects of soil creep. (After Bloom, 1968)

Mass movement by flow

A slow, downhill movement of regolith, known as soil creep, occurs on gentle-to-medium gradient slopes where the weathered mantle is deep, and is particularly common when it is also fine-grained. It can be shown to be active by reference to the piling up of soil behind hedges or stone walls and other similar features (Figure 6.3). The movement rate is dependent to some degree on the angle of slope, but above 30 degrees movement is often too rapid for vegetation to grow. Some form of disturbance is, however, necessary for particles to be moved in this way. Thus alternate expansion and contraction of surface particles will lead to heaving, in which particles are thrust out at 90 degrees to the slope but on contraction fall back along the vertical: this results in downslope movement. In humid temperate areas like Britain the wetting and drying of clay minerals in the soil combined with freezing and thawing in winter cause such heaving, whilst the action of the soil fauna and plant roots assist the process. Movement rates of 1 mm/year in the uppermost 2–5 cm have been recorded. In the tropics the wetting and drying of clay is again important, but the soil fauna is also more active. Greatest movement is thus experienced on slopes with a high clay and water content, rather than on the slopes with the steepest angles. On scree slopes rock creep occurs as rock fragments are supplied or as temperature changes affect the stability of the slope.

Solifluction is a process occurring in cold lands where there is an annual cycle of freeze–thaw. During the freezing and melting processes, which take place repeatedly during the spring and autumn (whereas there may be no freezing in summer, or melting in winter), surface frost creep (like soil creep, but with merely the freeze–thaw rhythm) gives rise to movement. Thawing leads to extensive melting, and if the ground is permanently frozen below a metre or so(see p. 111) the surface zone loses cohesion and flows (gelifluction). Rates of up to 0.5-5.0 cm/year have been observed, giving rise to lobes of detritus on the steeper slopes which push through the surface mat of vegetation. The moisture content of the regolith is the main factor influencing such rates of movement and it is only where slopes exceed 22 degrees that steepness becomes as important. Fine clays and silts are more affected than sand or gravel. Such movement gives rise to definite small-scale features on slopes, including lobes and benches. Block fields form when massively-bedded and -jointed rocks break up and become involved with the surface flow. Many of the Dartmoor tors are surrounded by greater numbers of blocks spreading down the slopes in stripe-like areas: the total volume of these is often several times the volume of the tor feature.

Frost action in tundra areas will also give rise to patterned ground due to heaving on flatter surfaces, and to unusual features like pingos, where ground ice becomes concentrated and leads to local updoming (Figure 6.4).

More rapid types of flow are related to increasing supplies of water on steeper slopes. Mudflows occur where sufficient water is concentrated in the regolith at the head of a valley to

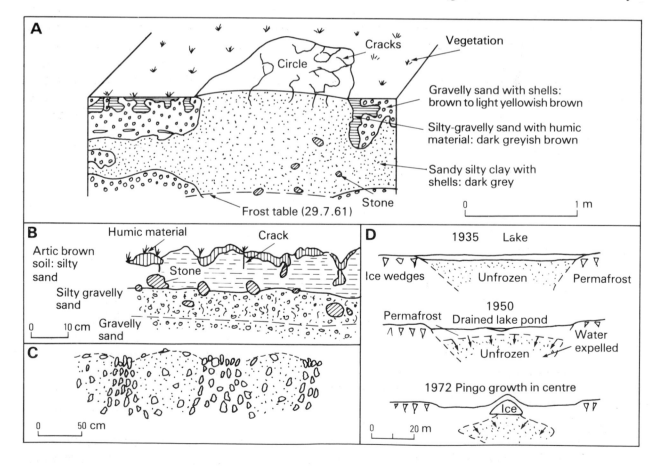

Figure 6.4 Features produced by frost action: patterned ground and pingo. (A) A cross-section of a nonsorted circle in north-east Greenland. (B) A cross-section of a small nonsorted polygon in north-east Greenland. (C) A sorted net. (D) The growth of a closed-system pingo in the Mackenzie delta, northern Canada. (After Washburn and Mackay, 1972, in Washburn, 1973)

overcome the internal cohesion, and they move at rates as rapid as river flow. They have a long, narrow track and spread out on reaching lower ground. This type of event is most common where vegetation cover is slight or absent, and so is frequent in semi-arid and mountainous regions. Earthflows occur beneath a vegetation cover, which may be scarcely disturbed apart from a crescentic scar at the upper margin. They are most common on clay soils in the tropics and have short flow tracks. All movements of this type are most common after heavy storms and affect the regolith and thick unconsolidated deposits of, for instance, glacial origin. They occur infrequently and often have the effect of exposing bare rock in their source regions and of assisting the headward extension of valleys. Debris avalanches may be viewed as the most rapid form of rapid flow, illustrated by the case of Huascaran with which this chapter began.

Mass movement by slope failure

When the moving mass shears from the stable rock along a slip plane it produces forms where the disrupted mass retains cohesion within itself (rotational slip, or slump), or those where a mass disintegrates (rock falls and rock slides). These all tend to be of catastrophic form, and have long posed problems for the engineers building railways and roads in mountainous areas. Shelters have been built over sections liable to rockfalls from slopes above, and recent road construction in the Alps (and also on the M5 south of Bristol) has left a gap between the road and the mountain side so that falling boulders will not litter the road or endanger traffic.

Rock falls and rock slides occur where a mass of rock is loosened on a steep slope or cliff. This may be associated with steeply dipping layers of rock, and especially with shale, and is common after heavy rain or during the melting season at high altitude. Slumps take place by failure along a rotational slip plane, a common situation for this being where massive permeable rocks overlie impermeable and weak clays (Figure 6.5). Many slumps in southern England seem to be inactive at present, and were probably initiated during a period of higher ground moisture concentration during the closing phases of the Quaternary glaciation.

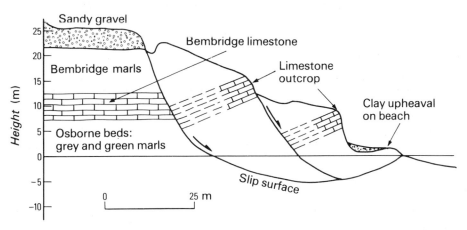

Figure 6.5 A rotational slip on the north coast of the Isle of Wight. Such slips are induced by an increase of water pressure in the rocks. (After Steers, 1962, in Pitty, 1971)

Mass movement by subsidence

Subsidence occurs when subsurface movement or rock material takes place. This includes the collapse of areas of the surface due to the underground enlargement of caverns in limestone areas (chapter 7), and surface downwarping due to the mining of coal or salt (chapter 11): certain areas of Cheshire and south Yorkshire are marked by water-filled subsidence hollows. Drainage of the Fenland peat soils has led to shrinkage and lowering of the ground level by 3-5 m.

The study of slopes

Slopes provide a series of intriguing problems for the student of landforms. Mass movement takes place over all slopes in the landscape — a contrast to the confined activity of running water in streams, of ice in glaciers, or of the sea around the coasts. With weathering it is thus the most widespread of all processes acting on the surface rocks. In addition, slopes are modified by other forces, including uplift and running water.

Slopes reflect the interaction of (a) the rate of uplift (or subsidence); (b) the resistance to weathering of the underlying rock-type; (c) the weathering and transport processes operating on the slope; (d) the effect of running water, glaciers or the sea wearing away or depositing at the slope foot (Figure 6.6). They thus pose a series of problems to which the answers are not to be found easily because of the multivariate nature of the factors. It is difficult to assess whether the inputs to the slope system (weathered sediment, rainwater, atmospheric conditions) balance the outputs (transported sediment and water), or whether there are definitive relationships between slope form, process, rock-type and time. This complex situation confronting the student of slopes has not been helped by the tendency in the past to base theories of slope evolution on vague, impressionistic descriptions instead of accurate measurements. Hence the study of slope evolution has been one of conflicting or complementary theories. This position is now changing, and a different approach to slope study has emerged: quantitative measurement of slope forms and processes are being carried out to test a range of hypotheses. Individual studies have begun to show the way ahead, but patience will be necessary before the data are sufficient to allow genuine synthesis and explanation. Slope studies remain a frontier of research into landforms.

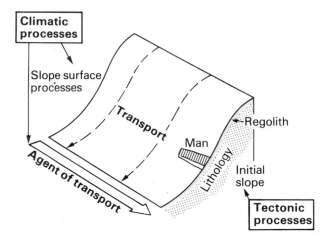

Figure 6.6 Forces operating on a slope.

The study of slopes based on field observation involves the measurement of process and landform. Only where this has been accomplished can theories relating to slope evolution be discussed meaningfully (Figure 6.7).

Slope processes

Slopes include all land surfaces, ranging from the horizontal to the vertical. They originate by a combination of tectonic and erosional activity: thus uplift or faulting will provide slopes, whilst valley sides, cliffs or depositional areas will have slopes of subaerial process origin. It is not correct to talk of an 'initial' slope since a variety of processes act on a slope to modify it as soon as it exists.

The processes acting on slopes include weathering (chapter 5) and mass movement, and in addition to these there is the surface and subsurface movement of water. The impact of rainwater on the surface will lead to the downslope movement of particles and in very heavy

Problem	Method of study	Type of study
The field evidence of a form or process which demands explanation.	The means of analysis: the factors and variables analysed and the methods used in explanation.	The general, common form for the approach and type of study.
Forms of slopes, e.g. convexo-concave.	Observed slope forms and processes in the field. The use of inference. The link between process and morphology. Process-response systems.	*Inductive:* The process-form approach.
Decline and retreat of slopes.	Evolutionary or sequential approach. Use of mathematical and geometric analysis of form.	*Geometric* The slope evolution approach.

Figure 6.7 Some approaches to slope study.

storms surface wash may be important over a complete slope, although it soon becomes channelled into gullies. This is particularly effective in semi-arid conditions. In wooded areas surface flow is almost unknown, and most water flows through the regolith, often concentrated in throughflow tunnels, which may collapse to initiate gullies.

Slope forms

It is necessary to provide a basis for process-form relationships by measurements of slope forms and the properties of slope regolith. The shape of the slope can be assessed from a vertical section (or profile) and from a map of slope facets.

Slope profiles

Slope profiles are surveyed lines across the ground following the maximum slope, and are usually carried out at points selected by sampling techniques. A slope profile may be divided into a series of slope units (Figure 6.8) for analysis.

Such studies add meaning to the vague statements often found in accounts of slopes. Thus the slopes of Chalk areas in southern England have been described in the following terms: 'the cross-section (of a Chalk valley) is usually a smooth curve'; 'the remarkably symmetrical double curve so characteristic of our Chalk regions'; 'the surface is billowy but not broken — the swells resemble Biscayan waves half pacified'; 'the whale-backed Downs'; 'the everrecurring double curve which was long ago styled by Hogarth in "the line of beauty"'. Measurement shows, however, that, instead of the convex–concave double curve, between 30 and 50 per cent of Chalkland slopes are rectilinear segments. Convexities and concavities tend to be smoother on massive rocks like Chalk and granite, whilst straight segments are most common where the rock type alternates frequently, as in the Millstone Grit of northern England. In Britain, convex slopes comprise 40 per cent of profiles on average, and locally as much as 80 per cent. If there is an outcrop of a rock layer in the slope, giving a cliff-like segment, the convexity above will be short and the concavity below will form up to 90 per cent of the slope. Common slope angles and profile characteristics can be correlated with features of the underlying rock-type, together with local relief and processes acting on the slope. The variables are so many, however, that it is difficult to determine answers to important questions — whether the slope profile form is related to local climate, geology and valley depth; whether it is related to past climates or periods of rapid erosion; or whether it is formed of units which are common to all slopes.

A

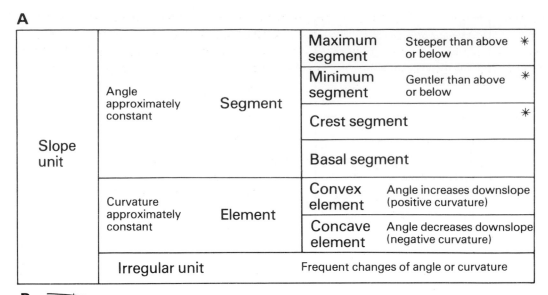

Slope unit	Angle approximately constant	Segment	Maximum segment	Steeper than above or below	✳
			Minimum segment	Gentler than above or below	✳
			Crest segment		✳
			Basal segment		
	Curvature approximately constant	Element	Convex element	Angle increases downslope (positive curvature)	
			Concave element	Angle decreases downslope (negative curvature)	
	Irregular unit		Frequent changes of angle or curvature		

B

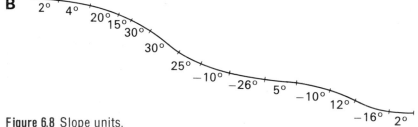

Figure 6.8 Slope units.
(A) The terminology used. A concavity includes all units where there is not increase in angle downslope; a convexity includes all units where there is no decrease in angle downslope; neither may include units marked with an asterisk.

(B) A slope profile: relate the units to terms used in (A). Numbers refer to curvature; angles have been measured for the segments. (After Young, 1972)

Slope maps

As the slope profile produces a two-dimensional vertical view, so the slope map provides a two-dimensional horizontal view. Since morphological maps recognise the areal extent of

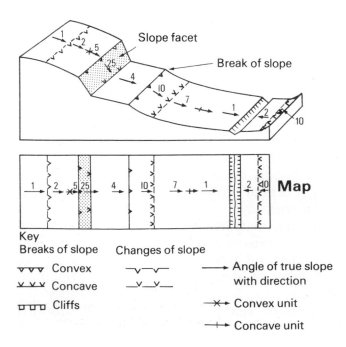

Figure 6.9 Morphological mapping units used on slope maps. It is common to shade the rectilinear facets. (After Curtis, 1965, in Young, 1972)

slope units, a similar series of conventional symbols are used (Figure 6.9). When the map has been prepared it may be difficult to interpret unless shading is used to identify slopes of varying steepness. As with the measurement of slope profiles, slope maps should initially keep to the evidence observed, rather than attempt to relate the slope form to its origin. Slopes can be grouped in a variety of ways for further analysis (Figure 6.10).

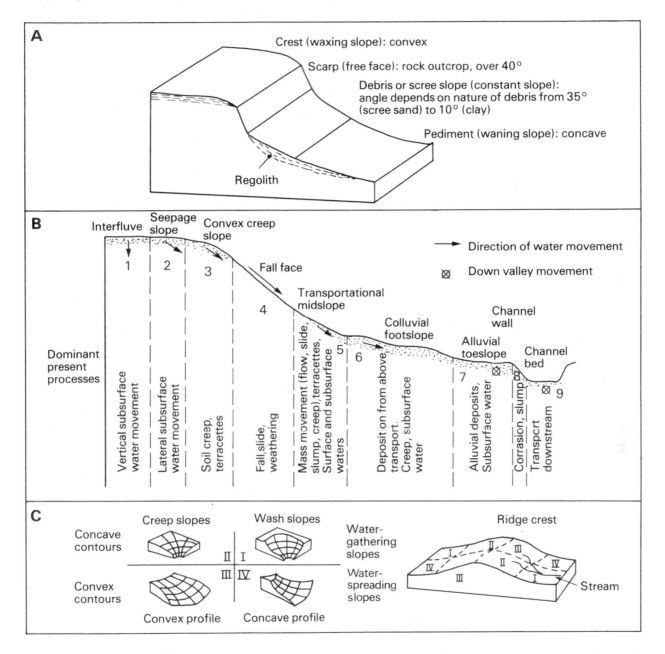

Figure 6.10 Slope classification: a variety of schemes.
(A) Slope facets. (After King and Wood, in Young, 1972)
(B) A nine unit model. (After Dalrymple, in Young, 1972)
(C) A classification of slope elements in a landscape. (After Troeh, 1965, in Bloom, 1968)

Slope regolith

The regolith on a slope is mobile: it is being formed by weathering and is being moved by solution, mass movement and by the flow of water through and over it. It is to be expected that there will be a close relationship between the slope form and regolith. Measurements show that the thickness of regolith often increases towards the foot of a slope (Figure 6.11), and that the material in the regolith becomes finer in the same direction. In the humid tropics the regolith is often over 10 m deep and up to 100 m, thinning near hill crests, but only the top one metre is mobile. In humid temperate lands, however, the regolith is much thinner, seldom reaching over one metre.

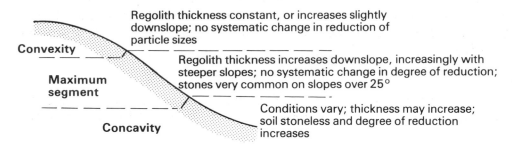

Figure 6.11 Regolith on slopes: a summary of observations. Regolith in the tropics is often over 10m thick and reaches 100m in many places, although only the top 1m is mobile. In humid temperate zones regolith is 1–2m thick at the most.

Regolith has a two-way relationship with its slope, since its protects the rock from weathering and thus affects the retreat of the slope. With fine solid materials (clays, silts) there is a greater proportion of heaving and more rapid creep; there is a lower permeability so that more water runs off the surface; and slope failure may become important on slopes over 10 degrees (cf. regolith accumulates on other materials up to slope angles of 30–40 degrees). Secondary hardening of sections of regolith occurs commonly in tropical laterites and bauxites, but also with the formation of iron concretions in temperate areas, and this forms a resistant layer within the regolith, known as a duricrust.

Process–form relationships

The correlation of process and form is a study beset by uncertainty, and is a matter of suggesting hypotheses which can then be tested against existing information and further measurement. These problems have occupied students of landforms throughout the history of such investigations, and it is instructive to trace the progression of ideas in this area. The

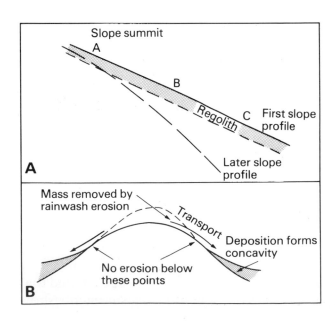

Figure 6.12 Some univariate approaches to process–form relationships on slopes.
(A) G. K. Gilbert. The regolith to be moved downslope increases from A to C: it will therefore need an increasingly steep slope for transport. The theory does not make it clear how this is achieved.
(B) A. C. Lawson. Rainwash erosion is most effective at the summit, and loses effectiveness downslope, since it becomes heavily loaded with sediment. Beyond a certain point there is no erosion, and then deposition occurs. (Diagrams after Small, 1970)

transition from the consideration of one factor to the realisation of natural complexity and the need for quantification reflects a general trend to be found again and again in this subject.

Univariate approaches

Early approaches to this subject tended to concentrate on one main process. Thus Fenneman (1908) discussed slope forms merely in terms of running water as an erosive agent; Gilbert (1909) considered the effects of soil creep in isolation; while Lawson (1932) looked at the effects of running water, but included its depositional role as well as the erosive factor studied by Fenneman (Figure 6.12).

Multivariate approaches

The univariate approaches are weak because slopes develop in a situation where there is a complex variety of interacting forces: they can provide only some slight indication of the contribution of individual processes to slope formation.

A Frenchman, H. Baulig, attempted to combine a consideration of both rainwash and creep. He regarded rainwash as limited in its erosional effect over the summit convexity because much of the flow occurs in the sheet form and is, in many cases, limited by the small volume of water due to a small catchment area. Erosion was seen as important at the slope base because of the greater volumes and velocities of water and the concentration of flow into channels. Creep tends to mould the summit but is not very effective in the lower sections of the profile. Both creep and rainwash are agents of regolith transport downslope and so they operate together. Through the summit area soil and regolith is removed by rainwash and creep. Although running water is active in channels over the basal concavity, creep may enhance the erosion process particularly if the soil is fine and moist. Field observation and measurement seem to suggest that in the upper slope area the increase in load is greater than the increase in efficiency of transport. Thus the gradient steepens and a convex slope emerges (Figure 6.13). In the lower slope area the increased efficiency of movement exceeds the increase in slope load and a concave slope results. Where a rectilinear section is present in the slope profile it would seem that the downslope increase in load balances the downslope increase in transporting power. A clear spatial relationship thus emerges between slope process, load and morphology.

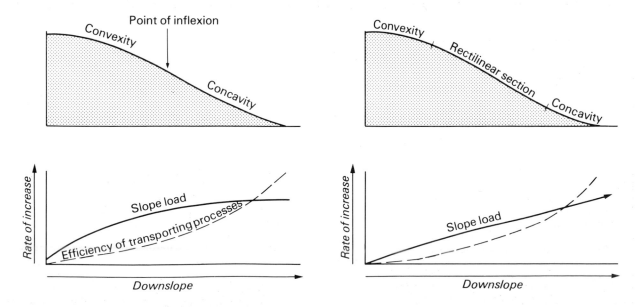

Figure 6.13 Relationships between slope processes, load and form. (After Small, 1970)

Slope evolution

The history of slope studies has dwelt less on the measurement of form and process followed by an analysis of their relationships, and more on a geometric view of the evolution of a slope profile from its initial state to the present. Theories of this type are related more clearly to a preconceived view of the way in which a landscape is worn down, than to a foundation of observation. Instead of an inductive approach, leading from observation and measurement to hypothesis and testing of the hypothesis by further investigation, such approaches begin with a hypothesis and are therefore deductive.

A number of major approaches have emerged.

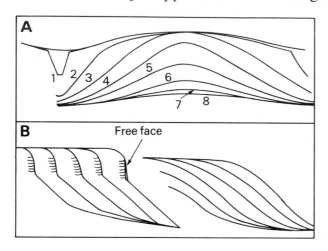

Figure 6.14 Hypotheses to explain the evolution of slopes.
(A) W. M. Davis: slope decline.
(B) L. C. King: parallel retreat, with and without a free face. (After Young, 1972)

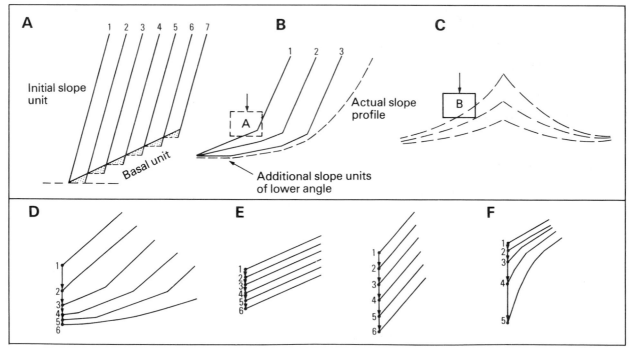

Figure 6.15 Slope evolution, after W. Penck.
(A)–(C) illustrate the development of concave slopes above non-eroding rivers.
(A) As the slope retreats (stages 1–7), a basal unit of transport is developed.
(B) The basal unit is modified and a concavity produced. (C) An overall decline of slope profile and a concavity is produced. (D)–(F) illustrate the relationship between river erosion and slope angle. (The numbers 1–6 represent successive river positions at equal time intervals.) (D) River is cutting down at a slowing rate: concave slopes develop above. (E) River is cutting down at constant rate: rectilinear slopes develop above. (F) River is cutting down at increasing rate: convex slopes develop above. (After Small, 1970)

1) W.M. Davis, an American geographer, put forward the concept of the 'cycle of erosion' at the end of the nineteenth century. In this he envisaged an initial landscape following uplift increasingly attacked by surface processes. This resulted in early, 'youthful' features, followed by maximum relief in 'maturity' and an overall lowering by 'old age' to a low relief landscape known as a peneplain (i.e. 'almost a plain'). Accordingly his ideas on slope evolution led to the smoothing of slope profiles and a lowering of slope angle with time (Figure 6.14 A).

2) L. C. King, a South African, favoured a different style of landscape evolution, whereby landscapes are reduced to 'pediplains' formed by back-wearing and parallel retreat of slopes (Figure 6.14 B). This was derived initially from studies on areas of tropical seasonal climate, but he extended its application (cf. discussion on tors, chapter 5).

3) Other studies have been more precise and detailed. The most important have been those of the German, W. Penck (1924). His ideas on slope evolution (Figure 6.15) were based on the assumption that the intensity of river erosion (itself related to rates of uplift of the land) determines the form of the valley side slope, rather than the slope processes themselves. He saw slopes as purely transportational features. This again emphasises the almost slavish devotion to an idea which the evolutionary approaches adopt: Penck's theory precludes the erosional activity of processes on the upper parts of slopes.

4) A. Wood, of Aberystwyth, attempted to overcome the problem left by Penck. His work (1942) assumed an initial cliff-like surface, which develops to a constant slope as the initial slope is weathered and the debris accumulates in a scree slope at the base (Figure 6.16). Convexity develops due to rounding of the uppermost angle with the plateau surface, and the lower part of the slope becomes concave as finer scree material is washed out: the convex and concave segments eventually merge. It is not clear, however, how this rather geometrical sequence is related to the actual slope processes.

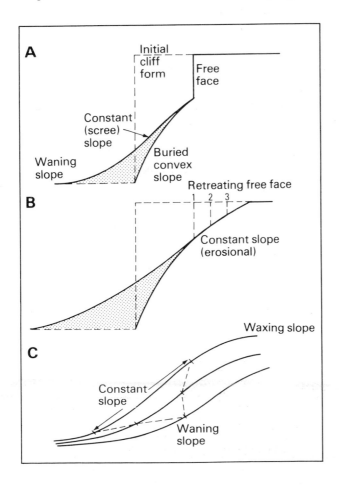

Figure 6.16 Slope evolution, after A. Wood, (A) A cliff face is weathered; fragments collect at the foot, protecting that portion from further weathering and reducing the free face height. A river begins to remove debris at the foot of the slope. (B) The free face section is eventually eliminated. (C) Later stages of slope evolution. (After Small, 1970)

Field investigations

The theories of slope evolution are not related to either observed form or process, but to some geometrical approximation of generalised conceptions of slope profiles. They are dangerous in providing an apparently clear-cut account of slope formation. Unfortunately at present the results from measurements and their correlation are at an early stage, but certain field investigations have shown some of the possibilities.

a) R.A. Savigear (1952) studied the cliff profiles of Carmarthen Bay in south Wales, where the development of salt marsh along a stretch of Old Red Sandstone cliffs offers the study of a slope which has been progressively protected from marine processes. The effect of slope processes alone has been to produce important contrasts in profile development (Figure 6.17). This example suggests that both parallel retreat and slope decline are possible in the same locality, but depend on the slope-foot conditions.

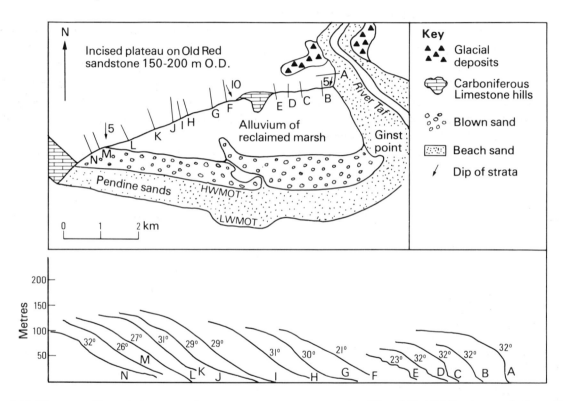

Figure 6.17 Slope profiles of the coast between Pendine and the Taf estuary in south Wales. The sand spit has grown from the west, giving increasing protection to the cliffs behind. Profiles A–C are those most recently abandoned by the sea, and still retain cliff at 80°. Middle slopes of all profiles are approximately 30°, and upper portions are convex. Basal concavities are formed by debris accumulation. (After Savigear, 1952, in Leopold, Wolman and Miller, 1964)

b) A.N. Strahler, an American, has carried out fieldwork (1950), measuring maximum angles on slopes in areas where the geology, climate, vegetation, soil and relief were as uniform as possible. He found that readings deviated only a degree or two from the mean and suggested that this showed that a state of equilibrium existed related to the steady removal of slope regolith — i.e. involving all the slope control factors. He also compared the angles of valley side slopes and gradients of streams in the valley. Where valley side slopes were steep, shedding debris into the stream more rapidly, the stream channel gradient was similarly steep. Thus debris supplied could be removed by the stream. Gentler slopes were associated with gentler stream channel gradients, except where they were unvegetated: the more rapid supply of debris in such cases was associated with a steeper channel gradient. This suggests that as a stream channel gradient is lowered with time, so the valley side slopes will decline in gradient.

Steeper slopes occurred also where a stream was flowing at the immediate base of the slope, whilst less steep slopes in the same valley could be found where deposition of slope debris was allowed to occur at the base. This situation is like that of the Carmarthen coastline, and suggests that both parallel retreat and downwearing are possible. Slopes may become steeper with time if the stream is cutting downwards very actively.

Slope development: a summary

The manner in which slopes develop through time, and the factors which influence the changes, are clearly complex problems to unravel. Slope decline or parallel retreat may not be due to a systematic series of events in terms of a rigidly followed cyclic sequence, but may be more closely related to a dominant local factor, which may change with time. Thus C.R. Twidale (1968) studied pediment slopes in central Australia. Those on granite and gneiss rarely exceeded 3 degrees, whilst those on schist were commonly between 5 and 8 degrees: he concluded that rock-type is the dominant control in that situation. Elsewhere it is clear that stream incision, marine erosion or tectonic activity may be dominant. There may also be the influence of changing processes evident in the same profile. Thus the slopes of southern England are related to the present processes acting on them, and to processes which acted on them during the tundra climates experienced during the Quaternary Ice Age. Earlier theories of slope evolution presented a simplistic view of the situation, but more recent investigations emphasise the complexities and the need for quantitative observation and analysis of the many variables.

Permafrost

Permafrost is permanently frozen ground, which is common in the tundra areas of the world (Figure 1), and gives rise to peculiar conditions of soil moisture. The permafrost zone is defined

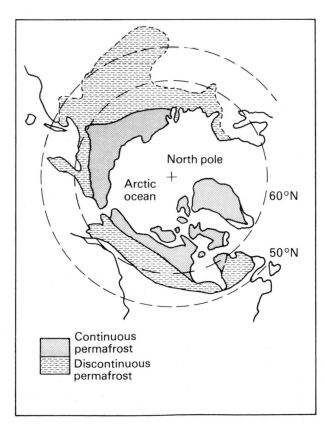

Continuous permafrost

Discontinuous permafrost

Figure 1 Distribution of permafrost in the northern hemisphere. The discontinuous zone has scattered areas of permafrost amid widespread areas of permafrost-free ground. (After USGS pamphlet 'Permafrost', 1973)

Plate 24 A building subsiding into thawing permafrost south of Fairbanks, Alaska. (USGS)

Plate 23 An ice wedge extending below the permafrost table in Alaska. (USGS)

Plate 26 Thermokarst pitting near Fairbanks; this is due to ice lenses in the soil thawing after clearance for farming. The soil had been cleared 13 years before 15 pits formed in one year. (USDASCS)

Plate 25 Ground subsidence along the Copper River railroad, Alaska. (USGS)

Plate 27 A pingo with a central crater in a valley floor on the Arctic Slope of Alaska. (USGS)

Plate 28 Polygonal soil patterns in a muskeg swamp area, 50 km north of Fort Yukon, Alaska: such patterns arise due to freezing and thawing of the ground, and are not covered by the meagre snowfall (0.5 m) in the area. (USDASCS)

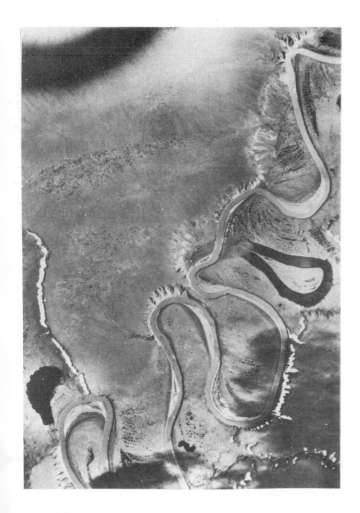

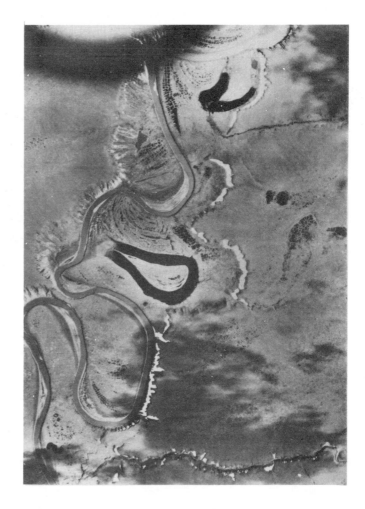

Plate 29 Stereo pair: the flood-plain of the Kogosokruk river, a tributary of the Colville river on the Arctic Slope of Alaska. Bands of ice-wedge polygons outline meander scars on the flood-plain, and thawed ice wedges give serrated edges to the terrace edge. (USGS)

as 'a thickness of soil or other superficial deposit (or solid rock) in which a temperature below 0° C has existed continuously for more than two years'. This occurs where the balance of winter freezing exceeds the summer thaw, and permafrost extends down into the soil until the freezing process is balanced by heat from the Earth's interior. Thicknesses of over 300 m may form over thousands of years, and will disperse only very slowly when temperatures rise.

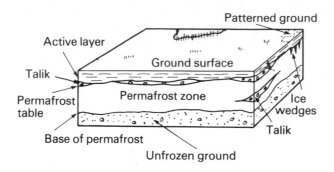

Figure 2 Features of the permafrost. The permafrost zone is up to 300m deep in North America and up to 600m deep in the USSR. The active layer has a maximum of 1–2m. (After USGS pamphlet '*Permafrost*', 1973)

The ground above the permafrost layer is known as the active layer (Figure 2), so-called because in it the ice melts in summer and re-freezes in winter. Permafrost may contain masses which do not freeze in winter (talik), and these may burst upwards under pressure and then flow out to form a surface layer of ice. The uppermost section of the permafrost layer contains masses of ground ice: this ice occurs as grain coatings, as individual ice crystals, veinlets, lenses and large ice wedges. The ice-rich permafrost is particularly common where the active layer is thin and the sediments are fine-grained and poorly-drained. These conditions are common on the North Slope of Alaska, where the ground ice makes up over half of the uppermost 3-5 m of the permafrost zone and the ground is honeycombed with a polygonal network of vertical ice wedges extending down from the permafrost table. Thawing may lead to subsidence along these lines, and patterned ground features develop. Streams may also reflect the patterns of subsurface ice wedges. Elsewhere thawing of ground ice leads to local subsidence in thermokarst pits, and once begun such thawing tends to extend the area to a thaw lake: these are often orientated to wind directions, since wind and wave action are concentrated.

Man's occupation of these areas has created many problems due to his ignorance of the delicately balanced environment: removal of the thin vegetation cover leads to local permafrost melting and subsidence. Vegetation tends not to be re-established, but the gaps created are extended. Roads, railways and buildings suffer subsidence because of thawing beneath them and need special insulation beneath so that heat is not carried down. This experience led to many of the objections made to the proposed construction of an oil pipeline across Alaska. Oil is transported through a 1.2 m diameter pipeline at 63° C (a little cooler than the temperature at which it comes out of the ground), and a buried pipe without insulation would cause undue permafrost melting. This pipeline would be 1260 km long, all but 40 km of it being across ground with permafrost, ranging from 650 m thick in the north to less than 1 m in the south. About half of the route is solid rock, or contains little ice, so that burial of the pipe will be possible there; elsewhere the pipe will have to be carried on piles or embankments, or be insulated under ground.

7

Running water

It is difficult to find an area of land surface in the world which does not show the effects of the action of running water. Only the permanently frozen central areas of the Greenland and Antarctic ice sheets are free from its influence, since the desert areas nearly all experience some runoff from rare precipitation (Figure 7.1), and running water is particularly active beneath the margins of glaciers and ice sheets (chapter 8).

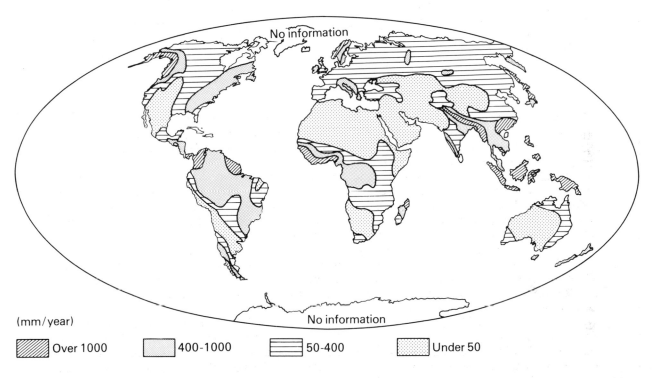

Figure 7.1 World runoff. Notice the correlation between areas of high altitude, high rainfall and high runoff. (After Lvovitch, 1958, in Gregory and Walling, 1973)

Water on the land surface sometimes occurs as sheetflow over the whole of a slope, but is commonly concentrated in stream channels, and the processes arising from the interaction of water with the underlying surface act largely within these channels. A drainage basin comprises the area drained by a trunk channel and its tributaries: the channels which drain a single drainage basin can be seen as a system. Inputs of matter (water and sediment) and energy (as provided by the Sun driving the atmospheric circulation and transformed by the local relief) can be related to the outputs of water and sediment resulting from work carried out in the channels (Figure 7.2). The study of drainage basins as systems helps to take account of the many variables acting in the natural situation; it emphasises the adjustments and relationships between landforms and processes; and it studies the total physical environment including the increasingly important part being played by man.

A drainage basin can be studied as a unit with well-defined boundaries. The inputs are transformed by the basin characteristics to provide outputs at the basin mouth: the relationship of landforms, rock and soil characteristics to the processes acting on them forms a

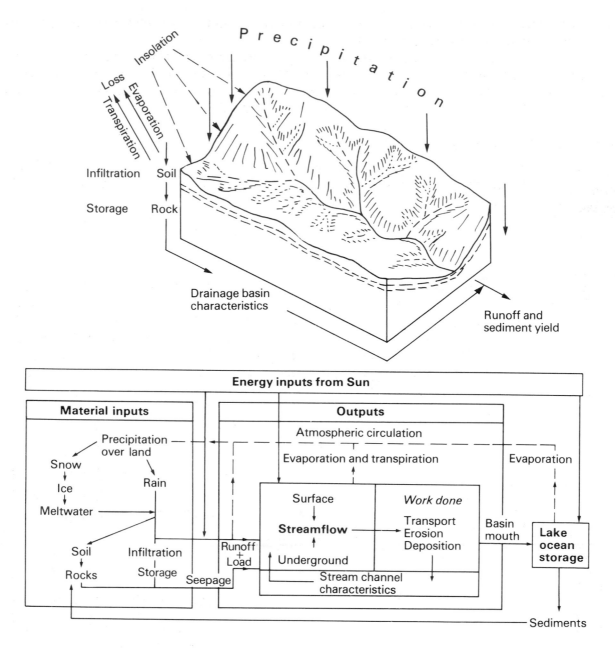

Figure 7.2 The stream system. At any point in the stream system the mass of water and sediment entering is balanced by that leaving: this is an open system (a closed system involves no transfer of energy or matter). (After Gregory and Walling, 1973)

process-response system. Alternatively the inner workings of the basin may be divided into a series of subsystems. Thus the slopes, the drainage network and respective stream channels can each be regarded as a separate subsystem. Each has particular characteristics of shape and composition and can be regarded as morphological subsystems. In each there are relationships with the inputs of rainfall and sediment, so that process-response subsystems develop. Sections of the stream channel can be measured and related to the varying passage of water through them. All the subsystems together may also be seen as a cascading system, where the output of one becomes the input of another. Outputs from the slopes become the inputs to the drainage network, and affect inputs to particular sections of the stream channels. Natural systems respond to natural processes, but in an increasing measure man is effecting changes of mass and energy outputs. Systems where a prominent variable, like man, are operative, are known as control systems (chapter 11).

There are two distinct branches to the study of water flow. Hydrology examines the relationship between the supply of water to the stream system (input) and changes in stream flow (output), whilst hydraulics is the study of the processes of water flow and the relationship of these to the work carried out by the stream.

Inputs to the stream system

Individual streams are supplied with water by tributary streams, by seepage via the valley side slopes, by water emerging from underground sources in springs, by rainwater falling on the stream surface, and by the melting of snow and glacier ice. Ultimately all can be traced back to precipitation as rain or snow, and the variations in the supply of the different forms of input are responsible for considerable changes in the work accomplished by streams.

Rainfall variations and runoff

Rainfall is periodic in occurrence: relatively short periods of rain are interspersed with longer, dry periods. This is the case even in climates which are humid throughout the year, and the effect of a long, dry season is to prolong the results of the rainless periods. There are also variations in total amounts of rain delivered within a particular time at a specified place, and in the intensity of fall: thus storms will deliver varying quantities over the areas they affect (Figure 7.3). All these characteristics of rainfall input are reflected in the flow of streams.

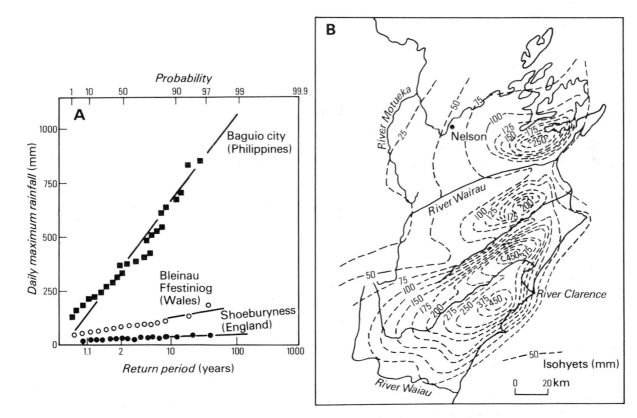

Figure 7.3 Variations in rainfall inputs to stream systems.
(A) *In time*. The probability of heavy falls of rain at three contrasting stations. Notice the difference between highland and lowland areas of Britain, and that between these stations and that in the tropics.
(B) *In space*. The localised pattern of storm rainfall in the north-eastern corner of South Island, New Zealand, in the 24 hours, 09.00 hr 23 May 1966 to 09.00 hr 24 May 1966. Suggest areas where the most intense flooding would take place. (After Rodda, 1970 and New Zealand Ministry of Works, 1966 respectively; both in Gregory and Walling, 1973)

When raindrops reach the ground the water will seep into the ground, be evaporated, or go directly into the stream channel. A simple equation is used to express this relationship:

Runoff in stream = Precipitation – Loss (Infiltration + Evaporation)

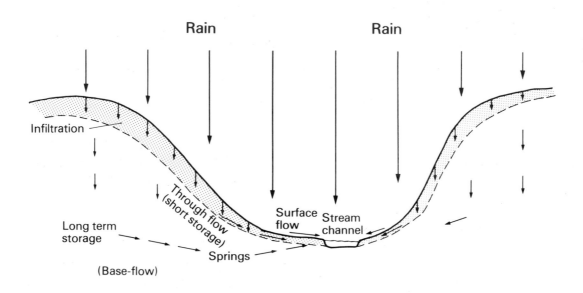

Figure 7.4 Water passage from rainfall to streamflow.

Stream channels cover a very small proportion of the land surface (Figure 7.4), and most rainfall infiltrates the soil and flows through it, often concentrated in small diameter natural pipes through the regolith until it oozes out into the stream channel at the slope foot. A proportion of this water percolates downwards into permeable underlying rocks, and is stored there until released to the stream system via springs. Some of these pathways lead to an immediate increase in stream flow, whilst others slow the passage of water and involve some degree of storage before the water enters the stream channel. Direct runoff takes up a greater proportion of the water supplied to streams when the volume of rain falling is greater; the intensity (mm/hour) is greater; the soil is already saturated after a period of wet weather; the slopes are steep; the soil and rock do not allow infiltration; the vegetation cover is poor; the evaporation rates are low (Figure 7.5).

There are thus two interacting sets of factors: a variable distribution of rainfall volumes and intensities in time and space on the one hand, and a variable relationship between precipitation, runoff and evaporation/infiltration loss on the other. This interaction largely determines the supply of water to rivers in most parts of the world. It can be illustrated first by examining the result of a particular rainstorm (Figure 7.6), which demonstrates the lag in the supply of water to streams. Secondly, a comparison can be made of flow patterns over a year in different areas (Figure 7.7) and these can be related to rainfall regimes, storm type, relief and geology. In the longer term also stream flow must be affected by changes in rainfall totals (Figure 7.8).

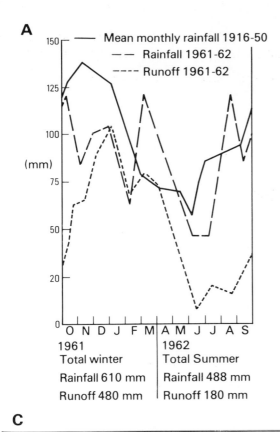

A

1961
Total winter

Rainfall 610 mm
Runoff 480 mm

1962
Total Summer

Rainfall 488 mm
Runoff 180 mm

B

75,etc Mean annual runoff
= 70-80% mean annual rainfall

C

Evergreen rain forest Brazil	Evaporation from tree crowns: 20% Running down trunk: 46% Penetrating to rain gauge: 33%	9.2% evaporation 9.2% absorbed by bark 20.7% absorbed by roots 6.9% reaches water table
Grasses Bunch grasses	(Both at rainfall of 25.9 mm/hour)	50-60% interception 30% interception
Temperate deciduous trees Coniferous trees	20% interception (17% without leaves), of which 5 (10)% stem flow 30-35% interception, of which 2-4% stem flow	
Cereal crops Clover	7-15% interception in growing season 40% interception in growing season	

Figure 7.5 Variations in the proportions of precipitation reaching streams to form runoff.
(A) Gunnislake gauging station on the river Tamar at the Devon–Cornwall boundary. The river here drains an area of 890 km². Note differences between (a) mean rainfall distribution through the months, and that for one particular year, and (b) summer and winter runoff. Account for these differences. Note that the 'water year' used by hydrologists runs from October to September so that summer and winter seasons are seen as a whole.
(B) How do relief, underlying rocks and evaporation rates affect the differences in proportions of runoff throughout Britain? (After Hanwell and Newson, 1973)
(C) Interception of rain by different types of vegetation cover. (From various sources, in Gregory and Walling, 1973)

Floods

Flooding rivers are major natural hazards. Two accounts of large floods are included here to demonstrate the effects on man's activities and the catastrophic effects of such rare natural events.
a) **The Lynmouth Flood, in north Devon, on 15-16 August 1952.** 'A disastrous inland flood occurred during the night of August 15–16, 1952, in north Devon, at and around the seaside town of Lynmouth (Figure 1). The rain gauge on Longstone Barrow, Exmoor, the tableland above

30

31

Plate 30 Raindrop impact, leading to the downslope movement of soil particles. (USDASCS)

Plates 31/32 Variations in flow at Great Falls on the river Potomac above Washington DC. (USGS)

32

Plates 33/34 Flooding on the river Wey, Surrey, England, in September 1968. These views show the effects in the wider flood-plain near Shalford, south of Guildford, and in the narrower zone in the city itself. (Aerofilms)

33

34

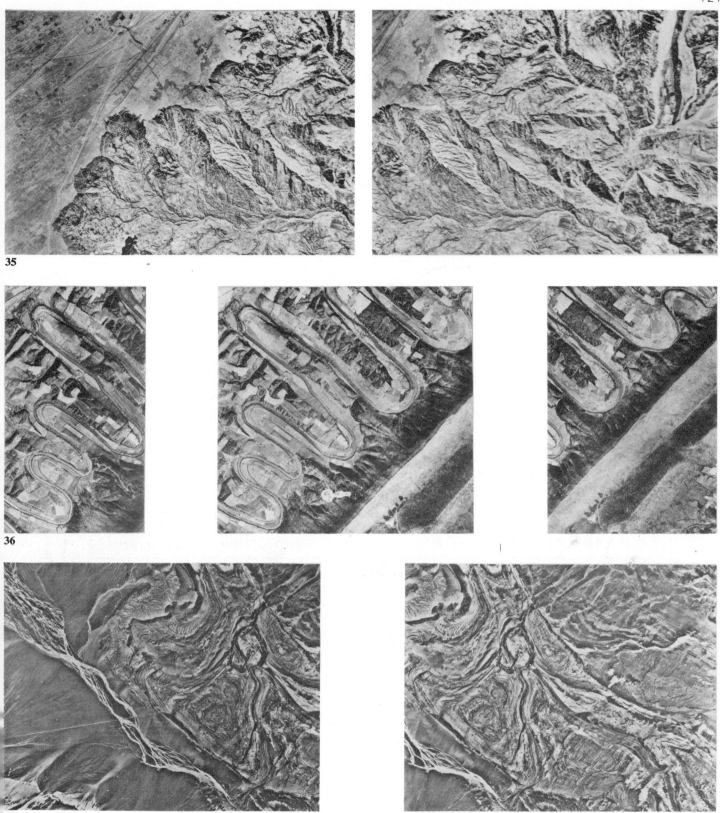

Plate 35 Stereo group: the Andes, south of La Paz, Bolivia. The Rio Mallasa, a tributary of the Rio La Paz, has headwaters cutting into the edge of the Altiplano (3500 m). Notice the extent of the drainage basins and the way in which the streams have cut deep valleys. The Achocalla mudflow resulted in the tumbled relief of the eastern part of the Mallasa basin: it descends 600 m in 10 km, and originally ponded back the rivers to form a lake in which 45 m sediment was deposited.

Plate 36 Stereo group: meanders in the North Fork Shenandoah River, Virginia. These meanders are incised 30–50 m below shale limbs, and have sinuousities of 3.2. The ridges to the south are of quartzite. (USGS)

Plate 37 Stereo group: folded rocks in the Atacama desert of northern Chile (Antofagasta Province), with a braided stream bed crossing a basin of sedimentation, filled largely by mudflow deposits. (USGS)

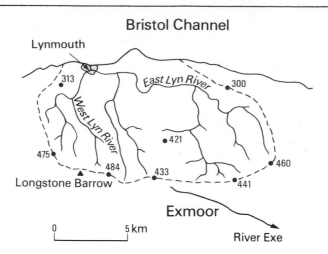

Figure 1 The Lyn catchment and Lynmouth, north Devon.

Lynmouth, recorded 230 mm of rain during the 24 hours. Such rainfall has been exceeded only four times since official records began in Britain 100 years ago. Remember that ordinary steady rain for 24 hours means less than 25 mm in a rain gauge. 230 mm represents over a quarter of a million tons of water per square kilometer. Throughout the whole area rainfall was about 95 mm, weighing about 90 million tons. The heaviest local fall was probably 100 mm in one hour.

The state of the ground made conditions worse. It was already waterlogged, and just below the surface a layer of rock prevented any appreciable percolation of water. Moreover, the East and West Lyn rivers, and their tributaries, fall 500 m through funnel-like gorges to Lynmouth in less than 7 km. Down these gorges on that terrible night pounded millions of tons of flood water. In dry summer weather the rivers have a depth of a few centimetres; now, at times, a solid wall of water nearly 15 m high raced down to the sea at 30 km per hour. Such a torrent is irresistable to everything except the heaviest and most solidly based objects. The water gouged out huge rocks and boulders — some weighing 15 tons — and carried them to the shore. Telegraph poles and motor cars followed. Trees, felled by earlier gales, and others washed out by the roots, were swept into the sea. The next morning, half a mile out to sea, hundreds of trees, presumably weighted down by rocks and soils entangled in their enormous roots, had their upper branches showing above the waves — a fantastic sea forest of stunted trees.

The flood waters dug deep into the earth. Road surfaces were scoured away, and the soft earth of the verges was gouged as by a giant excavator, some gullies being 7 m deep — right down to bare rock. The Lynmouth sewerage system and water mains were wrecked. A vivid illustration of this gouging effect of the flood occurred at a Lynmouth garage. Here petrol tanks were scoured from their foundations and swept away without trace.

When dawn broke the scene on the shore was fantastic. It was littered with the debris of scores of wrecked homes and buildings; smashed cars; telegraph poles; tree trunks, branches and complete trees; the smashed and mangled remains of the undergrowth from the surrounding countryside; some 200 000 m³ of silt, mud, gravel and stones, in some places massed 8 m high; some 40 000 tons of rocks and boulders; iron girders and bridges; broken masonry; and the bodies of animals, birds, fishes — and people.'
(From *'The elements rage'*, by F. W. Lane)

b) The Virginia Flood of 19 August 1969: the effects of Hurricane Camille in the James river basin. The James river drains an area of central Virginia extending back from the Atlantic coast into the Appalachian ranges (Figure 2). The floods of 19 August 1969 were due to a rainfall which is not likely to recur more than once in 1000 years. The combination of moist, thundery weather and the stirring action of the late stages of Hurricane Camille gave rise to totals of up to 675 mm in 8 hours — almost the maximum theoretically possible in that part of the world (cf. the previous state record of 205 mm in 12 hours). Totals of over 100 mm fell over the whole basin area. When this water reached the ground it was channelled in the steep-sided headwater valleys, and the flow of the river at Richmond reached 6092 cumecs. This will not be equalled more than 10 times in 1000 years (i.e. what is known as a 100 year flood). The damage caused to houses, and particularly

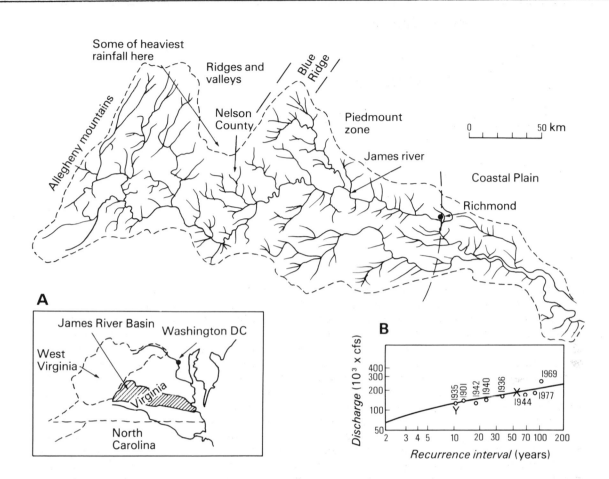

Figure 2 The James river basin, Virginia
(A) Maps showing the main features of the basin.
(B) Flow frequency curve at Cartersville. A peak of 125 × 10³ cusec will be exceeded at intervals averaging 10 years; a flow of 180 × 10³ cusec (X) will be exceeded at intervals averaging 50 years. (After Kelly, 1971)

the loss of life (152 people died) were due largely to the facts that the deluge came in the early hours of the morning and that it was not expected: it had seemed that Hurricane Camille was dying out during the previous day.

The water moved vast quantities of mud and boulders down slopes and down the valley. The areas of highest rainfall were scored by erosion scars and landslides, and deposits up to 10 m thick resulted. The local rocks are impermeable and the sheer volume of water caused the soil — and the vegetation growing on it — to lose cohesion. Much deposition took place at points where steep, narrow ravines opened into wider valleys: one of these piles contained boulders up to 3 m across and the mass was over 30 m wide by 60 m long; most were fairly thin and a typical fan-shaped deposit was 250 m long, 30 m wide and 0.5 m thick (i.e. 7000 tonnes of sediment). Valley floors were strewn with deposits: a stream which normally flows in a channel 1.0 – 1.5 m wide and in which the water is less than 0.5 m deep, deposited sand 2 – 3 m thick over an area 150 m long (14 000 tonnes).

Damage was immense and costly, especially in terms of farmland, stock, crops and business premises. In addition over 300 km of highways needed rebuilding and 94 bridges were destroyed. The Highway Department expected to take 2-3 years to repair this damage.
(Details from an account of the Virginia Flood of 1969, prepared for the US Geological Survey and the Department of Conservation and Economic Development in Richmond, Virginia by D. Kelly.)

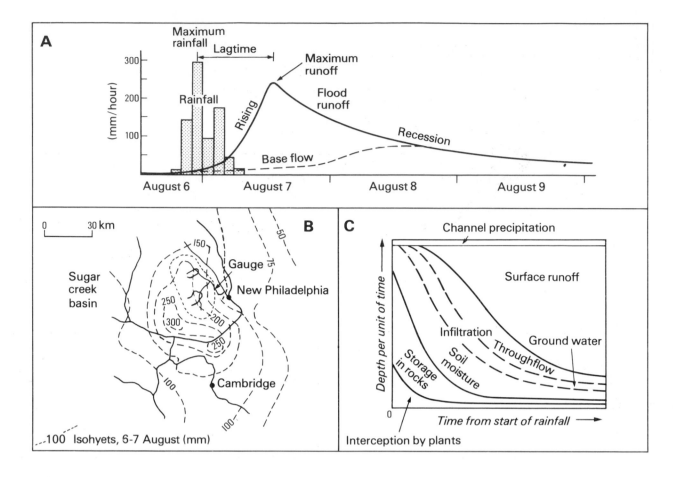

Figure 7.6 Aspects of runoff following a storm.
(A) Flood hydrograph. This shows the effect of a storm on flow in Sugar Creek, Ohio, in 1935. The stream basin drained is 805km². Notice the time lag between the maximum rainfall and maximum runoff, and the time taken for the stream to pass most of the water falling over its basin.
(B) Rainfall distribution over the Sugar Creek basin. The average total for the storm was 160mm, but the Sugar Creek stream discharge only accounted for 75mm of this: more than half was 'lost'.
(C) The changing pattern of runoff supply following a storm. Relate this to the information supplied in (A).

((A), (B) after Hoyt and Langbein, in Strahler, 1969; (C) after Linsley, Kohler and Paulhush, in Morisawa, 1968)

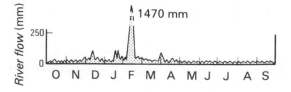

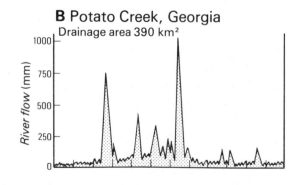

Figure 7.7 Annual hydrographs for three streams in the USA. Relate each to local rainfall regime, relief and rock-type where possible. Thus Florida is an area of limestone and low relief. (After Foster, in Strahler, 1969)

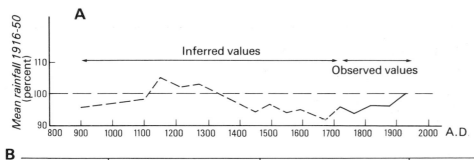

Present climatic type (and change)	Mean annual temperature (°C)		Mean annual precipitation (mm)		Values for changed climate (present = 1)		
	Present climate	Changed climate	Present climate	Changed climate	Mean annual runoff	Mean annual sediment yield	Mean annual sediment concentration
Central European (glacial)	10°	5°	500	750	5	0.5	0.1
British Isles (glacial)	10°	5°	750	1000	3	0.8	>0.2
Semiarid (glacial)	15°	10°	250	500	>20	>2.0	–
Tropical (glacial)	15°	10°	1000	1250	2	0.9	>0.5
British Isles (interglacial)	10°	12.5°	750	625	0.5	1.5	3
Tropical seasonal (interglacial)	15°	17.5°	750	625	0.3	1.7	7

Figure 7.8 Variations in climatic conditions over time.
(A) Rainfall variation over England and Wales since 900 AD. Before 1740 AD information is inferred from botanical and historical data. How would such changes affect stream flow? (After Lamb, 1966)

(B) Possible effects of climatic changes upon stream outputs.
(After Schumm and Dury, in Gregory and Walling, 1973)

Meltwater

Glaciers form important storage areas within the hydrological cycle: they hold approximately 75 per cent of the world's fresh water, equivalent to the total of 60 years' precipitation over the entire globe. Streams fed by glacial meltwater have a maximum of flow in the summer months, with distinctive diurnal variations during that season (Figure 7.9). There is no close relationship with input of snowfall for the same year, since some glaciers are retreating due to an excess of melting over supply, whilst others are advancing (chapter 8). At times glacier ice may block a valley exit for meltwater from other nearby glaciers, and the sudden release of this dammed water may cause floods.

The supply of sediment

Apart from water, the other major input to the stream system is sediment in solid and solute form. The various agencies producing and transporting this to the stream have already been discussed (chapters 5 and 6).

The particular ions and their relative proportions in the total solutes entering a stream will be controlled by the rock-types underlying the drainage basin (Figure 7.10). The solid sediment includes clay, silt, sand, gravel, pebbles and boulders.

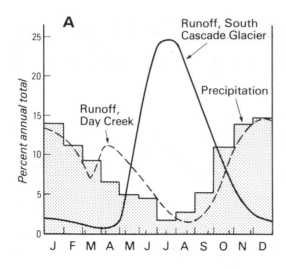

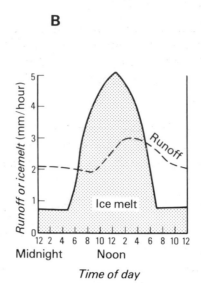

Figure 7.9 Meltwater flow.
(A) Runoff from Day Creek, in the foothills of the North Cascade Range, Washington, USA: this stream has no perennial ice in its basin. How is this yearly pattern, or regime related to precipitation and local snowmelt? How does it contrast with runoff from the South Cascade Glacier?

(B) Ice melt and runoff during the course of a day at the South Cascade Glacier (average 6–11 August 1961). Ice melt is measured over the glacier surface; runoff in a subglacial stream.
(After USGS pamphlet 'Glaciers, a water resource')

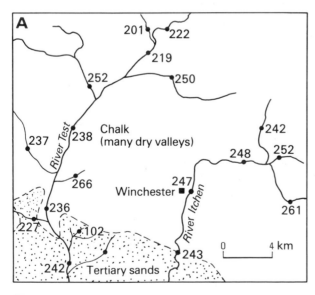

Figure 7.10 The chemical composition of stream water.
(A) Changes of chemical quality in streams in the southern Hampshire basin. Numbers refer to calcium carbonate in solution (parts per million). The streams crossing the Chalk have high values, which are maintained as they leave that rock; those rising on the Tertiary sands have much lower values. (After Pitty, 1971)
(B) The influence of rock type on the chemical composition of stream water in Bohemia, Czechoslovakia. Contrast the importance of each ion, and of the total soluble load, for each rock type. The Cretaceous rocks include limestones, sandstones and shales. (After Clarke, 1924, in Gregory and Walling, 1973)

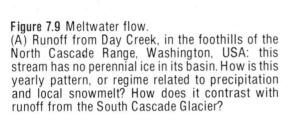

Rock type	Ion concentration in stream water (mg/l)						
	Calcium	Magnesium	Sodium	Potassium	Bicarbonate	Sulphate	Chloride
Phyllite	5.7	2.4	5.4	2.1	35.1	3.1	4.9
Granite	7.7	2.3	6.9	3.7	40.3	9.2	4.2
Mica schist	9.3	3.8	8.0	3.1	48.3	9.5	5.4
Basalt	68.3	19.8	21.3	11.0	326.7	27.2	5.7
Cretaceous rocks	133.4	31.9	20.7	16.4	404.8	167.0	17.3

Outputs of the stream system

The interactions of the inputs of water and sediment with the characteristics of the drainage basin (i.e. shape, size, relief, rock-type) give rise to particular patterns of water flow, stream channel shape, and the transportation and deposition of sediment. A particularly interesting relationship exists between the inputs and rates of denudation.

Stream channels

Water which does not sink into the soil or evaporate flows over the surface. At times this may occur in a sheet-like mass, but is normally channelled into the lowest depressions of a slope and thence to the valley bottom. Stream channels range in size from the tiniest rills and gullies on slopes to major rivers. These channels carry nearly all the water supplied by precipitation but not lost by evapotranspiration or deep storage in rocks, together with the debris resulting from weathering and slope processes. Overbank floods, demonstrating that the input of water is too great for the channel to cope with, occur relatively infrequently. Most stream measurements to date have been made in the humid temperate regions (Figure 7.11), and in these areas it is plain that rivers flow quietly within their channels for a large proportion of their time. The water level in their channels rises and floods over their banks on few occasions — less than once per year for most temperate rivers.

	River Tamar Gunnislake, Cornwall	Seneca Creek, Maryland, USA
Area drained	896 km²	256 km²
Daily discharge giving bank-full condition	225 cumec	84 cumec
Occurrence of bank-full and overbank floods	1 per year	1 per 2.3 years
Mean discharge	26.5 cumec	3 cumec
Number of days per year when mean discharge exceeded	110	125
Maximum (estimated) discharge	453 cumec	455 cumec
Minimum recorded discharge	0.75 cumec	0.05 cumec

Figure 7.11 Some figures of discharge relating to steams in the western part of England and the eastern USA. Note relationship of mean discharge to bank-full discharge (bank-full is where the channel is just brim full), and the proportion of time when low-flow conditions obtain.

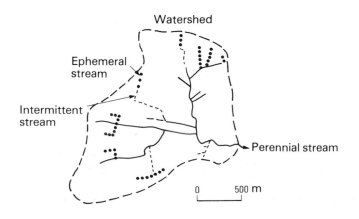

Figure 7.12 Changing patterns of stream flow in valleys. The drainage network functions in relation to the input of water. Perennial streams flow at all times, with a stable source of groundwater. Intermittent streams flow for only part of the year, when the water table is high enough. Ephemeral streams flow only after heavy rains for a short period or direct runoff. The network shown is a small drainage basin in east Devon. (After Gregory and Walling, 1973)

Many of the variations in water inputs to the drainage basin are accommodated by temporary extension of the drainage network. Some streams flow only in time of heavy rains (Figure 7.12).

Stream channels fall into three types in plan view: all may be found along the course of one stream.

a) The **straight channel** is not very common, except in artificially regulated drainage systems. In natural systems there may be stretches of 100 metres or so at the most, and these can be divided into alternate shallow and deeper sections — named riffles and pools after terms used by fishermen (Figure 7.13). The origin of the deeper sections may be due to the main flow of water being concentrated along the thalweg, preventing deposition. If this is so it would suggest that water will tend to adopt a sinuous path of flow even when its initial path is straight.

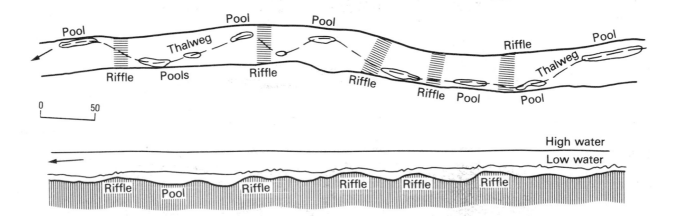

Figure 7.13 Features of a straight section of stream channel: alternating riffles and pools. The pecked line in the upper diagram, joining the pools, is the thalweg, which is the line through the deepest points in the channel. Notice how the thalweg wanders from side to side of the channel.

b) **Meandering stream channels** are the most common pattern, and occur anywhere along the course of a stream. Definite relationships have been established between such factors as meander wavelength and stream width (Figure 7.14); and, less clearly, between meander wavelength and stream discharge. This suggests that meandering channels are the normal and stable patterns related to river flow, rather than an accident due to special conditions (e.g. an obstacle deflecting the flow; or a course with a particular gradient). It is nevertheless true that meandering sections are particularly common in areas where fine silt or clay compose the dominant load, or where there is a moderate gradient of the stream channel. Flow of water in a meandering channel takes on a spiral course, like a screw thread, and this tends to maintain the meander form, but it is not known whether the flow pattern causes meandering, or vice versa.

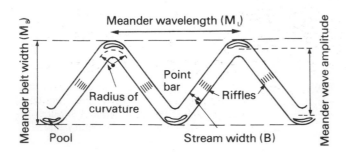

Figure 7.14 General features of meandering channels. Certain relationships have been established from measurements: thus $M_L = 6{-}10B$; $M_B = 14{-}20B$.

c) **Braided stream channels** are associated with coarser bed load and/or steeper gradients than are normally found with meanders. The stream divides and recombines as for most of the year it must find its way through or around piles of boulders or pebbles, or banks of sand, changing its route as one way becomes blocked, but never occupying the whole floor of the stream channel. Braiding is particularly common in situations where there is a marked season-al variation in stream discharge. This occurs below a glacier, around the margins of an ice sheet, or in semi-arid areas. The load is extremely coarse, giving rise to a high proportion of bed load which can be moved only during a period of maximum stream discharge. Alpine streams are typically braided, due to their steep gradients and maximum runoff following snow melt, but braided sections also occur on British streams. In cross-section stream channels tend to have steep sides and flat beds — rectilinear, rather than rounded in cross-section.

d) **Anastomosing channels** also divide and recombine, but on a larger scale, with hills between the branches. They are thought to result from the change from arid to humid climate (chapter 13).

The occurrence of the different types of channel forms is the result of response to stream processes. Elements of seasonal water input regime, the nature of sediment input, and channel slope are involved (Figure 7.15).

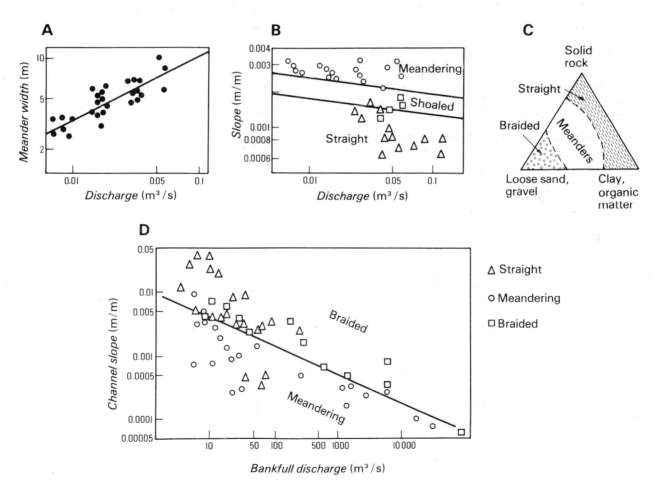

Figure 7.15 Relationships of channel form with other stream variables. (A) Meander width and discharge. (B) Slope and discharge, measured in laboratory flume studies. (After Ackers and Charl-ton, 1970)

(C) Channel patterns and local materials. (After Tanner, 1968)
(D) Slope and bank-full discharge, related to braided and meandering stream channel patterns. (After Leopold and Wolman, 1957) (All in Gregory and Walling, 1973)

Water flow in channels

The water in channels flows downhill under gravity. At very low velocities in artificial channels water may flow in laminar fashion, in which one layer moves smoothly over another, but in natural streams the water flows in turbulent manner with many eddies due to the irregular nature of the channel and boulders on the bed. At a particular point water within the stream may be flowing upwards, downwards, downstream, upstream, or across the channel. The maximum flow occurs just below the stream surface and above the deepest part of the channel: movement at the sides of the channel and along the bed is retarded by friction, and there is also a slight frictional effect between the water surface and the air above (Figure 7.16). The flow of water in a river is expressed in terms of the volume flowing past a point in a given time. This is known as the discharge, and is calculated from measurements in the stream channel:

$$\text{Discharge} \quad = \quad \text{Velocity} \times \text{Channel cross-section area}$$
$$Q \qquad\qquad V \qquad\qquad\qquad A$$

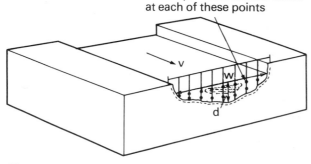

Velocity measurements taken at each of these points

Key

------- Wetted perimeter

Maximum flow

v   Velocity: speed of water flow (m/s or feet/s)

w   Width of stream

d   Depth of stream

Figure 7.16 The stream in its channel

w x d = A(Cross-section area of channel occupied by stream)

V x A = Q(Discharge of stream, measured in cumec (m³/S) or in cusec (ft³/S); 1 cumec = 34.3 cusec)

The cross-section area is calculated after detailed measurement or, approximately, by multiplying the depth of water in the channel by the width of the occupied channel. The velocity is measured by a current meter at selected intervals across the channel so that the mean may be calculated.

The discharge of a stream is related to other aspects of water flow in channels: as the discharge increases, so do depth, velocity and width of the water, in a regular relationship. This means that, once a series of measurements have been taken at a point on the stream, and the relationship between them established, future measurements can be made by reference to just one of the parameters. It is common to compute the discharge of a stream by measuring its height on a permanent calibrated stage pole and relating this to a graph (Figure 7.17). Continuous monitoring of river flow is carried out by recording the variation in water level, and converting this to discharge figures.

Similar relationships between discharge and stream channel parameters exist as the water flows downstream. Thus in humid climates tributaries join the main stream and increase the discharge, which is also associated with increased width and depth of the channel, plus — surprising to some — increased velocity (Figure 7.18). This fact is a part of the changing relationships within the stream system (Figure 7.19) and the interacting nature of the factors involved.

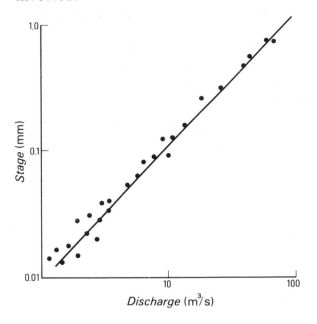

Figure 7.17 A streamflow rating graph. The dots represent measured values. The line (regression curve) averages them. These values are for the river Axe at Cheddar. (After Hanwell and Newson, 1973)

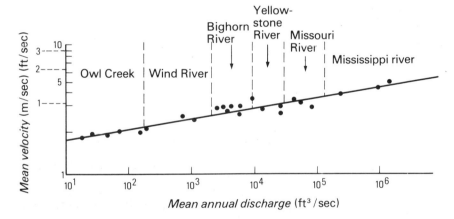

Figure 7.18 The Yellowstone–Missouri–Mississippi system. Velocity related to discharge: the dots represent gauging stations. Velocity increases downstream. Both scales are logarithmic. (After Leopold and Maddock, in Bloom, 1968)

Stream load and transport

The water flowing in a stream channel may be clouded with sediment. The movement of sediment by a stream is a sign that work is being carried out. If it is flowing through an area of clay rocks and soils, or if it is viewed in a tidal section, this may be a permanent feature, but where the rocks are sandy. or where they break into larger fragments, the river may remain clear for long periods and it is seen that the river is doing no work with the visible load, although the solution load is always being moved. The work carried out by a river is related to the energy available to it. Each stream has a certain quantity of potential energy, determined by the height of the source region and the volume of water entering the system following circulation through the atmosphere. This energy is converted to kinetic energy as it moves through the system, and then to heat energy by friction along the channel margins and by internal turbulence, this heat being lost into the atmosphere. Such a transformation is demonstrated by the observation that meltwater streams cut channels in the surface of glacier

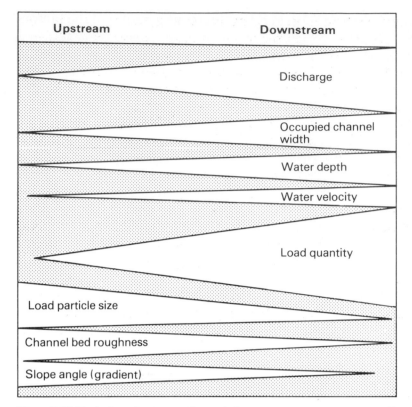

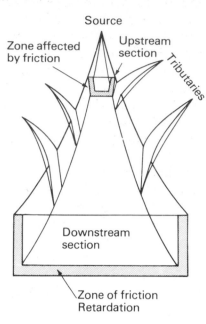

Figure 7.19 Variations in stream channel characteristics from the upper reaches to the mouth. How far are these interdependent? What is the explanation for increasing velocity downstream?

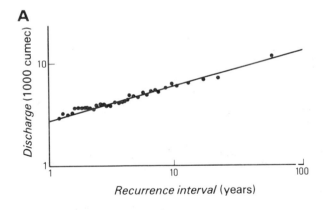

Figure 7.20 Frequency of work accomplished by rivers.
(A) A flood frequency graph for the Licking river, Ohio. (After Alrymple, 1960, in Morisawa, 1968)
(B) The time required to transport various percentages of total suspended load (After Leopold, Wolman and Miller, 1964)

B	Drainage area (km²)	Percent of total annual suspended load carried during			Days/year required to transport 50% load
		Maximum day	Maximum 10-day period	Events which recur 1 day/year	
Colorado River, Arizona	352 768	0.5	4	92	31
Rio Puerco, New Mexico	13210	5	31	82	4
Cheyenne River, South Dakota	22426	5	28	78	4
Niobrara River, Nebraska	7680	2	7	95	95

ice. It has been suggested that up to 95 per cent of the energy available is transformed to heat during the normal flow of streams. During floods, however, the discharge may be 10-100 times the average flow, frictional effects are lowered for the water body as a whole, and much more energy is available to carry out work. It is at this time that a stream is most active (Figure 7.20). The fact that floods occur so rarely makes it seem that river action takes place in a catastrophic series of events rather than in a slow, inexorable fashion. This view, however, is related to a man's life-span, rather than to geological time, which involves thousands and even millions of years. If a major flood occurs every 300 years, that is 33 000 major floods and 3300 catastrophic floods every million years.

The load of a stream consists of sediment particles, and can be divided into three main fractions.

a) The **bed load** includes the largest sediment grains or boulders, and is moved by sliding, rolling, or bouncing (known as saltation) along the bed. It is moved along on, and supported by, the bed of the channel. The bed load is moved only occasionally: normally it lies undisturbed on the stream bed and protects it. Considerable energy may be required to set it in motion, but once in motion less energy is needed to keep it moving. It then becomes the stream's tool for erosion (page 137). Sand particles (up to 2 mm diameter) may be moved more often than the pebbles (up to 10 cm), which will be moved only at bank-full stage — and these move more often than large boulders. Boulders a metre or so in diameter may be moved only in the largest floods.

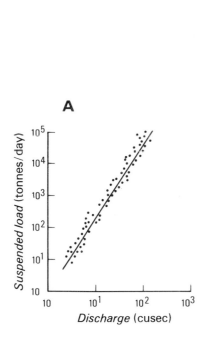

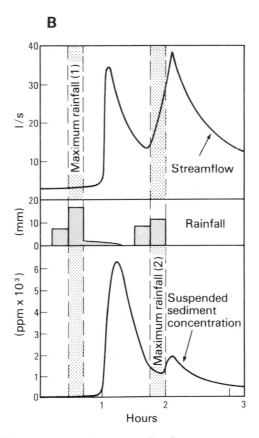

Figure 7.21 Discharge and suspended load. What do these diagrams tell us?
(A) Sediment rating curve for Powder river, Arvada, Wyoming. (After Leopold and Maddock, 1953, in Bloom, 1968)
(B) Response to two closely spaced small storms. The suspended sediment moved during the second storm was lower due to the effect of the first storm. (After Gregory, in Brunsden and Doornkamp, 1974)

134

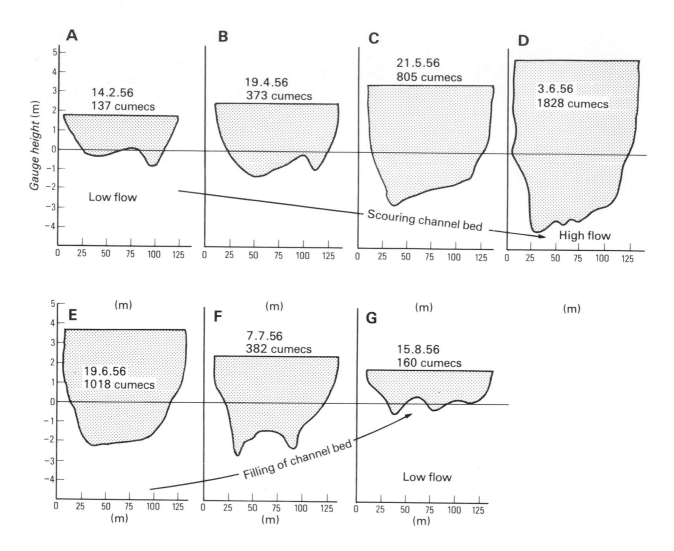

Figure 7.22 Scour and fill during the annual flood passage at Lees Ferry, Arizona on the river Colorado. How does the gauge height increase with discharge? (After Leopold, Wolman and Miller, 1964)

b) The **suspension load** consists of silt and mud which are carried in times of moderate/high flow, and may travel up to several kilometres at such times. The particles are carried (i.e. suspended) within the water and become part of the fluid mass: measurements of suspended load concentrations are shown as parts of sediment mass per million parts of water volume — $g/10^6 cm^3$ (ppm). At the highest rates of flow sand particles may enter this suspended part of the load. At all stages the river discharge will determine the competence (i.e. the largest grain size which can be carried), and the capacity (i.e. the total mass of the load) in the stream. The suspended load capacity increases rapidly as a flood discharge moves through a stream (Figure 7.21), and the increasing discharge may lead to an increase of the suspended load by a power of 2–3. Thus a small increase in discharge leads to a great increase in the suspended load. Large increases in discharge cause so much sediment to be taken up by the stream that the channel bed deposits may be scoured deeply (Figure 7.22) and the underlying rock may be eroded.

The highest values of suspended sediment concentration have been recorded at Lees Ferry in Arizona, where a mean daily value of 411 000 ppm and a maximum of 646 000 ppm have been recorded on the river Colorado. This water has been described as 'too thick to drink and

too thin to plough'. High concentrations of over 400 000 ppm have also been recorded in the Yellow river, China, and in the river Esk in New Zealand after an extreme flood in 1938. Proportions vary widely. Thus in the USA mean annual values of suspended sediment concentration range from below 200 ppm in the east and north-west (where rainfall is highest) to over 30 000 ppm in the semi-arid lands (Figure 7.23). A decrease in rainfall is related to less vegetation cover and increased sediment supply. Maps of world suspended sediment yield have been produced (Figure 7.24), showing where the highest rates of erosion are to be found.
c) **Wash load** is a term which includes the matter in solution and very fine particles of mud. These may enter the stream where it rises and flow right through the system without any intermediate stops. The quantity and nature of this fraction varies with the stream and the rocks in its basin, but may constitute a large proportion of the total load moved by the stream, since the bed load and suspended load are moved only from time to time (Figure 7.25).

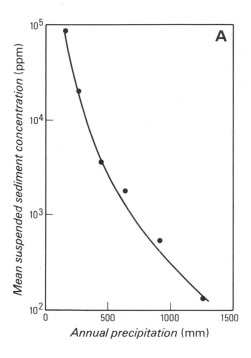
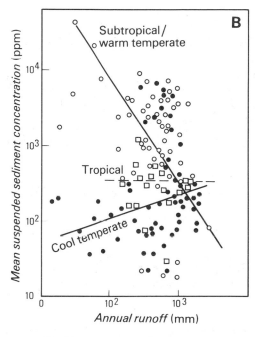

Figure 7.23 Suspended sediment concentration.
(A) The relationship between annual precipitation and mean suspended load concentration. Contrast this with Figure 7.21A, and explain the seeming contradiction. (After Leopold and Schumm, 1958)
(B) Annual runoff and mean suspended sediment concentration in terms of major climatic zones. Attempt to explain the differences. (After Fournier, 1969) (All in Gregory and Walling, 1973)

Transport of the load is related to the discharge of the stream. As the discharge increases, so does the turbulence in the water and the power to lift material off the channel bed. Whilst the wash load is carried at all times, increasing with discharge, the suspended load is picked up

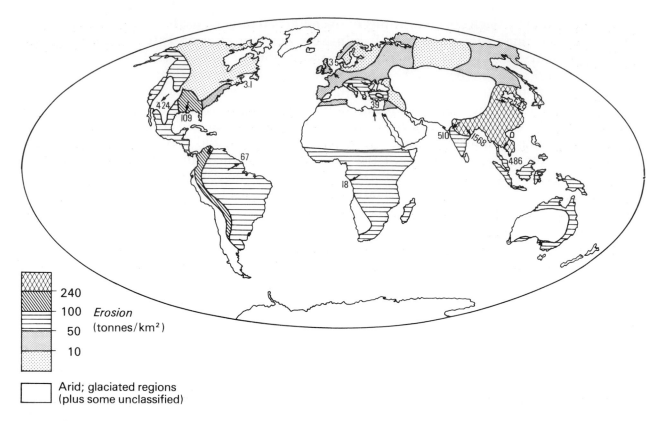

Figure 7.24 World suspended sediment yield. The arrows indicate readings for particular rivers and the numbers by them show the annual average suspended sediment yields for those rivers (tonnes/km²). (Map after Strakhov, 1967; figures after Holeman, 1968; in Gregory and Walling, 1973)

Area	Drainage area (km²)	Total stream discharge (m³/s)	Dissolved solids		Suspended solids		Dissolved: suspended solids
			(Tonnes per km²)	(Tonnes × 10³ per yr)	(Tonnes per km²)	(Tonnes × 10³ per yr)	
Atlantic seaboard	740 480	10 180	45.9	34 000	17.3	12 886	73:27
Mississippi	3 267 700	18 400	48.4	142 000	76	222 186	37:63
Total Gulf of Mexico	4 504 528	25 120	36.6	166 000	76	343 080	33:67
Pacific seaboard	1 638 000	14 130	24.2	39 300	54.9	89 872	30:70

Figure 7.25 A comparison of dissolved and suspended sediment loads of streams in the USA. (After Curtis et al., 1973, and Leifeste, 1974)

more seldom and dropped more easily, whilst the bedload is rarely moved at all. There is thus a process of 'leap-frogging' involved, with very slow movement of large particles, which are soon passed by the finer fractions.

It has been observed that the larger boulders often occur in the headwaters of streams, and the finer material in the lower courses. This has been attributed to the progressive breakdown of rock fragments as they are transported by stream action. It is possible, however, that the large boulders are moved only a few metres in rare floods and it is unlikely that they will reach the river mouth. The river Plym flows off Dartmoor in a series of steeper and more gentle sections, and the channel contains sections where large boulders are common and others (generally in the more open, low-gradient sections) where there are only large pebbles. There is no correlation of large boulders with headwater areas, or fine deposits with the lower sections of the Plym course (except for the muddy tidal section). Large boulders have entered the stream channel from the valley sides, or from quarries working close to the channel edge, and have seldom been moved more than a few hundred metres.

Stream erosion

Erosion is defined as the active wearing away of the stream channel wetted perimeter — as opposed to the picking up of loose material which has been dropped temporarily by the transporting stream. Few stream beds in Britain today are being eroded in solid rock: most, even in upland areas, have a thick covering of pebbles and boulders between the channel and solid rock; in lowland areas the alluvium is often 30 metres thick. The explanation for these facts is related to changes of climate and base-level in the past (pages 208-14). As with its transporting work, the erosional work of a stream increases with discharge. A wide variety of processes are included in stream erosion.

a) **Corrasion** is where a stream uses its load to scrape away its bed. In steep, confined sections of stream channels this may lead to fluting and pot-holing of the solid rock and the lowering of the channel (Figure 7.26).

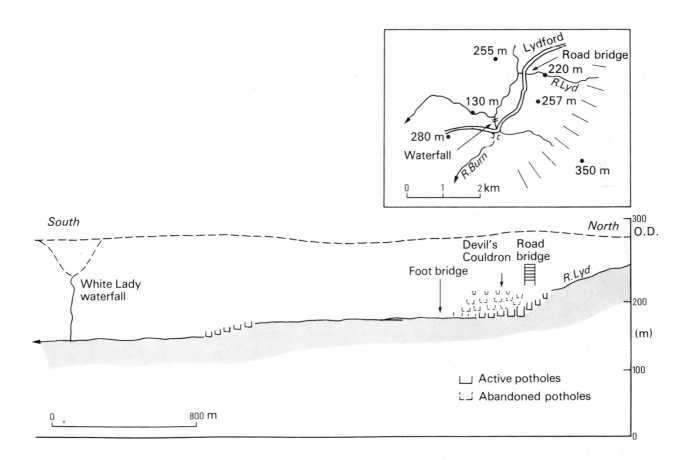

Figure 7.26 Lydford gorge, west Devon. Erosion is taking place at two sections, concentrated in groups of circular potholes which are cut into the slate rocks. This is an unusual situation, related to river capture, which has resulted in a very steep local stream channel gradient. The river Lyd formerly flowed south along the course of the present river Burn, but was diverted to the west.

b) **Corrosion** is solution by the stream water, which is important in limestone areas, adding to the solution carried out in weathering.

c) **Hydraulic action** occurs where the flow of water is responsible for undercutting banks of unconsolidated alluvium.

d) Other effects include the sucking effect of vortex-like **turbulence** within the river, and **cavitation**, a pressure effect at high velocities. In cavitation, bubbles form in narrow sections and collapse in wider sections, leading to shock waves which deliver hammer blows to the channel sides.

e) **Abrasion,** sometimes known as attrition, of the load, and particularly of the larger fractions, takes place and leads to rounding of pebbles and boulders.

f) Water may erode more powerfully when it flows under pressure in **confined conditions.** This is important in limestone areas (page 178), and in the case of meltwater being channelled through and beneath glacier ice. Water is effectively squeezed through the ice or rock, and this produces steep-sided, smoothed or fluted rock surfaces.

The erosion accomplished by these methods, which often work together, is related closely to the discharge, load and gradient of the river: the load is involved in corrasion and is itself produced partly by erosional processes. The overall denudation at the present time in humid landscapes is due to the combination of channel erosion and slope processes.

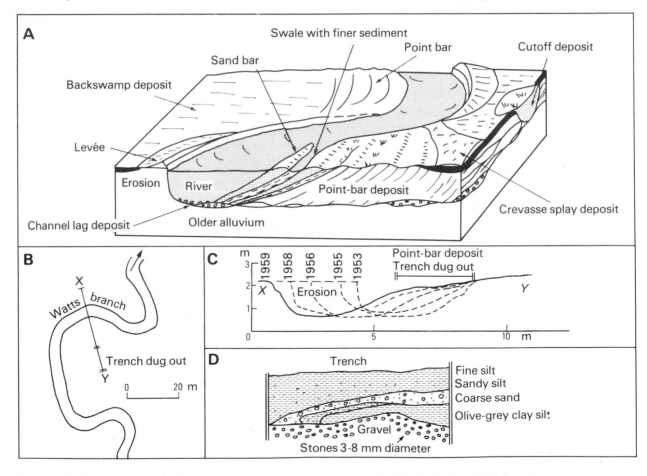

Figure 7.27 Flood plain deposits.
(A) The main types of deposit associated with meandering streams. Where do erosion and deposition take place? Crevasse-splay deposits occur during flooding, when the stream breaks through its banks and spreads gravel over the section of flood-plain near this outlet.

(B)–(D) Watts Branch, Maryland: a stream which has been studied in detail.
(B) A map of one meander. (C) A section through X–Y on (B), showing erosion and deposition over 7 years. Relate this to (A). (D) Deposits in a trench dug along part of the X–Y section. How are these related to the point bar deposits of (A) (After Leopold, Wolman and Miller, 1964)

Stream deposition

Deposition of the load occurs where the stream's capacity or transporting ability is reduced. As the stream's ability to increase its load is related closely to discharge, so, when the discharge is reduced, the stream will be unable to carry all, or part of, its load farther and deposition will take place.

a) **Flood-plain deposits.** Flood-plains are areas of relatively flat land on either side of the stream channel at bank level. The name comes from the fact that these areas are flooded as soon as the stream water overtops the banks of the channel. Flood-plains may be found throughout a stream valley, even extending in restricted fashion into upland areas. The stream usually follows a meandering course across older alluvium, carrying out both erosion and deposition (Figure 7.27). During normal (channel-confined) flow the stream flows in spiral fashion, eroding the outer banks, and depositing on the inside banks, of the next meander downstream. Pebbles accumulate in the deep channel, whilst sand-grade sediment builds up on the inside of a meander in a bar, behind which the finest sediment can settle out of suspension in a swale where it is soon stabilised by plants. These deposits build up laterally. In time of flood the river spreads beyond the channel, dropping immediately any coarser load around the outsides of meanders to form levées, and then spreading out to form a thin layer of water from which the mud settles, giving rise to back-swamp deposits: these deposits are built up vertically. Abandoned sections of channels (ox-bow lakes) also fill with sediment of finer grades, assisted by aquatic plants like reeds.

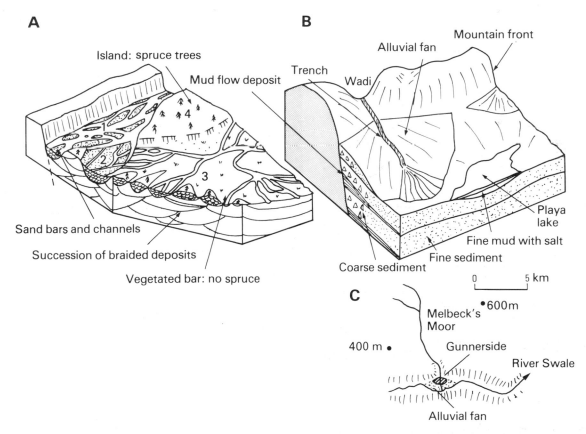

Figure 7.28 Deposits of braided channels and alluvial fans.
(A) A braided river floodplain with four levels of dissection (1–4): the lowest levels are most active, whilst the higher levels are disturbed only by the most extreme floods, and are stabilised by vegetation. (After Williams and Rust, 1969, in Blatt, Middleton and Murray, 1972)
(B) Alluvial fan deposits filling a basin to form a bahada.
(C) The alluvial fan at Gunnerside in Swaledale, north Yorkshire.

The deposits of flood-plains thus form when the stream velocity is lowered on the insides of meander bends, or when flood waters are spread across the valley floor. It seems that flood-plain deposits begin with vertical accretion at confluences, at meander crests and in abandoned channels, and that these deposits are then distributed in lateral fashion by a meandering stream and cover the whole valley floor with alluvium.

b) **Alluvial fan deposits.** Alluvial fans form most commonly where a stream leaves a confined valley and enters a flatter region. They may occur where a mountain front gives way to lowland, or where a side valley enters a main valley. The former situation is common along the margins of faulted blocks in western USA, where adjacent fans coalesce to form bahada deposits (Figure 7.28). There is an alluvial fan where a tributary enters upper Swaledale, Yorkshire.

Deposition takes place when the flow of water and debris spread out on reaching the open area of lesser gradient, increasing the width and decreasing the depth of flow. The flow may be a stream of water, or a mudflow: the stream flows typically in a braided channel and may disappear if the fan material is coarse and permeable. The fan is composed of a pile of interleaved mudflow and stream-laid sediments, and when it has been raised by deposition the flow will be diverted from one direction to form a radial series of depositional lines. The

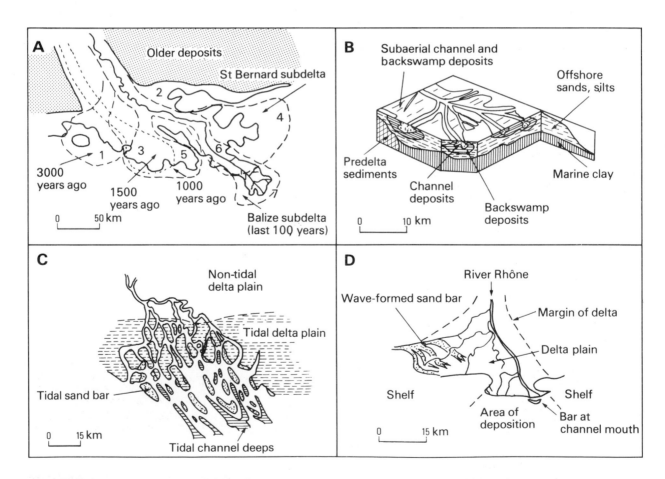

Figure 7.29 Deltaic deposits and landforms: a variety of types. (A) and (B) The Mississippi delta. Deposition by the river is dominant, leading to rapid extension out into the Gulf of Mexico along the distributaries.
(C) A delta in the Gulf of Papua, characterised by strong tides and the removal of sediment, plus gouging of tidal channels offshore.
(D) The Rhône delta, southern France. Wave action has affected the coastal margins of this delta.
((A), (B) and Frazier, 1967; (C), (D) after Fisher, 1969; all in Blatt, Middleton and Murray, 1972)

uppermost fan may be trenched by the shifting courses, and these trenches encourage mudflows to travel farther by channelling them. This also leads to greater variations in particle size and the masking of sorting processes initiated by the stream.

c) **Deltas** are formed where streams carrying heavy loads enter a lake or the sea: stream flow is checked rapidly and the load is jettisoned. They are thus most closely associated with high discharge streams which are able to carry the heavy loads. Around the Gulf of Mexico streams with less than 15 cumec average discharge never support deltas, but those with over 1500 cumec discharge always do. Deltas form where there is sufficient load brought by the stream to build its mouth out to sea. The detailed shapes of deltas are determined by a variety of factors: the depth of water which the stream enters; the tidal or nontidal characteristic of this water body; the importance of offshore currents; the density of the river water/sediment mixture as compared to that of the water body (fresh/salty) entered; and the effect of local earth movements leading to elevation or subsidence. Each delta is thus unique in form and rate of extension out to sea. The Mississippi delta (Figure 7.29) has been the most fully studied. The Ganges delta is very different in shape, being over 300 km across at the mouth and 300 km from the head of the delta to the sea. It is subject to greater tides than the Mississippi delta, and although the drainage basin of the Ganges is only one-third of the area of the Mississippi basin, the water volume flowing to the mouth each year is greater. This increased volume flow brings down six or seven times the load mass of the Mississippi each year, mainly as clouds of fine mud, which are deposited out to sea as well as on the surface of the flooded delta.

The features of these modern deposits have been used by geologists to interpret sediments formed many millions of years ago (Figure 7.30).

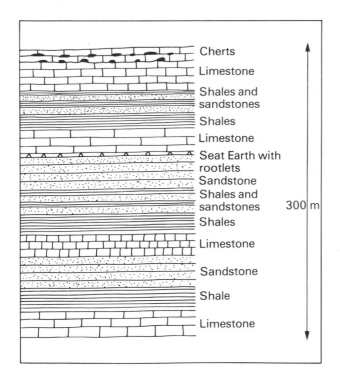

Figure 7.30 A sequence of rocks, formed 300 million years ago, and interpreted as having been deposited in a delta. Suggest how these layers could have accumulated, using the information supplied in Figure 7.29 A, D. The limestone layers formed some way in front of the delta, where the finest mud was being deposited in only very small quantities. This sequence occurs in Swaledale, northern Yorkshire.

d) **Glacial outwash (fluvioglacial) deposits.** Meltwater streams flowing on, in, or beneath the margins of ice masses, as well as across areas beyond their margins, transport and deposit sediment. Deposits formed in this way are different from those left by ice itself (Figure 7.31), because they are often stratified, sorted and contain rounded fragments; they also have distinctive characteristics when compared with other fluvial sediments, owing to the wide

Glacial till deposits **Fluvial outwash deposits**

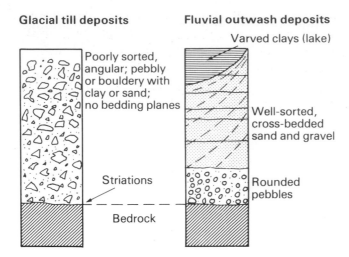

Poorly sorted, angular; pebbly or bouldery with clay or sand; no bedding planes

Striations

Bedrock

Varved clays (lake)

Well-sorted, cross-bedded sand and gravel

Rounded pebbles

Figure 7.31 Glacial and fluvioglacial deposits. The former are unstratified and the latter clearly stratified. (After Press and Siever, 1974)

fluctuations in discharge (Figure 7.9), combined with the large quantities of sediment provided by the melting glacier.

Two distinct depositional environments can be distinguished — the ice contact zone and the proglacial zone. In the ice contact zone the advance and retreat of the ice front results in many complications and re-working of deposits, with extensive changes over short periods of time, but two basic types of depositional feature can be recognised. Eskers are ridges composed of rounded gravel and cobbles with some silt and sand, and they are laid down in

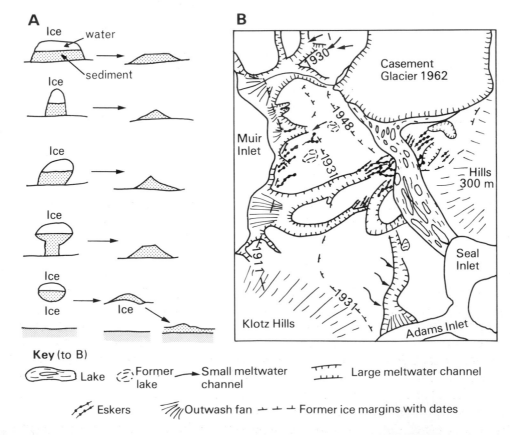

Key (to B)

⬭ Lake ⟨⟩ Former lake ⟶ Small meltwater channel ⊥⊥⊥ Large meltwater channel

⫰ Eskers ⫰⫰ Outwash fan ⊣⊢⊣ Former ice margins with dates

Figure 7.32 Fluvioglacial deposits: eskers.
(A) Forms produced when the ice around subglacial and englacial streams melts.

(B) The Casement glacier, south-east Alaska: eskers can be seen in relation to other outwash features and the retreating front of the glacier. (After Price, 1973)

subglacial or englacial meltwater channels (Figure 7.32 A). They may range from a few metres long up to 400 km and the height of the ridge is commonly related to its length (e.g. those 50-200 m long are 10-100 m high). Eskers are found in meandering, tributary and distributary patterns, and seem to be formed mostly by a mass of stagnant and decaying ice (Figure 7.32 B). Kames are mound-shaped deposits of sediment which have accumulated in water at the margins of the ice mass, and then collapsed when the ice melted (Figure 7.33). Like the eskers,

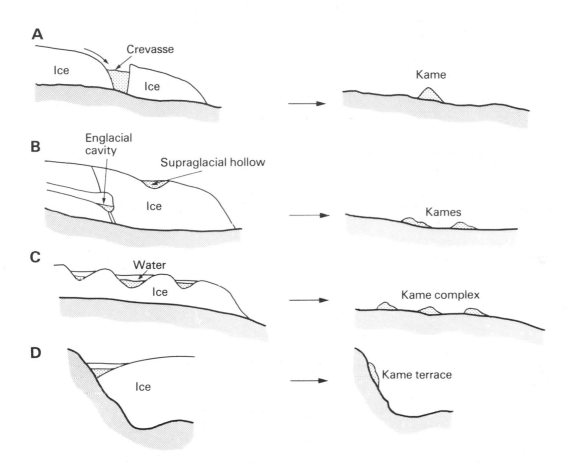

Figure 7.33 Types of kame formation. The sediment accumulates in water trapped around and in the ice, and forms mounds when the ice melts. (After Price, 1973)

kames are usually composed of gravel and sand, and are formed mostly at a decaying ice front. They are associated with kettle-holes, left by the melting of detached blocks of ice.

In proglacial zones the fluctuations of discharge result in overbank floods and the deposition of fan-shaped spreads of gravel, sand and silt, known as sandar (singular: sandur). In valleys the confinement of the side slopes leads to the concentration of deposition in a single zone, but in open plains wide areas are covered, broken by the flat-floored braided channels of streams flowing in the season between high and low water (Figure 7.34). The finer material washed from the glacier may be carried by a stream into a lake, where it is deposited in laminated, seasonal varves, or is taken into the sea.

Both the ice contact deposits and the proglacial sediments are found intermingled with deposits left by the decay of glacier ice, to provide problems of interpretation for the student of those areas (chapter 8).

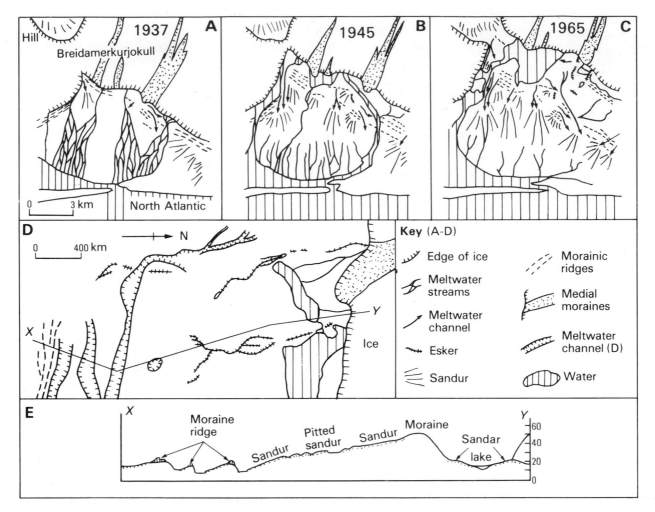

Figure 7.34 Sandur and proglacial features.
(A)–(C) The retreat of Breidamerkurjokull, Iceland, 1939–65, exposing glacial deposits (moraines) and fluvioglacial features (lakes, channels, eskers, sandur).

(D), (E) An enlarged map and section of the 1965 outwash plain, with channels cut into sandur and moraines modified by fluvial activity. (After Price, 1973)

The stream long profile

The long profile of a stream is often depicted as a simple, concave-upwards curve, steeper in the headwater area and flattening towards the stream mouth. This is a model which is broadly correct, but it is rare to find a natural stream with such a simple profile. The river Plym profile, in south-west England (Figure 7.35), is composed of a number of short concave segments, and many British rivers have this type of profile — although the breaks between the sections may not be so clearly marked. The sections of steeper gradient are associated with narrower channels and experience greater flow velocity, erosion and transport during periods of high discharge. During normal, low discharge all sections are shallow, boulder-strewn channels, and the steep gradient overcomes the loss of energy induced by the increased turbulence.

Debate concerning the significance of the long profile in the evolution of streams, and of the associated landforms, has continued through the twentieth century. It is another interesting example of the ways in which the study of stream processes and landforms has changed.

W.M. Davis, the American geographer whose ideas have been discussed in connection with slope evolution (chapter 6), saw the development of the stream long profile as an integral part of the evolution of the entire landscape. His idea was that a landscape developed from an initial uplift of land into which rivers cut steep-sided valleys: this was the 'youthful' stage, during which the stream long profile would be irregular with waterfalls and lakes. The load

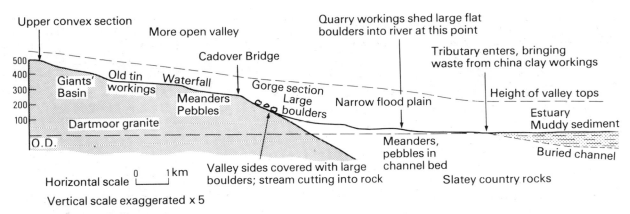

Figure 7.35 The long profile of the river Plym. Notice how the largest boulders (diameter 1–3m) are found only in, and at the foot of, steeper sections. What does this suggest about erosion and transport by this stream; about the contribution of valley slopes to the stream load; and about man's effect on the stream channel?

carried would increase in quantity and coarseness as valley deepening progressed. This stage would be followed by 'maturity', in which the stream profile would attain a slope of equilibrium where capacity to do work (i.e. erosion, transport) would be balanced by the quantity of work to be done. The load would still be increasing, but not necessarily in coarseness. This state of equilibrium was known as the 'graded profile' and was seen as a smooth, concave-upwards curve related to the base-level (i.e. the level below which the stream could not deepen its channel, normally sea-level). The graded state was seen by Davis as beginning in the lower sections of the stream and working up to the source. Downcutting would continue as a gradual process during which the valley side slopes were reduced in gradient and the velocity-load relationship altered: the final state of 'old age' was designated a peneplain, with low-angle slopes and the lowest stream profile possible to carry water and sediment from source to mouth.

This entire scheme was envisaged at a time when few measurements of stream processes had been made, and whilst Davis mentioned different sizes of load he did not attempt to quantify. Criticism of the idea of the graded profile came when it was found that the balance between erosion and deposition was not achieved, and that the type of load did not equate with the other features of a particular stage. The features of the Plym channel (Figure 7.35) demonstrate the problems of applying the concept of grade to real streams, and it is now regarded as having little contribution to make in the study of processes or landforms.

A more realistic approach has resulted from the quantitative measurement of streams. This suggests that the average system of stream channels develops and produces an approximate state of equilibrium ('quasi-equilibrium') between the channel characteristics and the movement of water and sediment through them. Downstream changes occur which can be related to the channel slope, but this is determined by a variety of factors including the tectonic and climatic environments of the area, which may also change over a period of time. The channel features are the result of the interrelationship of many factors (Figure 7.19), which vary with passage downstream. Many of these interrelationships are still not completely understood.

This approach changes the emphasis from the purely theoretical to a more firmly-based concept. It can also be related to the systems approach. In this a stream is viewed as a working system from headwaters to mouth, with no section awaiting a particular 'blessed state', and where quasi-equilibrium between form and process is soon attained whatever the previous history. The inputs of water and sediment at a particular time can be related to the outputs of channel characteristics in each sector of the profile. This is a more satisfactory and practical view of the forces at work in determining the profile. In addition, the stream channel and valley slopes are now seen as two subsystems rather than an integral part of a single process model, and are studied separately (chapter 6).

38

39

40

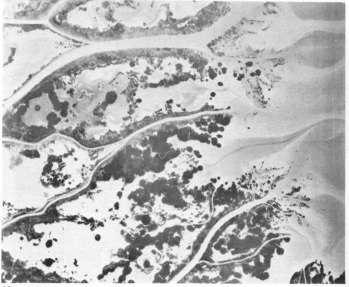

41

42

Plate 38 Stream-flow measurements in Rock Creek, a tributary of the Potomac which flows through Washington DC. The line from bank to bank allows measurements of stream velocity to be taken at regular intervals across the channel. (USGS)

Plate 39 Erosion of the outer bank of a meander on Satsop river, Washington, USA. Eight acres were lost between 1941 and 1957, but in 1957–61 a further 12 acres were removed by accelerated erosion. (USDASCS)

Plate 40 The distribution of overbank flood sediment on the flood-plain of the Washita river, Oklahoma. (USDASCS)

Plate 41 Detail of the sedimentation taking place at the Dead Women Pass mouth of the river Mississippi. Most of the area shown has been deposited since 1900, and few coastal features have had time to form. (USGS)

Plate 42 The snout of Mammoth Glacier, Wyoming, showing the contrast in size between the end moraine boulders and the gravel in the stream bed. (USGS)

Plate 43 Dendritic drainage lines form after heavy rains on a field in California. (USDASCS)

Plate 44 A dry climate in Arizona: the stream channel here is often without water. A small storm has just passed over the basin. (USDASCS)

Plate 45 Preacher Creek, Alaska: a meandering stream with many cutoffs in a tundra region. (USDASCS)

46

47

Plate 46 Stereo group: features and deposits in the river Mississippi flood-plain north of Memphis, Tennessee. Abandoned meanders and older point bar scrolls can be mapped. (USGS)

Plate 47 Stereo pair: alluvial fans at the abrupt scarp face of the Black Mountains overlooking Death Valley, California, and its salt pan floor. The largest fan is at the mouth of Copper Canyon. (USGS)

The drainage basin

Streams act as a system for channelling water precipitated over a particular area across and away from it to lake or sea storage. The area drained by a stream network is its basin, bounded by a watershed (Figure 7.36 A). The drainage basin is a natural, fundamental unit for the study of landscapes, and it is significant in terms of water supply, drainage, irrigation and lines of communication. The settlement of Britain by the Anglo-Saxons, and the later colonisation of North America, were by way of river systems. Recent stream basin development schemes like the Tennessee Valley Authority and the Mekong Valley Project emphasise the continuing dependence of man on water resources.

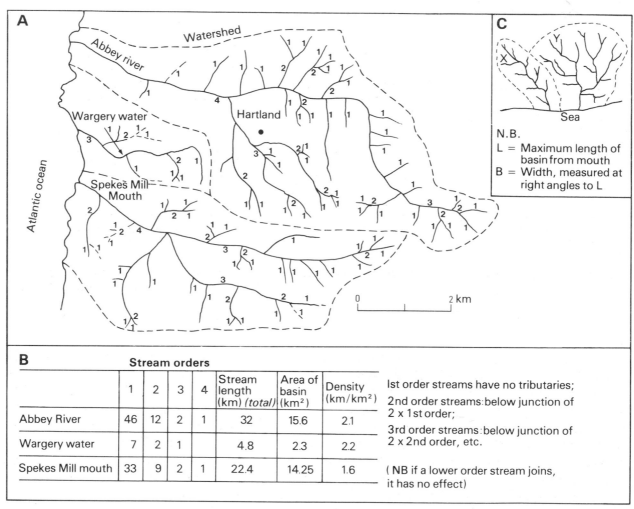

Stream orders

	1	2	3	4	Stream length (km) *(total)*	Area of basin (km²)	Density (km/km²)
Abbey River	46	12	2	1	32	15.6	2.1
Wargery water	7	2	1		4.8	2.3	2.2
Spekes Mill mouth	33	9	2	1	22.4	14.25	1.6

1st order streams have no tributaries;
2nd order streams: below junction of 2 x 1st order;
3rd order streams: below junction of 2 x 2nd order, etc.
(NB if a lower order stream joins, it has no effect)

N.B.
L = Maximum length of basin from mouth
B = Width, measured at right angles to L

Figure 7.36 Drainage basins and stream networks. (A) Three small basins in north-west Devon. The numbers refer to stream orders. The underlying rocks are sandstones and shales, highly folded along east–west lines, and cut by NE–SW and NW–SE faults.
(B) A comparison of numbers of streams, and their order, in the three basins, together with calcula-tions of drainage density and the method of work-ing out stream orders.
(C) Two drainage basins: compare the order of each basin (i.e. the highest stream order present), and compare the effect of heavy rain over the whole basin on a hydrograph drawn at each of the stream mouths.

Characteristics of drainage basins

The area and shape of the drainage basin are important characteristics, both for comparisons between different basins and for implications in terms of runoff patterns (Figure 7.36 B).

The geological characteristics of the basin are also important in relation to ground water storage, slope angle and type of sediment load produced. Vegetation types within the basin affect the interception of rainfall. Such properties of a drainage basin can be analysed in terms of their response to inputs and outputs of water and sediment, and this will have further significance in relation to the evaluation of facets of a landscape in terms of possible types of land use (Figure 7.37).

Class	General use	Soil conditions and slopes	Usage and conservation practice
I	Arable cultivation	Deep, easily worked, nearly level; little erosion.	Good farming land, maintained by use of rotations, fertilisers.
II	Arable cultivation	Moderate depth on gentle slopes; require drainage and subject to moderate erosion.	Require water control, special tillage and rotations.
III	Arable cultivation	Shallow soils on moderately steep slopes; liable to more severe erosion.	Restricted types of crops grown; special measures needed to protect soil.
IV	Arable cultivation	Shallow soils on steep slopes; severe erosion likely.	Kept in hay/pasture with one ploughing every 5-6 years.
V	Special problems	Stony or wet soils on nearly level ground.	Best left without interruption of vegetation cover.
VI	Grazing land	Shallow soils on steep slopes; liable to wind erosion.	Grazing and some forestry only with proper management.
VII	Forest land	Shallow soils on steep and rough land, dry or swampy.	Severe restrictions for use: care needed for any use.
VIII	Non-agricultural land	Poor for all uses.	Wildlife, recreation, watershed.

Figure 7.37 The Land Capability Classification of the US Department of Agriculture. This is based on soil and slope characteristics, and is related to the measures needed for preservation of the land in productive use. (After Hudson, 1971)

The stream network

Stream network analysis (Figure 7.36 A), provides another basis of contrast and comparison between basins (Figure 7.38). Drainage density can be calculated by measuring the lengths of all the streams and dividing the total by the area of the basin. Fieldwork can attempt to correlate particular orders of stream valleys in an area, and to set up models for the landforms associated with a particular order of valley. Drainage network patterns can often be related to the underlying rock structure, and certain patterns occur often enough to be given definite names (Figure 7.39). This type of evidence is important in the interpretation of aerial photographs, since a knowledge of the stream pattern/rock-type associations often leads the interpreter to suggesting the type of rock which occurs in the area.

Changes of base-level

Base-level is the lowest point to which a river can cut its channel. This is usually the same as sea-level, and the two are taken to be synonomous in this section, but it may be a more local phenomenon such as lake-level or a band of resistant rock in a stream course. Ideas of landscape evolution by rivers have normally depended on a stable base-level, but it is clear that the sea-level has varied from 200 m above the present to 150 m below the present during the last 1 million years, whilst calculations of rates of denudation (e.g. 1 m/12 000 years for Britain) suggest that it could take several million years for any landscape to be worn right down.

Streams on Dartmoor granite

Cherry Brook			Walla Brook			River Swincombe	
Order	No		Order	No		Order	No
1	42		1	38		1	55
2	12		2	9		2	16
3	3		3	2		3	3
4	1		4	1		4	1

Drainage density (Dd) = 3.6 miles per mile² 　　Dd = 2.6 　　 Dd = 3.3

Streams on Devonian rocks south of Dartmoor

Gatcombe Brook			Dittisham Creek			Badda Brook	
Order	No		Order	No		Order	No
1	43		1	57		1	51
2	9		2	17		2	14
3	2		3	5		3	2
4	1		4	1		4	1

Dd = 4.7 　　 Dd = 4.6 　　 Dd = 4.8

Streams on Culm rocks north of Dartmoor

River Yeo			Shippen Brook			Dorna Brook	
Order	No		Order	No		Order	No
1	44		1	45		1	38
2	11		2	12		2	9
3	2		3	5		3	3
4	1		4	1		4	1

Dd = 6.0 　　 Dd = 7.7 　　 Dd = 5.2

Figure 7.38 A comparison of stream orders and drainage densities for stream basins on three types of rock in the Dartmoor area of Devon. The details were calculated from 1:25000 scale maps. What differences are there in the proportions of stream orders, and in the drainage densities on the three rock types? Carry out similar studies using maps of other areas and investigating small basins (i.e. 4–10 km²). (After Brunsden, 1968)

Changes in sea-level relative to the land are caused by a range of effects, worldwide and local.

1) On the largest scale they may be caused by tectonic processes, such as the deepening of ocean trenches or the uplift of mountain ranges. These may cause displacements of water which are effective on a worldwide scale, and such changes of level are termed eustatic ('eu' — widely developed; 'static' — level). It is often difficult to determine that particular sea-level changes are of this type, since so many local changes of sea-level relative to the land also take place and mask the worldwide effects over large areas.

2) Eustatic changes of level also occur when ice sheets form or melt, leading to falls or rises of sea-level respectively. These are known as glacio-eustatic changes, and their effects are more clearly found, since they are more recent in origin.

3) More local changes may be due to warping or faulting of a section of continental crust, or to the loading or unloading of the crustal rocks by masses of sediment or an ice sheet. The latter effect is termed isostatic, since it affects the crust-mantle balance (chapter 4).

For some time it was assumed that evidence for changing sea-levels was related to changes of the land, and that the sea-level was permanent. The pendulum of ideas then swung so that the eustatic theory gained prominence. Today it is realised that eustatic changes show up best around fairly stable coastal margins, as around the Atlantic Ocean, whilst they are less clearly marked around the active Pacific Ocean margins.

Most British rivers and their valleys show the effects of sea-levels which have changed recently — both upwards and downwards. Where rivers cut across folds in the rocks (Figure 7.40), for instance, there are two possible explanations. The river may have had this course before the rocks were folded, and it was able to erode downwards as rapidly as the uplift was taking place: this is known as antecedent drainage. Alternatively the rocks may have been folded first and the stream has cut into them from a course determined by other factors at a higher level before the landscape was lowered: this is superimposed drainage.

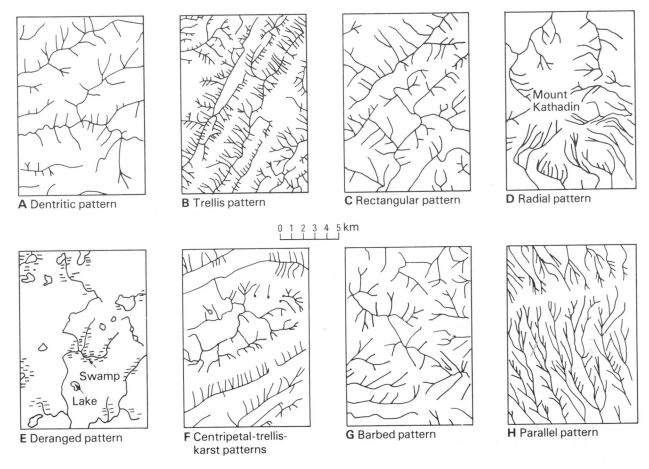

A Dentritic pattern   B Trellis pattern   C Rectangular pattern   D Radial pattern

0 1 2 3 4 5 km

E Deranged pattern   F Centripetal-trellis-karst patterns   G Barbed pattern   H Parallel pattern

Figure 7.39 Stream networks often reflect a relationship with the underlying rocks.
(A) Dendritic patterns occur where the rocks have uniform resistance to erosion (e.g. till).
(B) Trellis patterns occur with alternations of resistant and less resistant rocks.
(C) Rectangular patterns are often related to joint patterns in massive igneous rocks.
(D) Radial patterns occur around volcanoes and areas of domed uplift.
(E) Deranged patterns are the result of streams becoming re-established on a formerly glaciated terrain.
(F) Centripetal patterns may occur with karstland-forms.
(G) Barbed patterns occur with river capture, or with some joint patterns.
(H) Parallel patterns are found on steep slopes with little vegetation cover. (After Thronbury, 1954)

It is known that the rocks in one case (Figure 7.40) were folded a maximum of 30 million years ago, since rocks a little older than that date are involved in the folds. Deposits on the North Downs (the Netley Heath Deposits) show that the sea rose to 200 m above the present level nearly 2 million years ago, and it seems that the streams began to carve out their valleys as this sea receded. If this is so the antecedent explanation is incorrect, although it is possible for

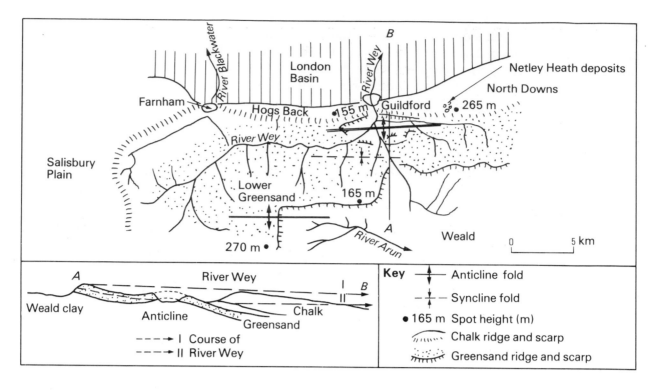

Figure 7.40 The course of the river Wey in south-west Surrey. It cuts across the Chalk ridge of the south: it thus has a course which is out of sympathy with both relief and geological structure. How could this occur? The superimposition hypothesis would suggest that the course I in cross-section A–B would have been an earlier one, cutting down to the modern course at II.

the folding movements or uplift to have continued far longer than is suggested here (see also discussion in chapter 12). The second explanation (superimposed drainage) suggests that the streams were once flowing at a height greater than, or equal to, the present height of the North Downs (200 m). If the sea-level was at this height the rivers would have drained to that base-level and have eroded the present valleys in response to falling base-level, leading to the superimposition of the streams across the underlying folded rocks.

The process of downcutting in response to falling base-level is known as rejuvenation, and has led to features such as the paired terraces along the middle and lower sections of the river Wey valley (Figure 7.41). These are remnants of older flood plains into which the stream has cut. The long profile of the river Wey also has a series of breaks in it, which may be knick-points related to the sequence of falls in sea-level — although such breaks may also be related to changes in rock-type (Figure 7.35). The process of downcutting has also led to river capture (Figure 7.40), since the river Wey has gradually abstracted more and more of the Blackwater headwaters and has left the latter as a beheaded, underfit stream trickling through a wide valley between Camberley and Aldershot.

The floor of the Wey valley, however, is not cut in bedrock for most of the river's length. There is an area of flood plain floored by alluvium of a thickness which increases towards its confluence with the river Thames. This suggests that there has been a recent rise in sea-level, leading to deposition by the river. The lower parts of the streams reaching the English Channel in Sussex — the Arun, Adur, Ouse and Cuckmere — confirm the latter observation. Each of these streams has carved a deep valley through the Chalk of the South Downs, and each has a wide flood plain beneath which the alluvium is up to 30 m thick. Thus the phase of deep valley cutting was related to a lower sea-level than the present, and the rise to the present sea-level (the most recent change) led to the filling of these valley floors, first with water and then with sediment.

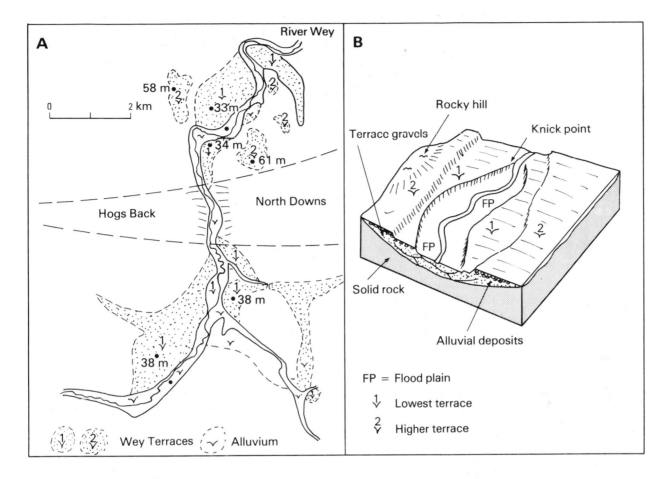

Figure 7.41 Paired terraces.
(A) Terraces along the middle course of the river Wey, north and south of Guildford, Surrey, as mapped by the Institute of Geological Sciences. The unshaded areas are solid rock. How do the heights of the terraces vary along the course of the river?
(B) Paired terraces in relation to knick points, flood plain and river valley deposits.

Changes in base-level lead to feed-back in the stream system. Gradients are increased or decreased, resulting in increased or decreased velocities, erosion or deposition. If rejuvenation and erosion are continued for long the superimposition of drainage patterns will take place, and there will be adjustments in river capture. Deposition will have the main effect of filling the area at a river mouth. If there is an alternation of periods of falling sea-level with periods or rising sea-level, the downcutting effect will be most obvious, since it will remove solid rock, whereas the intervening sea-level rise will lead to deposition of unconsolidated sediment which will soon be flushed out when the sea-level falls once more.

Changes of climate and drainage basins

Many of the features of present-day river valleys can be explained by referring to past changes of base-level; others are probably related to past climatic conditions. This much is clear where a small stream today drains a wide valley which was once filled by a glacier. In areas beyond the farthest extent of ice sheets during the Quaternary Ice Age lakes were more extensive in arid areas (e.g. Great Salt Lake, Utah, is a shrunken remnant of the large Lake Bonneville), which also experienced a greater influence of running water than at present.

In humid areas it is often found that the valleys meander, like the streams they contain, but in most cases the stream meanders are smaller than, and apparently independent of, the valley meanders (Figure 7.42). Such streams are known as underfit streams. Even where the streams

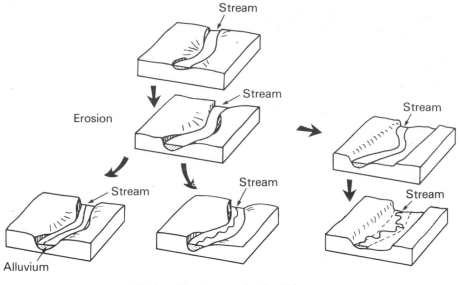

B Manifestly underfit (1)

A Underfit: Osage type   **C** Manifestly underfit (2)

Figure 7.42 Meandering valleys and underfit streams.
(A) Osage type: the stream meanders within the valley, but the channel does not occupy the whole width of the valley floor.
(B) The stream meanders within the meanders of the valley (cf. Figure 7.41A)
(C) The stream meanders on a wide flood-plain, but the meander belt also meanders within this framework. (After Dury, in Chorley, 1969)

meander with the valley, the distribution of pools and riffles in the channel is not as the laboratory flume experiments would suggest (Figure 7.14): the riffles come more frequently — another reflection of the stream being underfit. G.H. Dury has calculated that a period of increased storm activity and precipitation giving rise to an increase of 50-100 per cent over present rainfall totals and lower evaporation rates would be sufficient to account for a higher degree of erosional activity by streams (Figure 7.43) which could be responsible for cutting the valley meanders. There would be a great increase in the number of occasions on which the bank-full and overbank flood stages were reached. A situation of this type probably occurred about 10 000 years ago at the end of the Quaternary Ice Age. This also suggests that streams today are relatively impotent, flowing across masses of sediment they are unable to move except in restricted times of flooding, and in valleys they have had little to do with carving. It also emphasises the point that the state of 'grade', in the Davisian sense, could not have been attained by any stream because of the combined effects of sea-level and rainfall changes. Many landscape features refer to conditions which obtained in the past, and simple process – form relationships cannot be established.

Underground water

Much of the water which infiltrates the soil following a rain storm percolates directly downwards into the rocks.

Porous and permeable rocks

Rocks which allow water to pass through them are known as permeable: permeability is the volume of water passing through a measured area of rock in a given time under a standard pressure gradient (e.g. x litres/m² day under a pressure gradient of y N/m²). Porosity is a measure of the spaces between the grains in a rock compared to the volume of the whole sample. These two properties are not necessarily related. Some sand dunes and unconsoli-

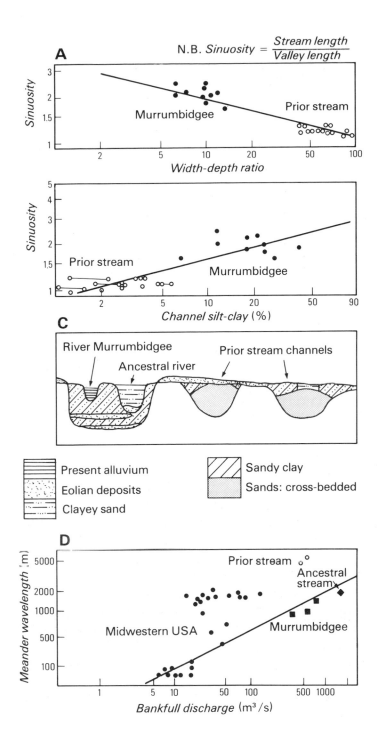

N.B. $Sinuosity = \dfrac{Stream\ length}{Valley\ length}$

Figure 7.43 Stream channels and climatic changes: the example of the Murrumbidgee river, Australia. The present river occupies a small channel and carries out little work. Depressions on the flood-plain are the sites of ancestral rivers (C), similar to the modern rivers in shape (width: depth ratio—(A)), sediment type and pattern, but larger. Prior stream channels are straight, wide and relatively shallow ((A), (B), (C)), and are filled with sands. Bank-full discharges of ancestral and prior streams were up to 5 times the present: ancestral streams were associated with humid conditions, but the prior streams experienced large floods in drier overall conditions. Similar features occur where Quaternary meltwaters caused massive runoff discharges. (After Schumm, 1968, in Gregory and Walling, 1973)

dated gravels have a high porosity, and are also permeable since the large pore spaces are interconnected and allow throughflow. Muds may also have a high porosity, but the pore spaces are so small and poorly connected that the water cannot move between them: they are impermeable. Compaction and the filling of pore spaces by a cementing medium decreases the porosity of a rock and may also lower permeability. Thus a limestone, originally formed on a shell-bank, will soon have the pore spaces filled by the solution and redeposition of the shell material, and may become impermeable. Other rocks may be permeable although they are not porous. Thus a limestone may develop joint cracks as it dries out or is folded, and these will give rise to lines along which water can percolate: such a rock is known as pervious, or as one having secondary permeability. Igneous and metamorphic rocks are normally impermeable, but some lavas and pumice are highly porous due to the gas cavities left in them, and joints in these rocks allow the water to circulate through them.

Springs and the water table

Other factors also influence the accumulation and movement of water underground. The water table is the line below which the rock is saturated; above it water moves downwards after rain, but may move upwards in dry seasons. The position of the water table in a rock is thus determined by the supply of water from rain, and hence varies with the annual regime. The geological succession of the rocks, the structure and the relief all influence the water-holding capacity of the rocks beneath London (Figure 7.44). Water movement underground is much slower than at the surface, unless it is able to move in open cave channels, or is extruded under hydrostatic pressure through a narrow passage. In passing through a permeable rock, water will move at rates of a metre or so per day to as little as a metre or so per year.

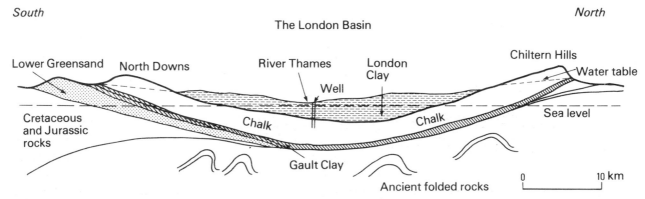

Figure 7.44 The London basin. Describe the succession of rocks beneath the London basin: which are permeable, and which impermeable? How has the relief and structure affected the accumulation of water in the Chalk? Why should water emerging from the well marked on the diagram do so under its own pressure (and therefore be an artesian well)? (N.B. Artesian conditions existed into this century, but excessive water extraction has now lowered the water table, rendering pumping necessary.)

As the water passes through the rock it will carry out solution and replacement of minerals in the rocks above the water table, but chemical reactions and changing pressure conditions will lead to precipitation of the ions carried in solution during a prolonged stay below the water table, where movement is extremely slow. This latter process leads to the increasing infilling of pore spaces and cementation of the rock.

Water may return to the surface after passage underground, and often enters a stream from a spring or line of seepage. Springs occur where the water table meets the surface — at the junction of permeable and impermeable rocks, or where the wet season water table rises to the level of a dry valley floor.

Karst landforms: water and solution in limestones

Underground water is particularly effective in the solution of limestones and produces a distinctive landscape together with underground caves. The term 'karst' is derived from words for 'bare, stony ground', and the area where it was first investigated in detail covers 20 250 km² of Jugoslavia bordering the Adriatic Sea. Limestone areas throughout the world possess generally similar features, although there are differences of emphasis according to the type of limestone, the nature of the relief and the local climate.

Limestones include a wide variety or rocks. Even in the small area of the British Isles these range from the rather soft and more evenly permeable Chalk of the Downs and Salisbury Plain, to the oolitic limestones of the Cotswold Hills, and the massive, older limestones of the

Pennines and central Ireland, which have become so compacted and cemented that any permeability they possess is due largely to joints and fissures cutting through the rock. The Chalk is up to 500 m thick, but the older limestones seldom reach this thickness in Britain. In other parts of the world limestones are often over 1000 m thick, and in the Jugoslav karst area they are 4000 m thick.

Limestones consist of at least 50 per cent calcium carbonate, together with varying amounts of mud and even sand. Some are also formed of dolomite (magnesium-calcium carbonate) which is thought to be the product of reaction between the calcium carbonate and mineralised fluids (e.g. seawater) moving through the rock after formation. Karst landforms develop on limestones where there is over 90 per cent of calcium carbonate, and on some dolomite limestones. Limestones on which karst develops may be open-textured (high porosity), allowing free contact with water, or may be of low porosity, but well-jointed and -fissured. It is essential, however, that the rock is sufficiently strong to support the formation of underground channels and caves. This is why neither the oolitic limestones nor the Chalk of southern England have many karst features, although springs issuing from them contain 230 and 260 ppm calcium carbonate respectively — evidence of more active solution than in karst areas on older rocks.

Limestones are dissolved by water, especially when it contains a high proportion of carbon dioxide in solution. Carbon dioxide is more soluble in waters of low temperature, which would suggest that limestones may be dissolved more rapidly in cold climates. It must be remembered, however, that such activity ceases during periods of freezing, and that heavier rainfalls in tropical areas can lead to greater rates of solution there. Solution of limestone under the action of weak carbon dioxide solution dissolves the carbonate as bicarbonate ions, forming temporary hard water. The carbon dioxide content of water increases with the acidity of rainwater (e.g. to the lee of industrial areas and on coasts). It also increases where there is microbiological activity in the soil through which it passes; where snow banks concentrate the carbon dioxide in their pores and thus in meltwater; and where two masses of water saturated with calcium bicarbonate meet and free some carbon dioxide on mixing. In addition to the carbon dioxide content of the water, other factors which assist the solution of calcium carbonate include water turbulence, the sheer volume of water passing over a rock surface, (little solution takes place in deserts), and the texture of the limestones (Chalk is very soluble).

When water containing little calcium bicarbonate reaches a limestone outcrop the rock is dissolved rapidly at first, the mineral matter being taken into solution until the water is almost saturated. After this initial burst the process of solution proceeds more slowly, associated with local additions of carbon dioxide to the waters and the moving in of less saturated water: there is a dynamic balance between solution and precipitation. The reaction in the equation: $CaCO_3 + H_2O + CO_2 \rightleftharpoons Ca(HCO_3)_2$ goes to the right as CO_2 increases, and to the left as it decreases. Some of the most spectacular features resulting from the solution of limestones occur near the margin of an outcrop, where uncarbonated water flows on the limestone. Rates of limestone solution and landscape lowering have been calculated: it has been estimated that they vary from 18 mm/1000 years in south Wales to 50 mm in north-west Yorkshire, about 300 mm in the southern Alps, and 72 mm in Jamaica. The differences can be related to varying types of limestone, relief and climate. The effect of seawater on the solution of limestone has been studied less than that of river water, and conflicting accounts have resulted (chapter 10).

Water flow in limestone regions has a distinctive pattern. There are three zones of water movement, all of which are underground (Figure 7.45). A characteristic of karst regions is the lack of surface drainage, and streams occur only where they have a sufficient discharge to maintain a flow which is greater than the rate at which water permeates the underlying rocks. Examples of this include the river Dove in Derbyshire, and the rivers Lot and Tarn in the Grand Causses area of central France. These are well-supplied with water from outside the limestone regions and have carved deep, gorge-like valleys through the rocks.

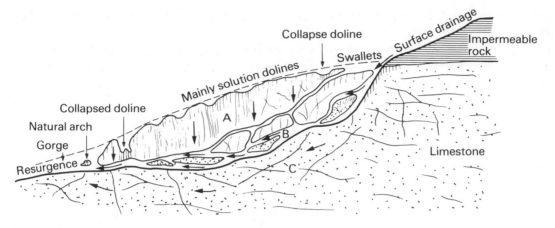

Figure 7.45 Underground water movement in limestone.
(A) The zone of percolation, solution and collapse.
(B) The zone of stream water corrosion and corrasion.
(C) The zone of phreatic water solution below the saturation level.
(After Williams, in Chorley, 1969)

Water falling on limestone areas passes into the surface rocks, especially where there is a high degree of fissuring. Humid tropical areas often experience such intense rainfall that there is also considerable surface runoff, but this seems to be peculiar to these regions. Once in the rocks the water will percolate downwards at rates depending on the permeability and interconnection of joint and bedding plane fissures. Rates vary according to the existence of a well-ramified system of these fissures. Measurements in one part of Derbyshire showed that it may take two days before precipitation falling on a rock layer 30 m thick will add to increased water flow in the stream issuing at the base. Some water enters the ground from surface streams, and may continue flowing in a stream-like form underground. Percolating water supplies additional volume to underground streams, which flow above the level of saturation in the rock. Thus the water flowing into the swallow hole above Malham Cove is only one-twentieth of the volume issuing at the foot of the Cove.

It is often difficult to locate the water table as a definite line in the limestones which give rise to karst features, since the main movement of water is concentrated along fissures. Three boreholes were sunk in Littondale (Yorkshire) at distances 100 m apart: one met water at 28 m depth, but the other two were dry although they pierced the base of the limestone formation at 120 m depth. This experience is also linked to the way in which water circulates in the permanently saturated zone. Movement is concentrated in the upper 10 m or so, and below this there is a very slow circulation, if any. Flow takes place under hydrostatic pressure, and is known as phreatic, or forced, flow (as opposed to vadose, or free, flow in the upper zones). Phreatic flow is determined by the head of water moving through the rock and is, in effect, a similar process to squeezing toothpaste out of a tube. As such, the water may be forced to flow uphill, and high discharges may pass through narrow passages.

Landforms produced in limestones may be related to these zones of differing water activity, and there is a major contrast between the features evident on the surface and caves.

Karst landforms: surface features
For long the surface features were regarded as the essential landforms, and it is only since the growth of cave studies and exploration (speleology) that the underground situation has been understood in as much detail. The surface karst has a chaotic, pitted appearance with little, if any, drainage or even valley forms. Rock exposed to rain, or affected by soil water movement, becomes fluted and pitted with 'microrelief' features known collectively as karren; a multitude of enclosed hollows of moderate dimensions, known generally as dolines, may dominate the

landscape in some areas; and there may be dry valleys, swallow holes and springs related to the restricted surface drainage

a) **Karren** are small-scale solution features a few centimetres across and up to a few metres long. They include flutings and chisellings on rock surfaces, together with small solution basins. Where the limestone is exposed at the surface, water running across the surface of blocks forms straight, rounded grooves with sharper ridges between: these are known as rillenkarren, which are common in Alpine karst and occur in larger forms up to 15–20 m long in the tropics. Rock fissures exposed at the surface are widened by solution into kluftkarren (or 'grikes' in northern England). These are related closely to jointing and narrow downwards: in Yorkshire they are 15–60 cm across and up to 3 m deep. The walls of grikes may be smooth or fretted, and it seems that they are formed after exposure to subaerial weathering and rain action. Deeper, corridor-like clefts occur in Jugoslavia and in tropical karst. Areas where limestone has been exposed from beneath a cover of soil and vegetation exhibit a pattern of rounded grooves and ridges due to the passage of water between soil and rock: small drainage systems may be integrated in response to the slope of the block into which they have been etched. Such rounded features are known as rundkarren, and occur on the blocks between grikes in Yorkshire (Figure 7.46). Limestone pavement is a common feature of this same area, composed of complex groups of karren phenomena. Separate blocks, or clints, with smooth or grooved surfaces, are divided by grikes. It seems that the limestone surface was scoured bare by ice movement, exposing the rock to atmospheric weathering when the ice melted. Striae would have disappeared soon (some scratched experimentally on such rocks vanished

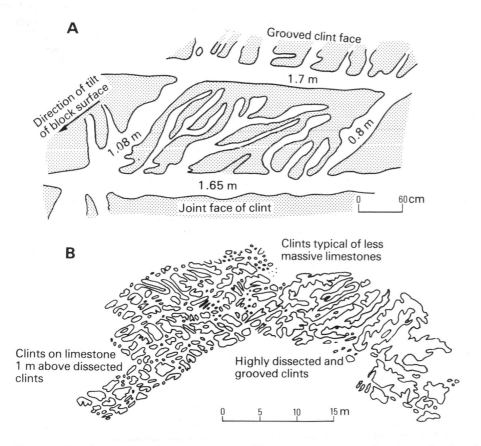

Figure 7.46 Karren and limestone pavement. The measurements in (A) refer to depths of the grikes (kluftkarren). Notice the patterns of rundkarren on the clint blocks. How does (B) demonstrate the influence of different lithologies within the limestone, and of joint patterns? Both diagrams are taken from surveys of Twistleton Scar, Yorkshire. (After Sweeting, 1972)

in 10 years), leaving the general pattern of relief resulting from glacial scour and patches of boulder clay between.

b) **Dolines** are often considered to be the essential features of true karst, and parts of Jugoslavia have over 60 per cent of the surface occupied by these hollows — up to 100-200/km². Some are bowl-like depressions, and some are cone-shaped; the sides are rocky or covered with vegetation; they may be 2-100 m deep and 100-1000 m across; and they may be isolated or in groups. Four main types of doline are found in temperate regions (Figure 7.47). Most are formed by solution and contain evidence of this in the facts that the bottom of the depression has limestone near the surface; any soil there is clearly residual with small fragments of limestone in it; and there are no signs of collapse. Such dolines are often formed beneath patches of snow in the Alps. In Britain the Mendip Hills have most — 566 were mapped in an area of 36 km², mostly in valley floors, but with no relation to the deep caves of the area. Few occur farther north in Britain, having been destroyed by ice erosion if they existed before the Ice Age. Recent studies of dolines have shown that the former conception of their irregular distribution in the landscape was incorrect. They are aligned along structural features in the rocks, or where surface runoff goes underground. Collapse dolines are particularly common in areas of large caves. Tropical solution hollows, known as

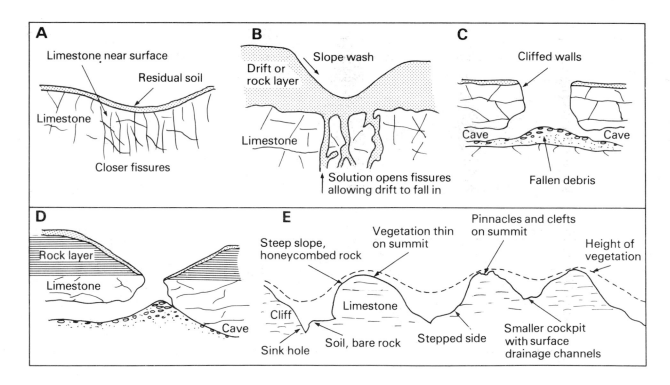

Figure 7.47 Enclosed hollows in karst. (A) Solution dolines. (B) Drift or subsidence dolines. (C), (D) Collapse dolines. (E) Tropical hollows, or 'cock-pits'. (After Williams and Aub, in Sweeting, 1972)

'cockpits' in Jamaica, are different in form due to the greater proportion of surface runoff (Figure 7.48).

c) **Swallow holes** (swallets) are where water in a surface stream sinks underground. This may happen without the formation of surface features as, for instance, in the bed of a stream, but there is often a small cave or vertical shaft into which the water disappears. Gaping Gill, on the slopes of Ingleborough in Yorkshire, has a round entry hole 13 m in diameter and drops of 120 m to cave systems below. Such holes (known locally as potholes) are common around the margins of patches of glacial drift in that area.

Plates 48–50 Images of the Washington DC area of the USA: one was taken at low altitude; another on infrared film by high-flying aircraft; and the third is a LANDSAT image taken from 900 km above the surface. Compare the areas covered and the information which can be extracted from each of the three images. In infrared images vegetation shows up in various shades of red; sediment gives streams a light blue colour. How can such information be of use to the student of landforms? These images depict an area which is largely built-up: what difference does this make to the action of natural processes on the Earth's surface?

48

49

50

51

52

53

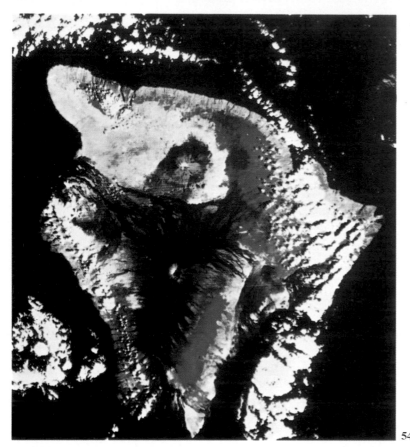

54

Plate 51 Peninsular India and Sri Lanka. Plate tectonics theory sees this landmass as having been carried northwards on a plate to collide with the Asian plate, raising the Himalayas. (NASA)

Plate 52 Basalt lava on the Mid-Atlantic ridge.

Plate 53 Glacial cobbles on the continental shelf off the coast of north-eastern North America. (52 & 53 USGS)

Plate 54 Hawaii. This is the largest of the Hawaiian Island group, and consists of five volcanic peaks. Recently erupted lavas show up as black (non-vegetated) around four of the five: Mauna Kea, north of centre; Mauna Loa, the largest mass; Hualala in the west; and Kilauea in the south-east. The fifth, Kohala Mountain, in the north-west, has not erupted in recent years. (NASA)

Plates 55/56 Lava in the 1969 Kilauea eruption. (USGS)

55

56

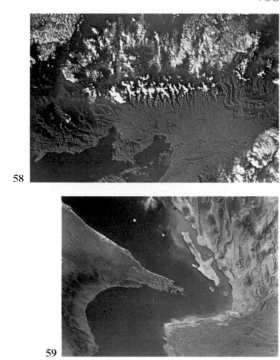

57

58

59

61

62

60

Plate 57 A Gemini IX photograph of the southern Red Sea, taken 750 km above the Earth's surface in September 1966. How does this suggest that the Red Sea might be the site of a new and widening ocean?

Plate 58 The Andes and Lake Titicaca in Bolivia: a Gemini V photograph (1965). Lake Titicaca stands on the Altiplano (3500 m) between the Andean ranges: the Cordillera Occidental (left) has a base of Mesozoic folded sediments covered by Tertiary basalt lavas and capped by volcanoes (up to 6000 m). The Cordillera Oriental rises to 7000 m with glaciers in the highest parts, and is formed mainly of folded Palaeozoic rocks.

Plate 59 The Persian Gulf (Gemini XII, 1966), with the folded ranges of the Zagros Mountains in southern Iran to the north of the Strait of Hormus. These fold mountains reach 1000–3000 m, and have been affected by the thrusting of the Arabian Peninsula (in the south of this picture). The dark patches in the Zagros Mountains are salt domes.

Plate 60 Mount Etna, Sicily (Skylab 3, 1973). This is the highest volcano in Europe (3100 m), and the plume of smoke issuing from the crest shows it is still active: recent lava flows show black, whilst older flows are red (having become vegetated). Contrast the features of the volcano with the Plain of Catania to the south, and with the surrounding hills. (57–60 all NASA)

Plates 61/62 Stromboli in the Lipari Islands to the north of Sicily. The views show the slope of ash and lava plunging into the sea and sulphurous gases rising to the surface. (A. J. Abbott)

Plate 63 Weathered debris at the foot of Ayer's Rock, Northern Territories, Australia. (Australian News and Information Bureau)

64

66

Plate 64 Exfoliation of granite (unloading) leading to the surface breaking into sheets: Bow river area, Western Australia. (C. R. Whitaker)
Plate 65 Surface flaking of rocks at Bulanik in eastern Turkey.
Plate 66 Spheroidal weathering of rock at Damghan, Iran. (65 and 66 A. J. Abbott)

67

Plate 67 Slope forms in Azerbaydzhan SSR. (A. J. Abbott)
Plate 68 The northern extension of the Gulf Coast plain of southern USA. The Mississippi river follows a meandering course west of the Yazoo plain (lighter coloured area, which it used to cross) with Memphis in the north-west corner. Streets of cumulus cloud cover the Appalachians in the north-east. The reservoirs of the Tallahatchie river system are situated in the Red Hills area on Tertiary rocks, well-dissected by the tributary streams. (NASA)

68

70

69

Plate 69 The upper Welschtobelbach near Arosa, Switzerland. The stream in summer winds a braided course through coarse sediment. (M. J. Bradshaw)
Plate 70 The meandering course of the Lower Mississippi, with the Ouachita and Red rivers joining from the west. Water carrying sediment shows up as blue, clear water as black, on this false colour LANDSAT image of 20 October 1972. (NASA)

71

72

73

74

75

76

77

Plate 71 The wide valley of the upper river Plym as it reaches the edge of Dartmoor at Cadover Bridge (cf. Figure 7.35). Here it meanders freely through the piles of rubble left by former tin mining operations; the company mining china clay on the edge of this valley has also altered the course by providing a larger car park for visitors to the moors.

Plate 72 The confined, gorge-like valley of the River Plym, 4 km below Cadover Bridge, near Shaugh Bridge. The large boulders have entered the stream from the valley sides and seldom, if ever, are moved by the stream. When floods occur there is no flood plain to accommodate the increased discharge, and the valley acts as a giant chute.

Plate 73 A section in alluvium at Plym Bridge, near the mouth of the river Plym. Here the flood-plain is 50 m wide and the deposits have a coarser lower section, related to the present channel bed deposits; the finer upper section was formed during floods. (71–73 M. J. Bradshaw)

Plates 74/75 Lydford gorge, West Devon (Figure 7.26). Evidence of erosion is seen in the waterfall from a hanging tributary, and in the potholes cut into slates. (H. Morris)

Plates 76/77 Vertical erosion in the Grand and Black Canyons in south-western USA (D. J. Griffiths and USGS)

78

79

80

Plate 78 The Nile delta. (NASA)
Plate 79 The delta of the river Maggia at the northern end of Lake Maggiore in southern Switzerland. (M. J. Bradshaw)
Plates 80–82 Limestone features of the Ingleborough area, north Yorkshire. Limestone pavement (80), with detail of karren (81). A disappearing stream cuts through the edge of a till deposit (82). (A. J. Abbott)

81

82

Plate 83 A skylab photograph of glaciers in the Himalayas (NASA)

Plate 85 The snout of the Black Rapids glacier, Alaska. (USGS)

Plate 87 Norwegian fjords. (R. K. Davies)

Plate 84 A cirque in Norway. (R. K. Davies)

Plate 86 An esker at Clonmacnois, Ireland, showing the nature of the debris deposited. (A. J. Abbott)

Plate 88 The Matterhorn, Switzerland. (T. R. Griffiths)

Plate 89 A cirque and glacial trough in Snowdonia, north Wales. (D. J. Griffiths)

Plate 90 Moffat Water: a glacial trough in Scotland. (Forestry Commission)

Plate 91 Langdale, English Lake District. The glacial trough floor in the foreground has been filled by alluvial and lacustrine deposits.

Plate 92 An erratic boulder in the Ingleborough district.

Plate 93 Glacial scenery in the Outer Hebrides: Benbecula. (91–93 A. J. Abbott)

Plate 94 Typical features of a glacial overflow channel (no stream, flat floor) at Wooler, Northumberland. (D. J. Griffiths)

Plate 95 Longitudinal sand dunes in the 'empty quarter' of Saudi Arabia. (NASA)

Plate 96 Sand dunes in Death Valley, California. (USGS)

Plate 97 Coastal dunes at Wendnine, Belgium. (A. J. Abbott)

Plate 98 Rock pedestal probably eroded by sand-blast in association with weathering, in the Palui Valley, Northern Territory, Australia. The bush in the foreground is less than 1 m high. (Australian News and Information Bureau)

Plate 99 Desert surface in Iran after finer particles have been winnowed by deflation. (A. J. Abbott)

Plate 100 Deflation of soil in the USA Dust Bowl. (USDASCS)

01

102

Plate 101 The Scilly Isles. This group of islands, situated off the coast of south-west England, is exposed to the full force of the North Atlantic waves: each has an edging of rocky foreshore. The islands shown here are the eastern group, including the Ganillys and Little Arthur. (G. White)
Plate 102 Storm waves attacking the rocks at Glencolumcille, Donegal, Republic of Ireland. (A. J. Abbott)
Plate 103 Heavy surf at Torquay, Victoria, Australia. (Australian News and Information Bureau)
Plate 104 Shore platform and cliff in Chalk dipping at 70 degrees to the north (right): Culver Cliff, Bembridge, Isle of Wight. (M. J. Bradshaw)
Plate 105 Shore platform and cliff with seaward-dipping Triassic red sandstones near Watchet, north Somerset. (W. Whitaker)

103

4

105

172

Plate 106 Hog-back cliff profiles on the north Devon coast near Lynmouth. (W. Whitaker)

Plate 108 Slapton Ley behind the bar which traps this freshwater lake in south Devon. (A. J. Abbott)

Plate 110 The Bahamas, with islands around the edge of the reef mound. (NASA)

Plate 107 A section of the strandflat in the Lofoten Islands, nor Norway. (R. K. Davies)

Plate 109 The coast of Long Island, New York, showing longshore drifting and the effect of groynes on the progress of debris. Seen on the approach to J. F. Kennedy Airport. (C. R. Whitaker)

Plate 111 An island in the Great Barrier Reef of Queenslan Australia. Notice the extent of the reef, compared with the size the island itself. (Australian News and Information Bureau)

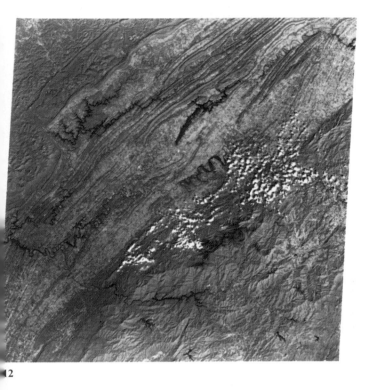

12

113

14

Plate 112 Man's use of the Tennessee valley and control of water flow in the river shows up on this LANDSAT image, where four major reservoirs are seen around Knoxville (blue area west of centre) in the ridge-and-valley province (folded and thrust rocks forming escarpments). The Blue Ridge lies to the south-east and the Cumberland Plateau is to the north-west. Contrast the three types of landform regions in relation to the scale of observation.

Plate 113 A view of Britain from Skylab 3. Skylab's orbit did not bring it north of 50 degrees north, and so this oblique view was the best it could do for the British Isles. Pick out major landform features, such as the Chalk outcrops of Salisbury Plain. (112 and 113 NASA)

Plate 114 Large erratics of Chalk in till at Runton, north Norfolk.

Plate 115 Glacial outwash south of the moraine ridge at Sheringham.

Plate 116 The Cromer Forest Bed at Runton. This is one of the uppermost 'crag' horizons, and is just below the older glacial tills. It is a peat deposit.

Plate 117 The interglacial Corton Sands, south of Lowestoft, sandwiched between two tills. (114–117 M. J. Bradshaw)

15

116

117

Plate 118 Ayers Rock, Northern Territory, Australia. (Australian News and Information Bureau)

Plates 119/120 Pediment in gneiss near the headwaters of the Ord river in the Kimberley ranges of Western Australia. (C. R. Whitaker)

120

Plate 121 Mountain desert: a section of the McDonnell Range, Western Australia. (Australian News and Information Bureau)

Plate 122 Calcrete in the Flinders Range of South Australia.

Plate 123 The southern edge of Lake Eyre. (122 and 123 C. Whitaker)

Plate 124 The Tasht and Ninz salt lakes in the Zagros Mountains in central southern Iran. (NASA)

Plate 125 A salt flat on the Turkey–Iran border, with Mount Ararat in the background.

Plate 126 A dry channel at Namakah, Damghan, Iran.

Plate 127 Stream channel deposits at Shahrud. (125–127 A. J. Abbott)

Plate 128 Sand dunes near Gacht Seran, Iran.

Plate 129 Stony desert at Agha Jari. (128 and 129 P. Boden)

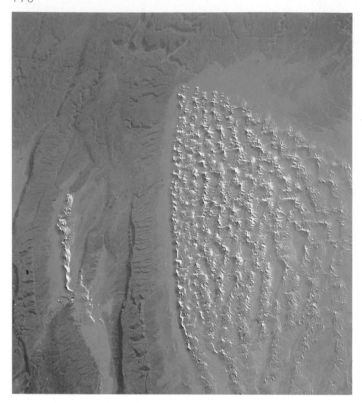

Plate 130 A Gemini VII photograph of dunes in southern Algeria. The dune area is quite distinct from the surrounding bedrock area, which shows clear marks of stream processes.

Plate 131 An Apollo 9 photograph of the south-eastern section of the Sahara. The Jabal al Uweinat is a granite intrusion in the Sudan, rising to over 2000 m; the small granite massif is in Egypt. Sands and dry channels occupy the areas between.

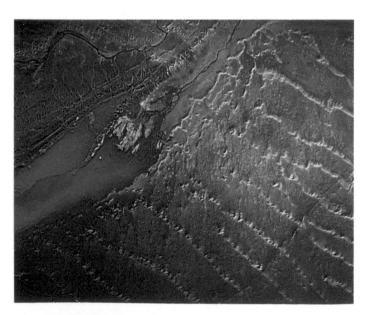

Plate 132 Dunes near the Ougarta range in southern Algeria. The Oued Asoura is full of water after a recent shower.

Plate 133 Libya: sand seas, rocky ridges looking towards the Mediterranean in the north. (All NASA)

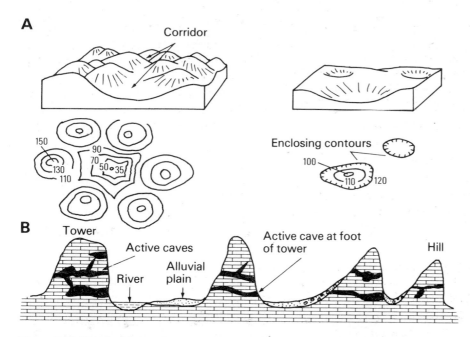

Figure 7.48 Tropical karst features.
(A) Cone karst (left), compared with a doline. Note the differences in three dimensions and on maps.
(B) Tower karst in southern China.
(After Williams (A) and Balaze, in Sweeting, 1972)

d) **Valleys** on limestones are normally dry, but rivers may occasionally flow across the region. The pattern of dry valleys in Derbyshire, or on Salisbury Plain, is similar to that developed by flowing streams, and demonstrates the existence of streams at the surface in these areas at one time. Watercourses are now absent or temporary. Various origins have been suggested for these valley systems, including the superimposition of a stream pattern from overlying impermeable rocks (Derbyshire) with an associated lowering of the water table, to the existence of surface streams in tundra (periglacial) conditions during the Quaternary Ice Age when the rock would be frozen, rendering such rocks as Chalk impermeable. Most limestone

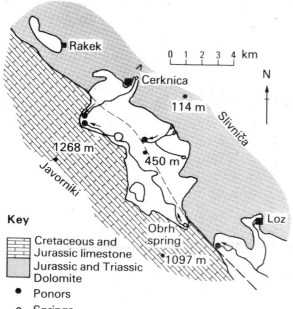

Figure 7.49 The Cerkniška polje, Jugoslavia. Figures refer to heights above sea-level (metres). Note the orientation of the depression and adjacent depressions, and their relationship to the boundary of permeable limestones and less permeable dolomites. This polje is the largest in Slovenia; the largest in Jugoslavia is 70 km long. It is thought that the floor of this polje was once a lake. (After Našě Jame, in Sweeting, 1972)

valleys in Britain show evidence of frost shattering, giving rise to rounded upper slopes. Many valleys in limestone areas end 'blind' in a cliff where water disappears, or reappears as at Malham Cove. The subject of dry valleys is discussed further on page 181.

e) **Poljes** are large depressions occurring mainly in the karst area of Jugoslavia. One or two have been described in southern France near Marseilles, and in Greece, but few elsewhere. In Jugoslavia they are depressions with flat floors, elongated along the north-west to south-east direction of folding (Figure 7.49). The floors are flat due to the deposition of alluvium, or less probably to the planation of limestone by lateral corrosion, and many poljes are flooded seasonally. Drainage may be by one or more rivers, which issue from springs at one end and disappear into swallow holes (ponors) at the other. Poljes are commonly situated near the contact of a permeable limestone with a less permeable rock. It is suggested that differential erosion at this contact led to the formation of basins in which alluvium collected. Shallow lakes in these basins corroded the limestone around the margins until the formation of ponors was sufficiently advanced to drain the polje. During the Quaternary period of colder climate large quantities of rock debris were washed into the polje floor, clogging the ponors and leading to further inundation and extension of the polje area by solution. The last hundred years have seen less and less frequent flooding, indicating that the ponors are being opened up once again. Problems follow for the local inhabitants, since the poljes ('fields') are their main source of cultivable land: flow across them and into the ponors leads to rapid erosion of the alluvium and exposure of the bare limestones. The essential feature of polje formation is impeded drainage leading to flooding and solution planation. It seems that this requires the rather special conditions of relief, drainage and history present in the Alpine-Mediterranean area, and that such features may not be typical of karst everywhere. Their study emphasises the fact that landform interpretation is a complex science.

Caves

The study of underground landforms is made difficult by problems of accessibility, which has led to much conflict of opinion in the past concerning the origins of caves. Many questions concerning the formation of caves still remain unanswered. If the water flowing through the rocks is saturated with calcium carbonate, how can it dissolve more? Yet for a considerable flow to take place cavities must exist in the rock — a chicken-and-egg type of situation. Recent studies by cavers have helped to resolve some of these questions, and it is now seen that caves develop in certain places at the junction of permeable and impermeable rock layers leading to a concentration of water flow at the margins, and undercutting; where there is sufficient relief to give a marked drop of water in flowing through the rock; and where there is a limited thickness of limestone (more caves occur in the thinner limestones of England than in the massive limestones of Jamaica or Jugoslavia).

Caves include large chambers and smaller channels, running both vertically and horizontally. In the deep karst of Jugoslavia they tend to have a vertical emphasis and are fewer in number, whilst in Britain the horizontal extent is most noticeable. Most caves display features indicating the activity of both phreatic and vadose water (Figure 7.50). Phreatic caves can be described as 'bore-tubes' through the rock formed by water flowing under pressure. They are circular or elliptical in cross-section, and may have uphill gradients in the direction of flow. The walls and ceilings are covered with honeycombed 'spongework' due to differential solution, and other small-scale features include fluting and scalloping. Where free-flowing water has been active there is evidence of erosion and corrosion in the cave floor, giving a more rectangular form to the cave, or even a triangular section with a flat base. Gorges may be cut in the bases of phreatic passages, and solution features are also common at the base of such passages. Cave roof collapse leads to enlargement, and to the eventual destruction of caves near the surface. Thus cavities and tubes created in the phreatic zone may become occupied by free-flowing streams when the water level falls, and further falls may lead to the cave being abandoned by running water. Groups of caves at different levels have been related to denudation of the surrounding area (Figure 7.51).

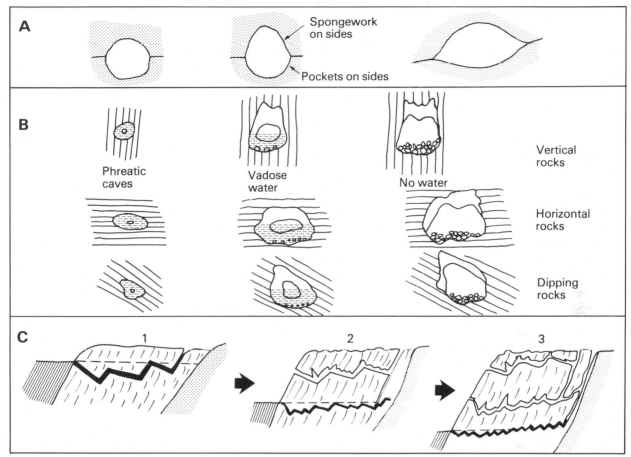

Figure 7.50 Cave forms in limestone rocks.
(A) Phreatic tubes. What is the characteristic shape?
(B) The transition from phreatic water to free flow and empty caves. How does this transition affect the cave shape in rocks of different dip?

(C) The development of caves in the Mendip Hills. Note the changes taking place as time advances and as base-level is lowered.
(After Ganes and Ford (C), in Sweeting, 1972)

Caves are also the scene of deposition. Many are clogged with fine mud which is able to enter through narrow joints with the movement of water, but masses of sand and pebbles may also accumulate. Some of this is brought in by running water, but may also be the result of roof collapse in part of the system. Many such deposits become cemented by the precipitation of calcium carbonate, so that cave passages may be blocked as well as opened. Calcareous deposits in caves include the precipitates formed by dripping water, or in areas of seepage by running water. Dripstones form in the humid cave air despite the fact that evaporation rates are low, as the precipitation of calcium carbonate is caused by another process. The water which percolates through the rock above the cave builds up a carbon dioxide content of between 25 and 90 times the normal atmospheric content. As the water enters the cave some of this is released and calcium carbonate is deposited on the cave roof. Stalactites begin by forming a straw-like deposit having a diameter equal to that of the water drop (approximately 5 mm). Later this becomes blocked and the calcium carbonate is precipitated in layers around it. Stalagmites, growing up from the cave floor, have more varied shapes, and form after further loss of carbon dioxide during the fall from the cave roof. A disc of calcium carbonate forms on the floor, and is added to rapidly. The Jockey Cap stalagmite in Clapham Cave

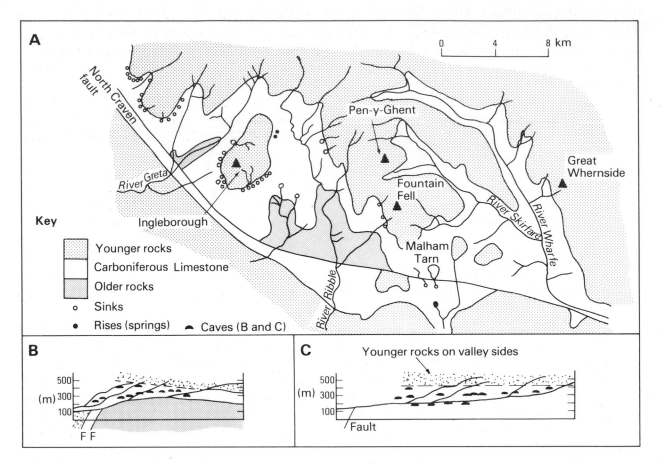

Figure 7.51 North-west Yorkshire: an area of karst features.
(A) A map of the region: note the relationship of sinkholes and springs to geological boundaries.

(B) The long profile of the river Greta and its tributaries, related to the geology and cave levels.
(C) The long profile of the Skirfare-Wharfe system.
(After Sweeting, 1972)

(Yorkshire) grew at a rate of 7.66 mm/year (both width and height) from 1839 to 1873, and since 1873 it has increased by 20 cm in girth and 2 cm in height. Other forms of deposition occur on the floors of caves (rimstone), and where seepage of lime-rich water may form curtain-like features.

Karst landforms and climate
Every karst area has unique conditions of limestone lithology, relief and climate, and thus of landform assemblage, but certain common types can be recognised.

a) **True karst** is that occupying much of Jugoslavia, where the thick limestones have been involved in mountain-building movements and form ranges over 2000 m high having a heavy rainfall (1000–1500 mm/yr). They are deeply fissured and deep cavities have been opened up in them with collapse features adding to the thousand of solution dolines. There are no surface streams or dry valleys in the central areas, where water circulation is at depth in patterns unrelated to the surface features. Poljes develop at the contact of limestones and impermeable rocks. Surface outcrops of rock with rillenkarren are common.

b) **Fluviokarst** is where a combination of fluvial and karstic processes are at work, as is the case over much of the limestone areas of western and central Europe and the USA. The limestones are less extensive and often thinner than in Jugoslavia, allowing streams to flow right across and water to circulate less deeply. River gorges and dry valleys are common, and many landforms relate to the disappearance and reappearance of surface water — swallow holes, blind valleys, springs. Caves occur most commonly in this type of karst, and may be found in series at particular heights related to stages in fluvial downcutting in the surrounding areas. Karst features are most in evidence between the main valleys with dolines and larger depressions combining several dolines (uvalas). Soil covering is thicker than in the true karst due to increased slope wash, and rundkarren form beneath this layer. Quaternary changes of climate and process are regarded as having affected these areas, especially if the valleys can be shown to be related to that phase. In Britain the Peak District of Derbyshire and the Mendip Hills have karst landforms of this type.

c) **Glaciokarst** includes areas on the margins of, or beneath, ice sheets, and many areas of very cold climate where snow gathers and melts regularly. Associated features are now exposed in areas once covered by ice, and these include the results of glacial scour (removing the soil and upper portions of weathered rock layers), and deposits. North-west Yorkshire and much of central Ireland are such areas. In north-west Yorkshire there are expanses of limestone pavement, but few dolines, together with areas of impermeable boulder clay surrounded by potholes. Near the margins of ice masses today the importance of meltwater can be seen in opening out subterranean passages. Tundra regions with permafrost, however, allow no underground circulation and have few, if any, caves.

d) **Tropical karst** is the most distinctive type of all, possessing few features similar to those of the karst in cooler regions. Rainfall and evaporation are both intense leading to surface sheet wash and gullying, but few cave systems underground. Cone karst is one typical form (Figure 7.48), based largely on solution hollows ('cockpits') with cone-shaped hills between. The relationship of rounded hills and irregular depressions contrasts with the rounded depressions (dolines) and irregular hills of cooler regions. Some of the hollows join to form corridors along faults or major joints. Tower karst occurs near major rivers where steep-sided, cliff-like hills are surrounded by alluvial plains where the rivers maintain the steepness of the hills by undercutting. These forms are common in south-east Asia, where there is the largest single area of karst — 60 000 km^2 in southern China — but occur also in the West Indies near extensive areas of cone karst. All the tropical karst forms emphasise the importance of runoff and intensive erosion which is prevalent in these areas. Even the occurrence of the wide alluvial plains is typical. Limestone dissolves rapidly and calcium carbonate concentrations may be up to 500-600 ppm in surface running water (i.e. more than twice the content emerging from Chalk springs in England). Caves, however, develop less commonly and may be filled by re-deposition of the calcium carbonate as rapidly as they are formed. This makes it difficult to calculate whether limestone is dissolved more rapidly in warm or cooler regions.

The dry valley problem

Normally streams are associated with valleys: a valley without a stream today suggests that changes have taken place since the formation of the valley.

Dry valleys include clearly defined valleys without streams, but also coombe-like forms on steep slopes and small-scale runnels known as 'dellen'. They may occur on areas of permeable rock, and many are on limestones, including Chalk, but others are found on sandstones. It is typical to find a comparison between stream density and valley density in such areas. Thus in the Dove basin of Derbyshire, on limestone, the stream : valley density is 1:1.5 and in south-east Devon on a variety of permeable rocks it is 1:1.7. Dry valleys are also found in areas of markedly seasonal or sparse rainfall, such as semi-arid, arid, tundra and seasonal tropical climates. In some arid areas the wadi systems have not contained flowing water in recorded history. Areas subject to tundra conditions during the Quaternary Ice Age also have dry valleys, since the freezing of subsurface water forced any water flow to take place on the surface. Mountainsides may be scored by short, steep, dry valleys.

182

134

135

Plate 134/135 **Plate 134/135** The Lynmouth floods of 1952, and the damage caused. (Western Morning News, Plymouth)

Plate 136 Mudslides in Nelson County, Virginia, resulting from the flooding (see p. 119). (USGS)

136

Plate 137 Stereo pair: sinkholes (dolines) in Kentucky, USA. There is a contrast between the area north of the road, where the sinkholes are broader and shallower in the St. Genevieve Limestone, and the south, where they are more numerous and deeper in the St. Louis Limestone. (USGS)

Plate 138 A drainage tube, rather like those occurring in limestone and ice due to water being forced through under pressure. This one formed by lava being forced out through older lavas in Hawaii: the man is some 40 m from the opening. (USDASCS)

Plate 139 Typical cone karst in northern Puerto Rico. (USDASCS)

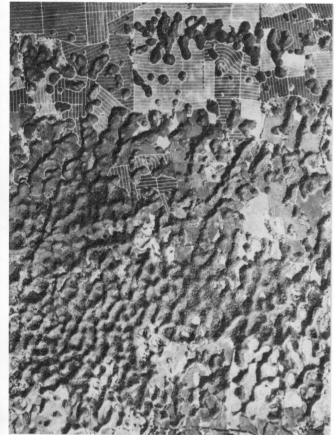

Plate 140 Stereo pair: tower karst in the northern part of Puerto Rico. The cultivated area in the north is on an alluvial deposit surrounding the individual limestone hills, known as mogotes. Many of these hills are steeper on the west than on the east. Farther south the more typical cone karst is found. (USGS)

183

A wide variety of causes have been responsible for the many dry valley forms. Thus mass movement on steep slopes may produce a valley-like scar which is seldom occupied by flowing water; solution collapse on limestone may result in a surface valley-like form; permafrost conditions also give rise to valley features by local subsidence along a line of ice-wedge accumulation; glacial meltwater under pressure beneath ice may gouge out steep-sided gorges which may not be followed by streams after the ice melts; and an area exposed from beneath the sea may have valleys which were once submarine canyons. All of these explanations are related to processes other than stream action. Many dry valleys may be occupied by streams in occasional seasons of unusually high rainfall, and the fact that they are normally dry may be related to the reduced rainfall that has been recorded in reference to underfit streams, leading to a fall of the water table beneath the valley floor. Certainly some valleys contain evidence of higher spring lines which are now seldom in action. The fall in water table height has also been explained in terms of a falling sea-level, or the recession of an escarpment (Figure 7.52). The reduction in stream network has been seen as due to the superimposition from an impermeable cover to a permeable layer.

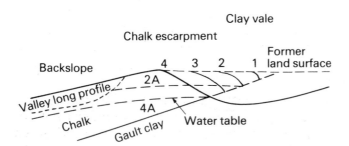

Figure 7.52 Fagg's hypothesis of dry valley formation.
The present water table (4A) is too low to supply a stream in the backslope valley. If the escarpment has been worn back with the erosion of the clay vale, the water table at 2A could have provided water for the backslope valley. The recession of the escarpment is thus responsible for the dry valley. (After Fagg, in Sparks, 1960)

Completely different systems of denudation may also have led to the present dry valleys. During the Quaternary Ice Age many of the desert areas had increased rainfall, and these 'pluvial' periods may have been responsible for many relict valleys there (chapter 13). Southern England, on the other hand, would have been dominated by tundra conditions, with permanently frozen soil (permafrost) and extreme solifluction effects, and perhaps the effects of increased runoff in the melt season.

Some of the most discussed dry valleys are those occurring on the Chalk of southern England. Many of the ideas mentioned in this section have been applied to this area. As more dry valley forms have been recognised, so the different ideas on the origin of Chalk dry valleys have been increased. It may be that several explanations could be true.

8

Ice

Ice, the solid form of water, produces landforms and deposits very different from those sculptured by running water. Like water it reaches the continental surface by way of the hydrological cycle, and then moves downhill under gravity and returns eventually to the oceans. Its solid nature, however, results in a different type of movement which is much slower, but takes place in a much larger mass. It reacts with the underlying rocks in ways which contrast with running water, and is particularly sensitive to temperature changes, which control the area covered by ice.

Ice accumulation

Ice forms on the land or sea surface where temperatures are low enough to cause water to freeze, or to allow the accumulation of sufficient depths of snow for the lower layers to be compounded into ice. The freezing of water surfaces occurs in the Arctic and Antarctic areas, and on lakes and ponds in tundra and midlatitude regions during cold winters. Such ice is of little effect on the form of the land surface compared with the ice formed by compaction of snow on the land, particularly at high altitude.

Snow accumulates when the air temperature is too low for it to melt. Ice will form only where the snow cover is thick and continuous and is not melted during a summer thaw. In mountainous areas the lowest level on slopes continuously covered by snow and ice is known as the Permanent Snow Line (Figure 8.1). At present it is estimated that 10 per cent of the world's land surface, and 7 per cent of the oceans, are covered by permanent ice.

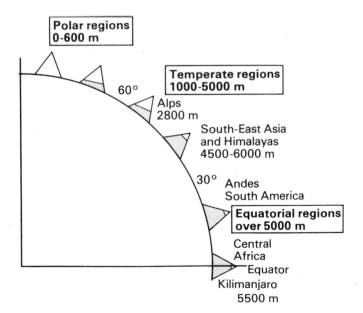

Figure 8.1 The snow line, above which snow does not melt completely in summer. The altitude varies with latitude in general, although there are local variations due to differences in precipitation and rates of melting.

Snow, firn and ice

The process of changing snow into ice is related to local conditions of temperature and pressure. As snow piles up, by actual precipitation or drifting, compaction of the lower layers takes place and the increasing pressure causes a slight lowering of the melting point of snow.

Meltwater percolates through the interstices of the snow mass and refreezes (a process known as regelation) around snow crystals to form ice granules. This process takes place more rapidly in temperate regions where the temperatures are closer to 0° C, and where there are longer periods of temperatures above freezing point, than in the colder conditions of the Arctic and Antarctic. Once the permeability of fresh snow has been reduced, so that the air pores are no longer connected, the mass is known as firn (or névé), and there is an accompanying increase in density from approximately 0.1 g/cm³ to approximately 0.5 g/cm³. Firn is a dull white, impermeable and structureless form of ice. Further pressure from the overlying weight of snow then converts the firn to ice (density 0.8–0.9 g/cm³) with few air pockets and a blue sheen. Such a change may take only 3–5 years in the Yukon, between 25 and 40 years in the Alps and up to 200 years in Greenland due to the increasing cold and lack of melting. Measurements in the Antarctic ice show that the snow-firn change takes place at depths of less than 10 m, but the firn-ice change takes place gradually down to 40–50 m. Ice accumulating in this way retains a solution of chlorides and other salts in a thin film around each granule. These salts are derived from the atmosphere.

The annual accumulation of ice can often be measured, since it builds up in the winter season, whilst summer melting and flow at the surface will often give rise to a 'dirt band', or a zone of discontinuity, in the recrystallised ice. In areas of heavy accumulation each layer may be over 2 m thick. In Antarctica, on the other hand, there is little summer melting, low accumulation and little atmospheric dust to give rise to such dirt bands.

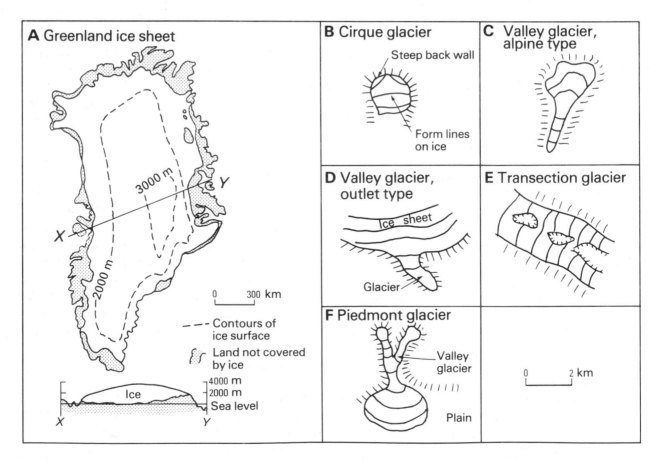

Figure 8.2 Ice masses.
(A) The Greenland ice sheet. The mass of ice is up to 3000m thick and has depressed the central area isostatically: this area would be drowned if the ice melted. Over 83 per cent of the ice is above 1400m, where ice accumulates. Much of the coast is ice-free.
(B)-(F) Forms of valley glacier.
(After Flint, in Strahler, 1969 (A); the rest after Price, 1973)

Types of ice mass
The form of an ice body resulting from the accumulation and burial of snow over periods of thousands of years will be related to the shape of the underlying land; the rate of snow supply and melting; the trend of climatic changes; and the length of time over which accumulation has taken place. Small ice masses may suggest that a cooling is beginning, or that a colder phase is giving way to a warmer, but the largest ice sheets have resulted from the persistence of extreme cold for thousands of years.

A number of ice mass types are recognised (Figure 8.2).

a) **Cirque glaciers** occupy depressions on mountain slopes: they are often at the highest altitudes, and are characterised by steep backwalls.

b) **Valley glaciers, Alpine type,** occur where ice fills a valley with several cirque glaciers feeding from the higher slopes of accumulation. The valley glacier will descend to heights determined by the rate of ice supply from above, and the rate of melting as it reaches warmer levels. They frequently descend below the regional snow-line. Valley glaciers are common in the young fold mountain areas such as the Alps, Himalayas and the Alaskan coastal mountains.

c) **Valley glaciers, outlet type,** are related to ice sheets or ice caps, rather than to cirque glaciers. They occur where the ice sheet becomes channelled between mountains, and are associated with nunatak peaks, which are the tops of mountains projecting through the marginal areas of ice sheets. Outlet type valley glaciers are common around the Vatnojökull ice cap in Iceland, and around the Greenland and Antarctic ice sheets.

d) **Transection glaciers** are valley glaciers which have become so thick that they spill over dividing ridges and join with other glaciers in adjoining valleys.

e) **Piedmont glaciers** are formed where a valley glacier extends over a lowland and spreads out horizontally. The Malaspina glacier of Alaska is the best-known example today.

f) **Ice caps** bury the relief in upland areas, as in southern Iceland, although they can occur at lower altitudes in very cold areas like Baffin Island. The ice is no longer confined by valley walls, and moves in directions determined by the gradient of the upper surface of the ice. Ice caps on lower land are usually smaller and slower-moving.

g) **Ice sheets** are the largest forms of accumulation, and at present there are two in the world. The Greenland ice sheet contains 11 per cent of the world's ice, whilst a further 85 per cent is in the Antarctic ice sheet. When ice bodies exceed a diameter of approximately 500 km the weight of ice becomes sufficient to cause isostatic downwarping of the Earth's crust. It is estimated that Greenland and Antarctica have subsided by between 25 and 33 per cent of the ice thickness, depressing the rocky surface of central Greenland below sea-level. Such large masses of ice also tend to create their own climates: air above their surface is cooled and flows outwards, repelling the advance of much moist air, which is forced to precipitate snow on the margins of the ice sheet only. The extreme cold spreads to surrounding waters where the ocean surface freezes around the ice sheet or outlet glacier for much of the year, so that the water becomes covered with a mass of floating ice. Ice breaks from the edge of the sea ice to form icebergs in warmer seasons or when waves attack the outer margin.

Glacier regimes
The input to the glacier system, in the form of snow and ice, is largely responsible for movement in the glacier and the changing position of the lowest point. Melting at the lower end of a glacier varies with climatic conditions and will be greatest in hot, dry summers. Most glaciers around the world have been retreating — albeit with shorter periods of readvance — during this century, but it seems that the large ice masses of Antarctica and Greenland may be approximately in balance, or even increasing (Figure 8.3).

The glaciers in areas of high relief and precipitation (e.g. south Norway; South Island, New Zealand; Alaska) experience the greatest variations. Those in Alaska are subject to sudden

Region	Mass balance	Trend
Antarctica	Accumulation difficult to measure due to drifting, but low: estimates 0.1-2.0 cm water equivalent per year. Loss includes iceberg calving, some melting: not so much.	Slight positive mass balance (e.g. 1.1×10^{18} g per year).
Greenland	Accumulation estimates = 300 mm water equiv. per year; loss normally balances this.	Approximately in balance.
Barnes Ice Cap Baffin Island	Accumulation and loss both small.	Slight overall loss.
North Sweden e.g. Storglaciaren	Accumulation (1946-62) 4.0×10^6 m³ water equivalent; loss 6.2×10^6 m³ per year.	Loss 55% more than net accumulation: rapid retreat.
Iceland Vatnajokull cap	Considerable variation from year to year but overall loss exceeds accumulation.	Slight loss.
South Norway Nigardsbreen glacier	Longer period of observation than most suggests broad periods of advance and retreat in this century.	Recent loss (to 1962).
Alaska Malaspina and Upper Seward glaciers	Variable patterns with occasional surges punctuating retreat phases. Overall excess of loss over accumulation.	Overall loss.
Alps e.g. Hintereisferner	Variable mass balance, but every year negative except one, 1952-61.	Overall loss and retreat 25 m per year.
Southern Alps New Zealand e.g. Tasman/Franz Josef glaciers	Loss exceeds accumulation west of Southern Alps: variable pattern in this century, but overall excess of loss over accumulation.	General loss.
Equatorial Africa e.g. Mounts Kenya, Kilimanjaro	Kilimanjaro has accumulation on lower part above snow line and melting near the peak; Lewis glacier (Kenya) retreated 60m, 1934-58.	General loss.

Figure 8.3 Glacier regimes: a world picture. Is there a general trend? notice that, because of its size, a positive mass balance in Antarctica will outweigh losses in all the other ice masses of the world. (After data in Embleton and King, 1968)

surges during which the glacier front will advance several kilometres in a period of months: the Black Rapids Glacier surged forward 6 km in 5 months in 1937, in between periods of retreat, and the Tweedsmuir Glacier on the British Columbia/Yukon border advanced in similar fashion in 1973. Such surging may lead to damming of the main valley into which the tributary valley glacier flows, and the ponding of a lake up to 20 km long. The release of this lake water when the glacier dam is lowered by summer melting could lead to catastrophic flooding. The evidence for such a flood associated with the nearby Lowell Glacier 200 years ago shows that the valley sides were stripped of forest up to 100 m above the present stream bed. Such glacier surges seem to be due to periodic increases in precipitation, which affects the down-glacier movement after a period of years. The increased accumulation leads to imbalance in the glacier ice, which may be released suddenly by a local earthquake or increase in subglacial meltwater and give rise to the surge. Once the surge has taken place a large mass of ice becomes stranded in the lowest portion of the glacier with no constant supply from above, and this ice slowly melts.

Ice decay

A glacier can be divided into an upper accumulation zone, and a lower ablation zone (ablation is the combined loss by evaporation and melting), divided by the firn line, below which no new firn is produced (Figure 8.4). In the zone of ablation, melting and evaporation increase

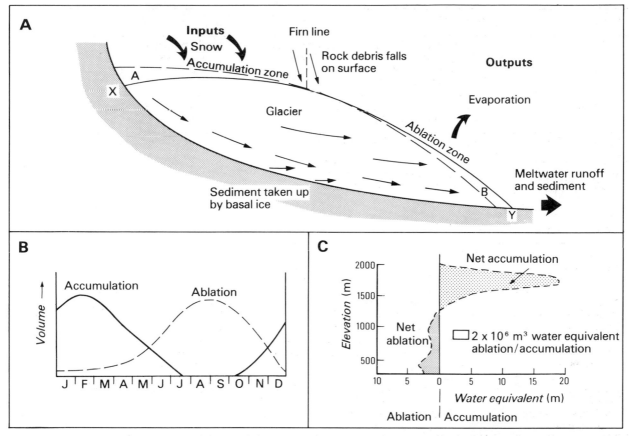

Figure 8.4 Accumulation and ablation of glacier ice: the glacier system.
(A) The glacier budget. X-Y shows a state of balance, with A representing winter accumulation and B summer ablation. If A exceeds B the glacier advances; if B exceeds A the glacier retreats.

(B) The seasonal distribution of accumulation and ablation in a northern hemisphere glacier.
(C) The mass balance of Nigardsbreen, a Norwegian glacier, for 1961-62: the surplus accumulation was 94.9×10^6 m³ water equivalent. (After Ostrem 1962, in Embleton and King, 1968)

towards the ice 'snout', or front, so that the ice becomes thinner and the proportion of rock debris to ice increases.

Ice masses will advance or retreat according to the balance between accumulation and ablation: these are controlled by the supply of snow precipitation and the temperatures in the ablation zone. Retreat of an ice front may be due to reduced precipitation, or increased temperatures, or both. It will be accompanied by thinning ice and extensive meltwater activity at the margin. Advancing ice fronts result from increased precipitation or cooling of atmospheric temperatures, or both.

Ice movement, transport and deposition

The study of glacial movement is still in its early stages, although increasing numbers of measurements on glaciers have been made since the 1950s. The difficulties of such studies are due largely to the fact that ice, as a solid, moves very slowly, and it is not easy to discover what happens at depth within and beneath the ice. Apart from the rapid surges (e.g. Black Rapids Glacier in 1937 moved at 75 m/day), most valley glaciers move at velocities of less than 1 m/day. The greatest rates over long periods are associated with glaciers possessing steep gradients along their courses and having large accumulation areas: some Greenland outlet glaciers have moved at 20 m/day. There is also a seasonal variation in glacier movement, with higher velocities in the summer season when meltwater and rain provide lubricating subglacial water — although higher winter velocities have been recorded above the firn line.

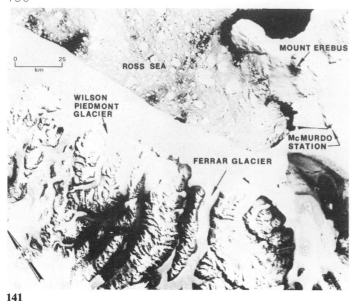

141

142

143

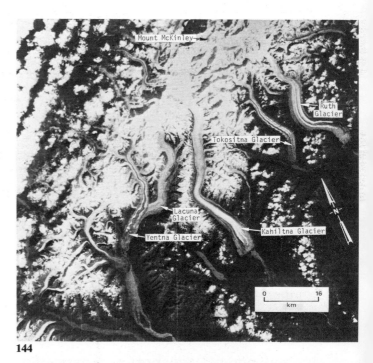

144

Plate 141/142 The edge of Antarctica. Two views of the Ross Sea and McMurdo Sound area. The 1973 LANDSAT-1 view shows the site of the US scientific station (McMurdo) with the channel cut by ice-breakers to it; Mount Erebus is the most active Antarctic volcano. The Ferrar glacier with nearby ice-free valleys can be seen clearly. The oblique aerial photograph, taken in December 1957, shows the snout of The Ferrar glacier and the adjacent ice-free Taylor valley with expanded-foot glaciers entering from a small coastal ice cap. The valleys are approximately 5 km wide. Old, partly-melted pack ice is frozen into younger sea ice in McMurdo Sound, the fragments being outlined by dust blown from the Taylor valley. (USGS)

Plate 143 Snow on slopes. An avalanche on a slope in the Canadian Rocky Mountains, with the results of an earlier avalanche to the left. Snow moves to fill hollows in this way, and by wind action. (National Film Board of Canada)

Plate 144 Valley glaciers in south-central Alaska near Mount McKinley, as seen on a LANDSAT image. Most glaciers flow 5–30 cm per day and have straight moraines; a few have alternating periods of stagnation (up to 50 years) and high flow rates (up to 1 m per hour for a year) and are known as surging glaciers, having zigzag patterns of surface moraines. (USGS)

Plate 145 Cirque glaciers at the foot of a steep ridge near the Finsteraarhorn in the Alps of the Bernese Oberland. (Swiss National Tourist Office)

Plate 146 The Yentna glacier, Alaska (see position on 144): a valley glacier with tributaries fed from cirques. (USGS)

Plate 147 Sea ice off Antarctica: ice berg and pack ice. (Australian News and Information Bureau)

Plate 148 Malaspina Glacier, St. Elias Mountains, Alaska— a piedmont glacier. The ice spreads out on a coast plain, and the moraines are compressed as the ice slows, as well as being dragged out sideways to give the complex patterns shown. (USGS)

Plate 149 Stereo pair: the Worthington glacier near Valdez, Alaska. Draw a sketch-map of this area, labelling the landforms and features of the glacier. (USGS)

Ice movement

Ice masses move by two main types of mechanism.

a) **Internal deformation** within the ice is caused by a combination of small-scale effects. These include the deformation of individual ice crystals by the gliding over one another of layers parallel to the basal plane of the crystal; and by the movement of individual grains along the direction of maximum stress facilitated by the early-melting saline film around each. Moving ice behaves as a plastic body, in which no deformation will occur until a certain stress level is exceeded, beyond which deformation will increase progressively. Stresses affecting ice masses are supplied by the increasing weight of ice accumulation, the increasing effect of gravity on steeper slopes, and the compression of ice at rocky obstacles or in narrowing channels.

b) **Basal sliding** may be the major mechanism of movement for many glaciers. This is induced by a combination of local melting with increased stress at obstacles where the ice is close to its melting point, and enhanced plastic flow near an obstacle. In addition, any meltwater reaching the basal zone of a glacier will increase the sliding velocity: this accounts for the increased velocities measured in summer seasons.

These mechanisms vary in proportion and effect in different ice masses. A most important distinction has been made between 'warm' glaciers — where the ice temperature is only just below $0°$ C — and 'cold' glaciers — where the ice is permanently at very much lower temperatures. The warm glaciers occur in temperate areas (the Alps, Scandinavia, Alaska), where summer temperatures rise to cause considerable melting and where the pressure melting point is reached throughout the ice thickness. The formation of firn and ice is thus rapid, carrying downwards the surface warmth in melted snow. Both internal deformation and basal sliding are encouraged, and the latter causes 50 per cent of surface movement on average. Cold ice masses occur in polar regions, and experience slow firn-formation and movement. Meltwater streams flow only on the surface and the ice is frozen to the bedrock.

Movement of ice in large masses, such as ice sheets, is particularly complex. In the Greenland ice sheet temperatures rise with depth (from $-24°$ C at 10 m to $-13°$ C at 1400 m). This rise is due partly to the insulating effect of the thick ice, and partly to the addition of small amounts of heat from the Earth's interior (156 joules/cm^2 year). This fact, together with the increasing pressure at depth, suggests that movement is greatest at depth. If the pressure melting point of an ice sheet is reached in the basal ice due to increased rates of accumulation,

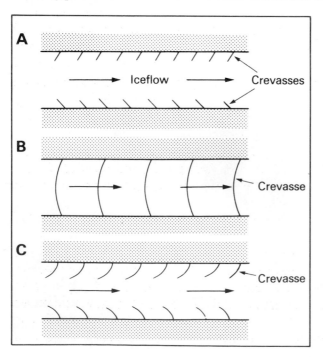

Figure 8.5 Crevasse patterns in valley glaciers. (A) Shear stress provided by valley walls only (i.e. the drag effect of the valley on moving ice). (B) Shear stress with extending flow (i.e. with ice flow velocity increasing down-valley). (C) Shear stress with compressive flow (i.e. with ice flow velocity decreasing down-valley). (After Paterson, 1969, and Nye, 1952, in Price, 1973)

there will be rapid expansion horizontally of the whole mass. This may be followed by shrinkage due to melting of the thinned ice mass in warmer areas. The movement of the ice is thus related closely to the interior temperature of the ice mass.

Movement of glacier ice is reflected in the surface pattern of crevasses. They form due to the longitudinal extension of the glacier ice, and their patterns are related to the rates of flow and subglacial topography (Figure 8.5). Crevasses play an important role as zones of penetration for meltwater and rock fragments in the ice. They are seldom more than 30 m deep, but some over 36 m deep are known in the cold ice masses of Greenland and Antarctica, where the lower temperature ice can support higher shear stresses.

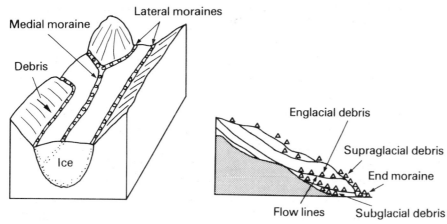

Figure 8.6 Sources of debris associated with a valley glacier. (After Price, 1973)

Ice transport

Ice masses are particularly effective agents for transporting rock debris. Their large mass, filling a valley or blanketing an entire landscape, together with their solid nature, enables them to transport enormous quantities of debris, and to include very large blocks: the largest of those found in Germany, for instance, measures 4 km by 2 km and is 120 m thick.

The input of rock debris to the glacier or ice sheet systems may come from the weathering and landsliding or avalanching of mountain slopes above the level of the glacier surface; from the entrainment of unconsolidated materials over which the ice mass advances; or from the erosion of rock by the ice mass. Thus the debris either arrives on the top surface of the ice, and may be taken down into the glacier via crevasses or meltwater, or it enters from the sole or base of the ice. Three distinct zones are recognised within the ice (Figure 8.6), but little supraglacial debris occurs in ice sheets unless nunataks project above the surface. The supraglacial debris occurs as relatively narrow surface bands (lateral or medial moraines) up to several tens of metres wide, but seldom more than one boulder thick. There may be wider spreads covering the entire surface of the glacier in the ablation zone as englacial debris is revealed by melting. This material in the ablation zone may be thick enough to be gullied by rain in the warm season, and to become virtually a landscape of running water.

Erratic blocks demonstrate not only the transporting ability of glaciers, but its uniqueness, since they may occur in places hundreds of metres above their original sources, and several hundred kilometres away. They have been used to trace the extent and direction of movement of former glaciers and ice sheets (Figure 8.7).

Depositional features left by the ice

Deposits revealed beneath retreating ice masses may be divided into those which are stratified (i.e. occur in layers), and those which are unstratified (Figure 7.31). The stratified deposits are those which have been redistributed by meltwater streams issuing from the decaying and retreating ice: they are restricted to sand and silt grades of debris, except for larger particles moved at the height of the melting season, and are described in greater detail in chapter 7. The unstratified deposits are dumped by the melting ice in a completely unsorted (i.e. mixture of

194

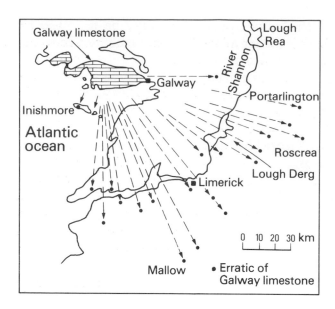

Figure 8.7 Glacial erratics in central Ireland. The Galway limestone has a limited outcrop, and erratic blocks of this rock have been recognised at the places shown by a cross. (After Charlesworth, 1957)

sizes) fashion. The stratified and unstratified deposits form a complex mixture near the margin of an ice sheet or glacier, reflecting periods of advance or retreat of the ice front (Figures 7.32 and 7.34). Ten per cent of the world's land surface lies within a zone where such associations of stratified and unstratified deposits formed near the margins of former ice sheets are common: these areas were covered by ice during the Quaternary Ice Age. In places like Lincolnshire and East Anglia much land would be submerged today if it were not for the glacial deposits.

Most investigators into the nature and origin of glacial deposits have been concerned with those which were left by old and long-disappeared ice sheets, and far fewer have looked at the less accessible rock debris beneath the glaciers and ice sheets of today. An understanding of the mechanics of deposition can be gained only by studying both.

In studying the unstratified deposits dropped by melting ice, it is important to use a clear terminology: 'till' is used here in reference to the nature of the deposits, and 'moraine' will refer to the landforms produced by deposition. Till deposits are sometimes divided into lodgement till, formed beneath the ice, and ablation till, formed from surface melting, but it may be difficult to distinguish these two varieties. Tills may form in both situations at once (Figure 8.8).

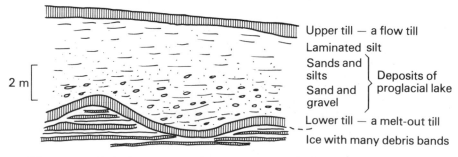

Upper till — a flow till
Laminated silt
Sands and silts
Sand and gravel
} Deposits of proglacial lake
Lower till — a melt-out till
Ice with many debris bands

Figure 8.8 A complex till, accumulating on the edge of Erikabreen in Oscar II Land (Spitzbergen). The upper till is thought to have flowed into a meltwater lake on the irregular glacier surface. The lower till is even in thickness, although irregular in section: this was formed by melting of the upper ice below the lake deposits. (After Boulton, 1970, in Price, 1973)

The essential character of a till is that it is formed of a mass of broken rock fragments, mostly angular or subangular, and of many different sizes and compositions. The term 'boulder clay' has been used to describe this variety of sizes contained, but is not a good general term since not all tills contain boulder-size fragments. The larger fragments in a till are usually related to local bedrock (cf. the 'Chalky Boulder Clay' of East Anglia), but the finer material often comes from farther afield. Stones in the tills commonly have their long axes orientated in directions parallel to the direction of ice movement. This has been attributed to different causes: it may be due to the gradual accumulation of debris beneath the glacier followed by this layer being moved with the glacier ice in a plastic state; it may be due to the orientation of the long stone axes in the direction of ice movement and subsequent deposition as the ice melts without disturbance of this pattern; it may be related to meltwater movements beneath the ice; or it may be due to the ice pushing or squeezing the underlying debris in particular directions. These different suggestions illustrate the facts that a variety of processes operates beneath the ice, and that little is known of their relative importance.

In general, two major mechanisms are responsible for the deposition of material carried by the ice. On the one hand there is the melting of ice from beneath debris carried on the surface, and on the other there is the melting of small quantities of ice within a rich debris layer either in the base of the ice or near the surface. It seems unlikely that ice could have actively plastered the underlying rocks with debris over large areas, since this would necessitate large quantities of basal ice debris together with movement of the ice during deposition, whereas most ice is 'dead' when deposition occurs. Most till seems to be derived from the more passive melting ice which may not move forward during this process; much comes from the basal zones, but ablation till may be more important where a cold ice mass, frozen to the underlying rock down to its snout, decays.

Moraines are the landforms produced by the deposition of till. Ground moraine is a low relief area produced by the deposition of till sheets, each a few tens of metres thick and perhaps up to over 100 m in total thickness. Older valley systems are filled and the relief masked in lowland areas, which are extensive in central North America and northern Europe including eastern England.

Drumlins are moraines which have an elongated, oval shape, and a steeper face ('stoss end') upstream. Their origin, however, has presented a puzzle, since their study has been restricted to older, exposed tills. They are often found on the ice side of an old terminal moraine, and tend to occur in groups (Figure 8.9); single drumlins are rare. Drumlins may be composed of clay till, a sandy or loamy till, or they may have a rock core or be formed from the re-working of older till. Those with a rock core occur by the side of other types. Some have

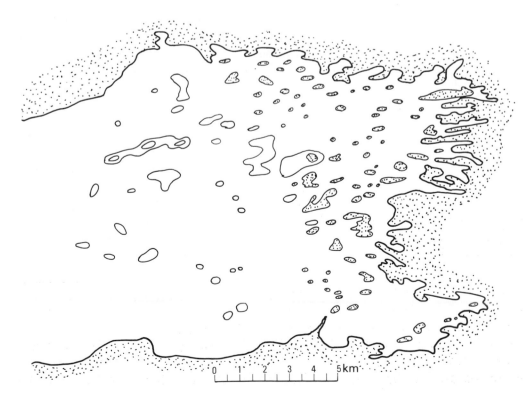

Figure 8.9 A drumlin field in Clew Bay, Ireland. The unshaded drumlin-like shapes are those covered by water on the bay floor. (After Charlesworth, 1957)

lenses or layers of stratified material; others contain material which has clearly been subjected to pushing. Theories concerning the origin of drumlins have included those which emphasise the erosion of a rock outcrop, or a previous drift cover: evidence such as the formation of solid rock drumlins or the pushing of older till suggests such an origin. Other theories, which are most numerous, look to depositional processes, mostly involving the subglacial accumulation of till in streamlined forms. The complex and varied composition of drumlins, however, shows that similar forms may well have different origins.

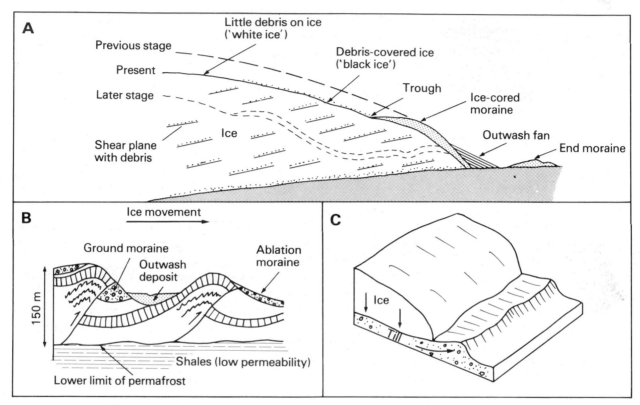

Figure 8.10 Moraine ridges at the fronts of ice masses. (**A**) The retreating margin of the Barnes Ice Cap, Baffin Island. Notice how a series of ablation moraines will become end moraines spread over the area once occupied by the retreating ice. The debris is brought to the surface along shear planes.

(After Goldthwait, 1951, in Price, 1973) (**B**) Ridges formed by ice-push. These incorporate moraines and underlying rocks subject to permafrost. (After Kupsch, 1962, in Pitty, 1971) (**C**) Morainic ridges produced by squeezing of till from beneath an ice front. (After Price, 1973)

Moraine ridges fortunately can be studied both in the process of formation, and in older deposits. Some are due to dumping following the melting of ice (Figure 8.10), either at the frontal zones of ice masses (terminal, or end, moraines), or along the margins (lateral moraines). Crevasse fillings may also give rise to short morainic ridges of this type. An advancing glacier may bulldoze a mass of unconsolidated material into a ridge known as a 'push moraine', and these commonly show evidence of such disturbance in internal folding or thrusting. Other ridges are formed where the excess of meltwater in the sediment beneath the ice enables it to be squeezed upwards or outwards. All these forms may be mixed in different proportions, and the larger morainic ridges accumulate where the ice front stays in place long enough for the ridge to be built up: this would appear to be an uncommon occurrence, since extensive, high ridges are unusual. Some end moraines are several hundred metres high (e.g. 430 m in Franz Josef glacier, New Zealand) and a thousand metres from front to back; those across Swaledale in Yorkshire are 10–20 m high and 50–100 m wide.

Deposition beneath an ice sheet is a complex process, involving the action of ice and running water. The two have been separated in this book, but normally act together. Most deposition occurs during the retreat of an ice front, leaving behind a landscape of ground moraine, drumlins and morainic ridges, together with sandar, kettles, eskers, kames and stream channels. Near the front of ice masses today such features change markedly over a period of a few years (Figure 7.31), but due to the recent predominance of glacial retreat over the last 150 years, they are usually obvious.

Erosion by ice

Whilst the processes of ice transport and deposition are now fairly well understood, the processes of erosion are much more difficult to observe at first hand. Most of the evidence available is in the shape of landforms which are presumed to have arisen largely by erosion. Tunnels have been cut to the base of glaciers, and the evidence resulting from such investigations has advanced the understanding of what happens enormously.

The mechanics of glacial erosion

Ice carrying no rock debris would result in little erosion when passing over an unweathered rock surface. Glaciers, however, normally carry considerable quantities of rock fragments in their basal sections, having moved across areas which have been subjected to weathering of the surface rocks, often in prolonged periglacial (tundra) conditions before the onset of ice. They erode, on their own, by two main mechanisms.

a) **Corrasion** takes place where the fragments in the basal ice wear away the underlying rock. This occurs in warm glaciers where basal sliding is important. If the fragments consist of silt or sand grains the rock beneath will be polished; if they are gravel or boulders the rock will be scratched, or striated, if it is softer than the fragments. Striations are usually found in groups, each scratch being between 1 cm and 1 m long, on surfaces which vary from level to vertical, but which most commonly were ascended by the ice in its passage. They give a general idea of the direction of ice movement, but other features have also to be taken into account in interpreting the past directions of ice flow.

Thermal zone	Subglacial erosion	Transport	Deposition
Zone A net basal melting	Subglacial streams may be very important; also ice with embedded particles: crushing and striations.	Melting and re-freezing at base, giving debris-charged regelation ice; also meltwater.	Subglacial lodgement till when frictional drag on particles moved over bed = tractional force exerted by glacier ice. More under greater ice thickness with rough bed, giving drumlinoid forms and filling hollows.
Zone B balance — melting = freezing	Similar to Zone A, but reduced unless down stream of Zone A.	Similar to Zone A, but basal debris layer of constant thickness on smooth surface	Rates higher than Zone A if no water.
Zone C Water freezes to moving glacier sole	Sole slips over bed as meltwater freezes to it. Crushing, polishing and plucking.	Water freezing in lower sections will add debris-rich layers to ice base	Net freezing tends to re-incorporate material dropped, giving net erosion and englacial debris.
Zone D Subglacial materials frozen	No slip: differential movement above ice/rock junction. Erosion where rock protrudes into glacier. No crushing of polishing.	Frozen material gained by plucking. Thrusting in Zone D below Zone C gives upward movement of englacial material.	Less subglacial deposition than ablation till. Some drumlins where rocks project.

Figure 8.11 The relationship of thermal conditions of ice masses to glacial activity. The zones may be combined in various ways along the length of a glacier, and one or all may occur in one glacier.

Notice the contrasts between conditions where the ice is frozen to the glacier bed, and where there is melting of the ice at this level. (Data after Boulton, in Price and Sugden, 1972)

There is still considerable debate as to the efficacy of corrasion in wearing away the rock beneath a glacier. It certainly seems to have been effective on limestones in northern England, since much of the limestone pavement has been attributed to the stripping activity of glacial corrasion. In a recent experiment on the Breidamerkurjokull glacier in south-eastern Iceland, blocks of marble and basalt were bolted to solid rock and left for three months for the boulder-studded sole of the glacier to pass over. During this time the glacier moved 9.5 m, and afterwards it was seen that the two surfaces were heavily striated; the surface of the marble had been reduced by 3 mm and the basalt by 1 mm.

b) **Plucking** is another mechanism of erosion by the glacier itself, but subglacial observation has shown that it is not caused merely by the freezing of ice around a protuberance of rock in the bed of the glacial channel, followed by a pulling away of much of the rock as the ice moves forward. It particularly affects the downstream side of outcrops of well-jointed rocks: these raise the ice base, leading to stresses in the base of warm glaciers and resulting in compression melting on the upstream side together with re-freezing on the downstream side. The water which re-freezes fills joints, leading to the breaking of blocks from the rock mass. These and other blocks of rock between the ice base and the underlying rock may be pushed along by fragments protruding from the ice base and then become entrained in the ice by regelation (i.e. re-freezing of temporary meltwater). They will then be carried along planes of weakness into the ice.

Many of the processes acting beneath an ice mass are more in the realm of weathering (chapter 5) than of erosion by the ice mass. There is also the subglacial movement of meltwater (chapter 7), and a range of water–ice–sediment mixtures under hydrostatic pressure, which are regarded by many as the most effective erosional processes taking place beneath the ice. Certainly it seems that little erosion can take place beneath a cold ice mass where the ice is frozen to the underlying rock. Ice erosion, like other glacial processes, is controlled largely by the thermal characteristic of the ice (Figure 8.11).

The actual mechanisms of ice erosion are thus poorly understood, yet the efficacy of the mechanisms are not doubted. Measurements of silt fractions in adjacent streams, one issuing from beneath a glacier, and the other not supplied by glacial meltwater at all, suggest that glaciers eroded from 10 to 20 times the amount removed by the stream on its own over the same period. The most diagnostic features of glacial erosion occur in areas of high relief and precipitation — surely the areas which endured the longest and most effective covering by ice in the past. It seems from this evidence too, that ice erosion would be most effective where the thickness of ice, local relief and the basal temperature provided maximum velocity, and where previous weathering had broken up the underlying rocks.

Landforms due to glacial erosion

The effects of ice erosion are seen particularly in upland areas of heavy precipitation, which were once glaciated. In Britain these include north Wales, the Lake District and north-west Scotland. In the Alps, Norway and Alaska there are still many glaciers filling parts of valleys which have previously been more completely occupied. Two associations of landforms dominate these areas.

a) **Cirques** are steep-sided semicircular depressions, often with their floors overdeepened to rock basins. In middle latitudes they occur in groups at relatively high altitude and above the level of major valley glaciers, at the heads of those systems, or along the tops of the valley sides. In and around Antarctica they may occur just above sea-level. They are most numerous where the amount and rate of accumulation could give the greatest intensity of erosion, i.e. where the alignments of high ground would permit surface shade from melting sunshine, and where snow accumulation and drifting would be encouraged. Those in the northern hemisphere are commonly on north-east-facing slopes (Figure 8.12), and those in the southern hemisphere on south-facing slopes. It is also clear that areas which have experienced the longest periods of glaciation have the best-developed cirques: those in Antarctica, the high

Himalayas and Alaska, still occupied by ice today, are much larger than those in deglaciated areas.

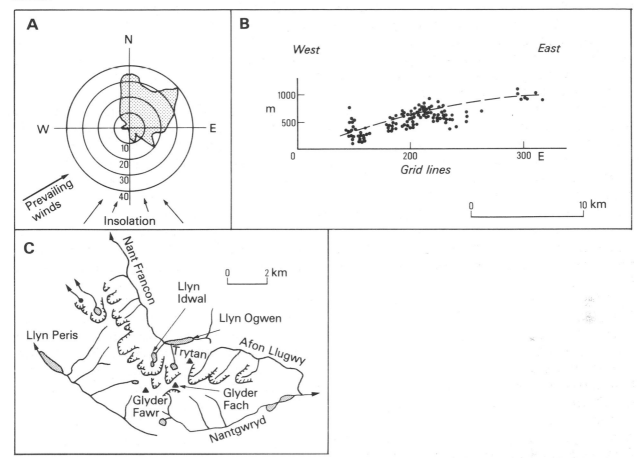

Figure 8.12 The distribution of cirques.
(A) Orientations of Scottish cirques: the scale shows the numbers of cirques in each 15-degree sector. (After Sissons, 1967)
(B) Altitude of Scottish cirques along a belt from Mallaig on the west coast where the cirques occur at a lower height. Farther east the mountains were protected from such heavy accumulation at low altitude. Almost 90 per cent of Scottish corries occur in areas now receiving over 2200 mm precipitation annually, and none are in areas with less than 1500 mm. (After Linton, 1959, in Sisson, 1967)
(C) Cirque distribution in the Glyder range, north Wales (After Sparks and West, 1972)

Cirques have been the subject of much speculation in relation to their origin, and, once again, little has been observed of the actual mechanics of erosion beneath cirque glaciers today. The steep backwall seems to be produced largely by the shattering of the rocks in freeze-thaw alternations. This process is effective above the ice, but then becomes less so until a zone is reached where the ice is at pressure melting point. It seems unlikely that the process has any effect in the bergschrund — or surface crevasse in the ice near the cirque headwall — since little melting occurs there. Such shattering by freeze-thaw action is thought to be responsible for the overall enlargement of the hollow. The overdeepened rock basins, on the other hand, are attributed to the effects of rotational movement which has been observed (Figure 8.13).

b) **Glacial troughs,** where most pronounced, are wide, straight valleys with steep, smooth sides and irregular long profiles in which there are rock steps and deep basins cut from the rocks. A preglacial stream system gave rise to the initial relief in many cases: weathering at that stage and immediately before glacial occupation resulted in the break-up of valley-floor rock (Figure 8.14). Whereas the stream occupied a narrow channel, however, the ice occupies much, or all, of the valley and erosional processes affect this entire occupied area. Protruber-

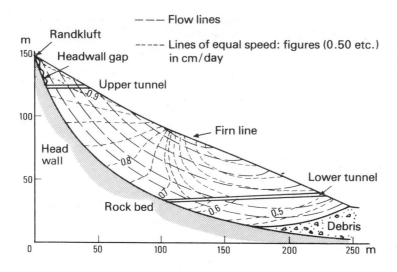

Figure 8.13 Rotational movement in cirques, studied from measurements made on the surface, and in tunnels into the Vest-Skautbreen cirque glacier in Norway. The randkluft is an alternative name for bergschrund, although at times the two are used where a double head-wall gap feature occurs. (After McCall, 1960, in Embleton, 1972)

ances, including rocky spurs on the valley sides, are removed by corrasion and plucking, and the larger volume of ice in the main trunk valley erodes more deeply to leave tributary valleys hanging above them when the ice melts. The rock basins result in lake-filled depressions in the main trough after melting of the ice.

Glacial troughs occur in a variety of situations (Figure 8.15). Those of the 'Alpine' type are contained within a preglacial stream network, separated from other troughs by high ground. Many of those in the Lake District, like the Ullswater valley, are of this type. The 'Icelandic' type is associated with a plateau ice-cap from which the ice discharges into troughs which may follow older valley lines, or be newly cut for the purpose. These are found in southern Norway, as well as in Iceland; troughs in the eastern Scottish Highlands and in the Southern Uplands of

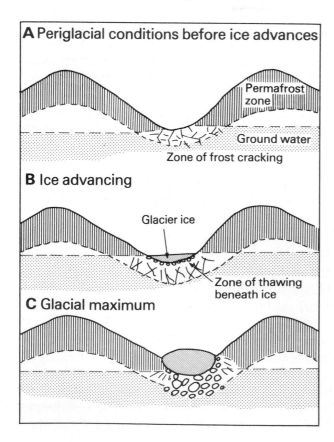

Figure 8.14 The effects of preglacial valley-floor weathering and of the presence of glacial ice on the break-up of rocks. (After Tricart and Caileux, 1962, in Embleton and King, 1968)

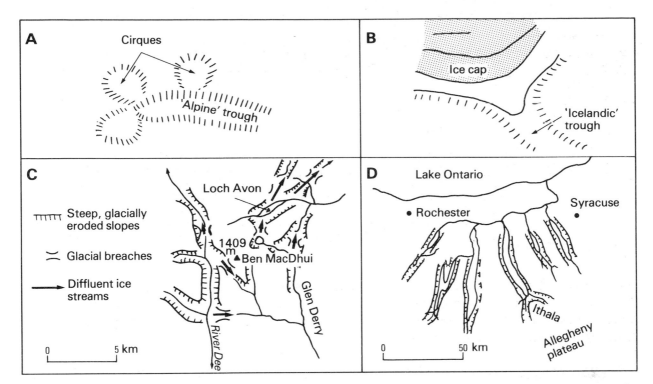

Figure 8.15 Types of glacial trough.
(A) 'Alpine' trough: a simple system fed by cirques.
(B) 'Icelandic' trough: the glacier is fed by an ice cap. (Both after Price, 1973)
(C) 'Composite' troughs due to watershed breaching: the Cairngorm area of north-east Scotland. (After Linton, 1949, in Embleton and King, 1968)
(D) 'Intrusive' troughs. The Lake Ontario ice mass

moved south carving deep and very steep troughs in the Allegheny plateau, which later gave rise to the New York State Finger Lakes. The ice boundary was just to the south of these lakes, on the New York/Pennsylvania state boundary. The lakes have short headwater streams. (After Clayton, 1965, in Embleton, 1972)

Scotland have been related to such a past association. 'Composite' troughs include sections of preglacial valleys, and also sections of trough formed by transection glaciers breaking through former drainage divides (Figure 8.16). Some of these arrangements are connected with a radial pattern of glacial troughs which is associated with a former ice dome centre of heavy precipitation (coinciding with present-day maximum precipitation in areas like the Lake District and western Scottish Highlands). 'Intrusive' troughs occur where ice was thrust against a highland area after a lowland path. This occurred in the Finger Lake area of New York State, and in the Midland valley of Scotland.

Fjords are glacial troughs on the coast (Figure 8.17) in areas of high land and high precipitation, normally in latitudes 50-65 degrees. The deepest known, Sogne Fjord in Norway, descends 1306 m below sea-level.

The erosive effects of ice are also found in lowland areas, although they may be difficult to distinguish from glacial deposition where unconsolidated deposits occur at the surface, since fluted and streamlined forms occur following the action of both types of process. Where rock occurs at the surface, however, it is clear that erosion is responsible for the formation of rock basins, 'crag and tail' features, parallel grooves and 'roches mountonnées'. The last is produced by plucking of the downstream side of a protruberant rock. All these features occur also in glacial troughs, but tend to be less noticeable because of the scale of the upland relief.

It seems also that the Chalk escarpment north of Hitchin has been lowered by as much as 75-100 m due to ice movement across it, which did not affect the Chiltern Hills south of this line. The tills to the east are full of Chalk boulders. Wide troughs in the clay rocks of the east Midlands of England extend below present sea-level, and also suggest that ice moved southwards concentrated in definite zones and gouged out such channels.

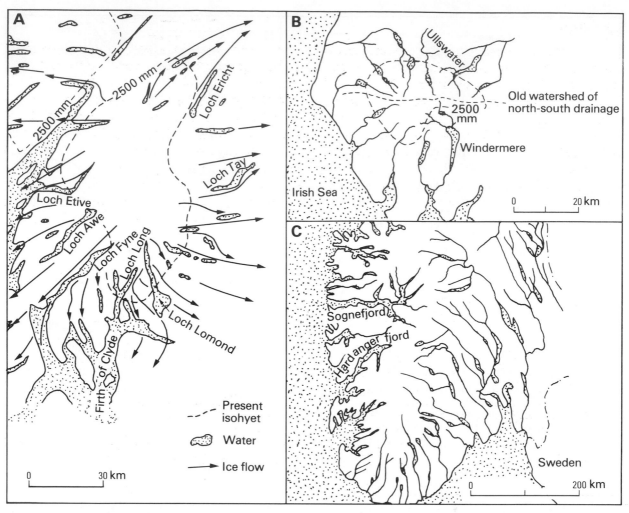

Figure 8.16 Glacial. breaching related to radial drainage patterns.
(A) South-west Scottish highlands: the pattern of radiating troughs is related to the zone of high present-day precipitation.
(B) The Lake District of north-west England: drain-age diverged from an east-west watershed before glaciation. The radial pattern was developed by ice.
(C) Southern Norway: a much larger area, in which several radiating centres can be recognised. (After Linton, 1957, in Embleton, 1972)

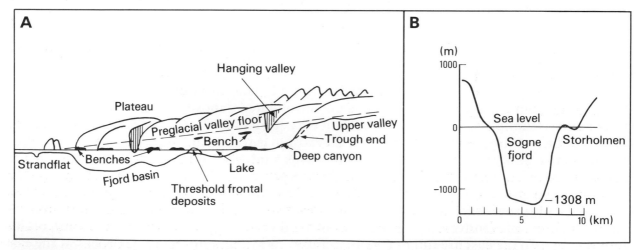

Figure 8.17 Fjord features. (A) A diagrammatic section along a Norwegian fjord. (B) A cross profile of middle Sognefjord, the deepest fjord basin in Norway. (After Gjessing, 1966, in Embleton and King, 1968)

Ice Ages

The abundance of features of glacial erosion and deposition in areas now experiencing temperate climatic conditions prompts the question of how an Ice Age is initiated. The glacially scoured valleys of the English Lake District, the northern Appalachians and the foothills of the Alps testify to the presence of ice in the geologically recent past. Millions of square kilometres of the surrounding lowlands are covered with till. Much of the local topography in the British Isles, northern Europe and North America shows the strong influence of glacial and periglacial activity. At present 10 per cent of the world's land surface is covered by ice. Evidence of past extensions of the ice sheets shows that uup to 33 per cent of the land surface (47 million km²) was covered during the Quaternary Ice Age: 18.5 km² in North America and 10.5 km² in Eurasia. This accounted for half (by volume) of the Quaternary ice: the other half was to be found in Greenland and the Antarctic (Figure 8.18).

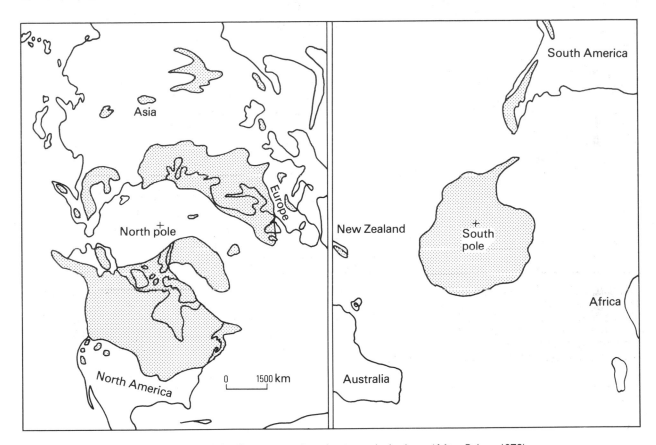

Figure 8.18 The maximum extent of the Quaternary ice sheets and glaciers. (After Price, 1973)

Evidence for past Ice Ages

When an ice sheet advances it sweeps away much of the unconsolidated surface material, so that it is difficult to know what the local preglacial relief features were. In some areas local features have protected till from the scouring of later glaciations which have themselves deposited new drift on the old. Evidence of glacial chronology (i.e. the sequence of events) is to be found in such areas (Figure 8.19). A study of the composition of several layers of till shows that each contains erratics from different sources, and differing stone orientations, indicating different directions and modes of ice movement. This, together with different degrees of activity, would suggest that there have been several advances of the ice, rather than a single glaciation. This hypothesis is further strengthened by the evidence of climatic change which is associated with the onset of Ice Age. Much evidence has been collected from borings into

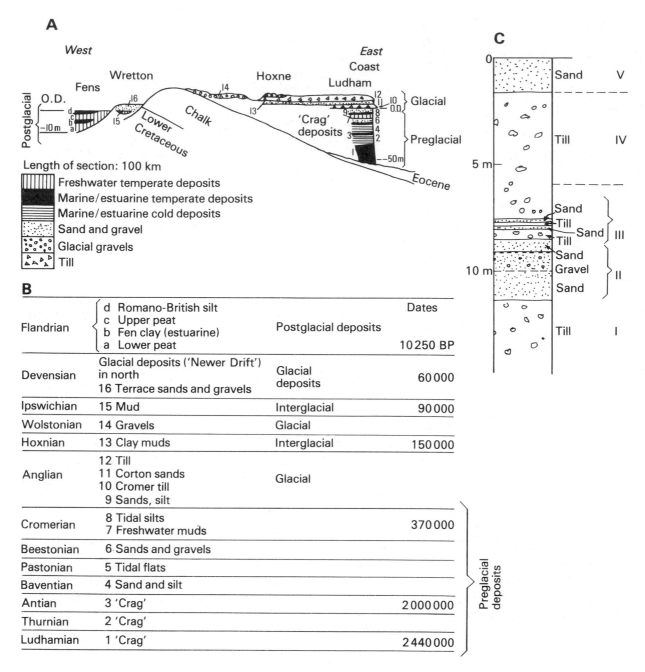

Figure 8.19 Glacial deposits and the sequence of events during the Quaternary. (A) A west-east section through East Anglia. (B) A Quaternary chonology related to (A). (Both after Sparks and West, 1972) (C) A section at Park Burn, Midlothian, Scotland, where there is an alternation of till and fluvioglacial deposits. It is now thought that I-III may be related to the same ice advance (cf. Figure 8.8). (After Kirby, 1969, in Price, 1973)

ocean-floor foraminiferal ooze. From careful examination of the ooze, and an analysis of those species indicative of temperature changes contained in this, it is possible to calculate the surface temperature of the oceans over long periods. Types of foraminifera (of which *Globigerina* is the main one found) indicative of polar and tropical waters occur in close proximity to each other, reflecting considerable fluctuations in climate over the last few ten thousands of years (Figure 8.20). Further studies into the nature of varves found in freshwater lakes adjacent to former ice masses provide additional proof of the changing pattern of climates in recent geological time. The glacial and interglacial periods lasted from several tens to hundreds of thousands of years, and the air temperatures would have been

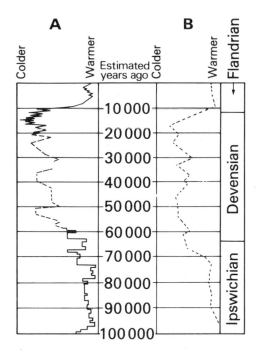

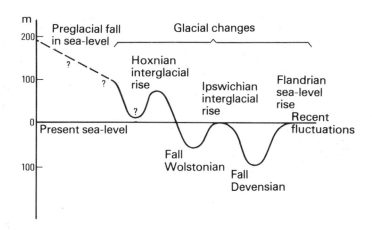

Figure 8.21 Quaternary changes in sea-level, as recorded in deep channels and raised beaches. These changes had important effects on inland and coastal landforms.

Figure 8.20 Climatic variations in the last 100 000 years, as measured by oxygen isotope ratios in ice deposits. (A) Camp Century ice core, Greenland, (B) Byrd Station ice core, Antarctica. (After Dansgaard, 1969 and Epstein, 1970, in Sparks and West, 1972)

reduced in temperate latitudes by 5–7°C below the annual mean for the present day. Temperatures in interglacial periods may have been a degree or so higher than those at present. Since the last ice left Britain only some 10 000 years ago it seems that the present climate may be that of the early part of an interglacial period. An alternative view suggests that pulsations of climate have taken place more often (every ten thousand years or so), and that the peak of the present interglacial warm period was in 1940.

Great fluctuations have also occurred in sea-levels in recent times. During some glacial periods sea-levels stood at over 100 m below the present, as a result of the vast quantities of water locked up as ice on the land (Figure 8.21). Some changes in local sea-levels are the result of isostatic adjustment. The amount of adjustment that has occurred is as much as 300 m where the British ice sheet cover was at its thickest. The speed at which the recoil following the melting of the ice takes place is fastest (approximately 1 cm/year) at the centre of the old Scandinavian ice sheet in the Gulf of Bothnia; the maximum uplift in Scotland is 3 mm/year.

The origins of Ice Ages

Most of the evidence for past Ice Ages comes from the most recent, Quaternary, extension of ice sheets over the northern continents. This also affected the southern hemisphere, but there is hardly any land in the relevant zone. Deposits resembling the tills of modern and Quaternary glaciers and ice sheets occur in the Carboniferous and Permian rocks (approximately 300 million years old) of the 'Gondwanaland' continent (chapter 4) and in the late Precambrian rocks (approximately 600 million years old) around the North Atlantic. Such ancient tills together with the evidence from the Quaternary Ice Age, especially for an alternation of glacial and interglacial periods, have demonstrated the existence of several Ice Ages, each of which may have had internal climatic fluctuations, but no explanation put forward so far for their nature and occurrence at such intervals in time has been completely satisfactory.

Two main groups of theories have been advanced. Those involving changes in the emission of solar energy suggest that variations in the output of energy from the Sun could affect the temperature of the Earth's atmosphere, and that a lowering could give rise to the extension of ice on the surface. Such theories are unsatisfactory, since the Sun, acting as a nuclear furnace, has a steady output of energy, solar flares are too short-lived to effect such major changes, and the events giving rise to major changes of solar mass would take place at wider-spaced intervals than the known occurrence of Ice Ages.

The second group of theories looks to changes at the Earth's surface, which may give rise to alterations in the pattern of atmospheric and oceanic circulation and thus tip the balance towards cooling in the middle latitudes so that ice could accumulate. Some of these suggestions are more speculative than others: thus ideas including variations in the Earth's orbit around the Sun, or the shifting of the Earth's axis, are difficult to prove — or disprove. Changes in atmospheric circulation would certainly have taken place with the late Tertiary uplift of the Himalayan ranges. The Gulf Stream ocean current was initiated following the uplift of Central America. This current directs tropically-warmed waters northwards, leading to a contrast in air conditions above it in the North Atlantic, and increased storminess and precipitation. The combination of mountain uplift and ocean widening has left the Arctic Ocean as an almost isolated body of water which receives little addition of warmer water, and can thus initiate a phase of intensive cooling.

None of these ideas by itself is adequate to explain either the initiation of ice age cooling, or the fluctuations within that event. It is possible that a complex coincidence of different factors could be necessary to account for these phenomena.

A lowering of temperature and increase in precipitation in the midlatitude areas is most important for the extension of ice masses, since the warm ice masses are the most active. A self-supporting cycle of glacial advance and retreat might account for the fluctuations experienced in the Quaternary Ice Age. A growing mass of ice reflects more solar radiation and becomes its own centre of atmospheric circulation; after a time less precipitation arrives, the ice mass spreads out and thins, with its margins retreating after a period of maximum extension; eventually most of the ice melts and atmospheric activity increases again; and

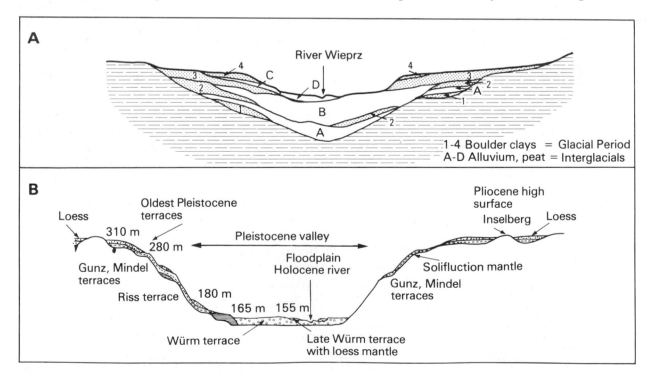

Figure 8.22 Glacial chronologies as revealed in sequences of valley deposits in central Europe. (A) The river Wieprz, a tributary of the river Vistula, in eastern Poland. The valley was carved before glaciation. (After Jahn, in Homes, 1965) (B) An idealised profile of valley features in central Germany. (After Budel 1963, in Derbyshire, 1973)

finally a new ice advance begins. It has been estimated that during the advance of ice sheets the net accumulation would be 20 cm water equivalent/year, requiring 15 000–30 000 years for the building of an ice sheet. This may have been more rapid once the build-up had resulted in the warming of a basal layer in the ice. Dissipation of the ice sheet could be extremely rapid due to high ablation rates as the temperatures rose.

The Quaternary glaciation

Once the Deluge Theory of Noah's Flood had been abandoned, geologists began to suggest that a possible multiple glaciation had occurred in both Europe and North America. Poly-glaciation was accepted as early as 1858 by some geologists and by 1877 Geikie had recognised four types of till in East Anglia. On the basis of this idea a tentative plan of glacial advance and retreat was put forward in 1895 to explain the evidence as found in Europe, but, as the field observation to support this was scanty, the theory was for long not widely accepted.

It was not until 1909 that A. Penck and E. Bruckner patiently worked out the succession of deposits on a river north-east of Lake Constance in an area of maximum Alpine glaciation. Here they found four clearly identified types of till, together with fluvioglacial outwash features, interspersed between other deposits not normally associated with glacial periods. These latter deposits, of varying thickness, revealed the remains of animal and plant life found only in temperate climates, providing proof that the particular location had seen four glacial and three interglacial periods before the present temperate conditions. Using the names of four tributaries of the river Danube — Gunz, Mindel, Riss and Würm — to describe these glacial phases, a comprehensive chronology of Alpine glaciation was developed (Figure 8.22). Many attempts were made to estimate the total length of time involved, and the figure of 600 000 years was favoured for a long time. Dating by radioactive methods of volcanic rocks formed before the first ice advance suggests a longer time-scale of up to 2.2 million years to contain all the phases of the Quaternary glaciation. The Quaternary is sometimes divided into the Pleistocene, which included all the glacial events, and the post-glacial Holocene. The Holocene, however, covers only 10 000 years, and can thus scarcely be afforded the dignity of being designated a geological epoch.

Problems arose when new research areas were correlated with this first established sequence. It soon became recognised that the details of the chronology of ice advance and retreat in one area did not necessarily correspond to those in another. While there is general acceptance that the Riss glaciation was the most severe, local advances and retreats in different parts of Europe and North America varied considerably. An earlier stage than the Gunz, the Donau, has been identified in some areas, while in parts of Switzerland the Great Interglacial that separates the Mindel and Riss appears to have been interrupted by shorter

Provisional numerical order	Great Britain	Alps	Northern Europe	European Russia	North America
Last glaciation	Devensian	Wurm	Weichselian	Valdai	Wisconsinan
Last interglacial	Ipswichian		Eemian		Sangamonian
Fourth glaciation	Gipping	Riss	Saale	Dnepr	Illionian
Third interglacial	Hoxnian		Holstein		Yarmouthian
Third glacial	Lowestoft	Mindel	Elster	Likhvin	Kansan
Second interglacial	Cromerian				Aftonian
Second glacial		Gunz	Pre-Elster	Pre-Likhvin	Nebraskan
First glaciation		Donau			Pre-Nebraskan

Figure 8.23 Various chronologies for the Quaternary glaciation from areas throughout the northern hemisphere affected by the ice sheets. (After Embleton and King, 1968)

glaciations (the Kaner and Glutsch). Evidence for the Donau and Gunz is absent over most of northern Europe, and a different classification has been produced by Polish scientists to suggest three major advances — the Elster, Saale and Weichsel, with the Warthe as a minor glaciation between the second and third. Where evidence exists in Scandinavia for a fourth, earlier, glaciation it is called the Elbe, and may be correlated with the Gunz. A different scheme is used in North America, and a new one has now been adopted for the United Kingdom (Figure 8.23). The pattern is generally similar in the southern hemisphere, although there is some disagreement over the earlier phases.

Changing climatic conditions did not therefore produce an immediate, identical series of ice movements, nor is it realistic to think that climatic change is necessarily simultaneous throught the world. While ice advanced in some areas it was possible for it to be stationary, or even retreating, in others. It is the major pattern of glacial and interglacial periods which appears to have been most widespread.

The Quaternary glaciation in the British Isles

The glacial chronology of the British Isles does not coincide directly with that of Europe. It is established from a great deal of evidence that as the Scandinavian ice cap developed and spread over what is now the North Sea into the North European Plain, other ice caps were developing in the western highlands and islands of Scotland, the Lake District, north Wales and in the mountainous areas of Ireland (Figure 8.24). As the ice accumulated so the individual ice caps merged with the Scandinavian, giving a continuous ice cover from the Atlantic coast of Britain into central USSR. Just as there were oscillations of the Alpine ice cap, so the British ice cap fluctuated as a result of climatic factors.

The net result of glaciation was the deposition of till in lowland Britain as far south as a line which for long was thought to stand approximately from the Severn estuary to just north of

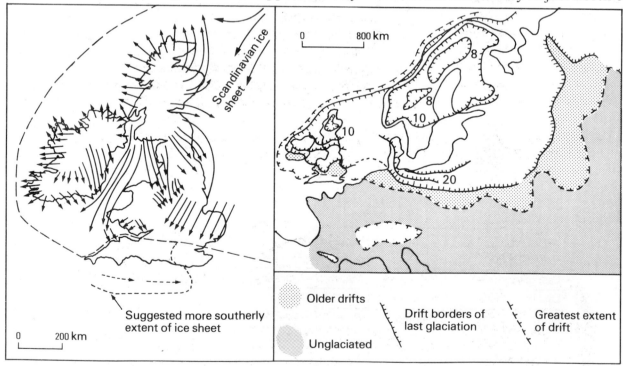

Figure 8.24 Quaternary ice sheets and the British Isles.
(A) Sources of ice and the greatest extent of its coverage. Some authorities would place the boundary even farther south, but the evidence of this is not completely convincing at present. (After Wright, 1937, in Sparks and West, 1972)

(B) Limits of ice advance, and later stages in retreat. The numbers by the moraines indicate an approximate date (10^3 yr ago) when ice stood at that position. (After Flint, 1957, in Embleton and King, 1968)

the Thames. Some geologists are now recognising tills south of this line in north Devon and south of Bristol. Numerous features of glacial erosion are to be found in the high country of the north and west, as well as many water-cut channels resulting from hydrostatic and meltwater overflow processes. The ice tended to follow well-defined courses, even in the lowland areas, but evidence from deposits shows that, at different glacial periods, the direction of ice movement changed. South of the ice sheets periglacial conditions prevailed. Permafrost affected the soils, the products of solifluction blocked the valleys, and the drainage system was adjusted to a base-level which may well have been at least 100 m lower than the present. This caused the rivers in the periglacial zone to incise steeply their courses, with results which are highly significant in the present topography of Britain (cf. Figure 7.35 and chapter 12).

The warm phases of the Quaternary — the interglacials — were accompanied by rising sea-levels and deposition of sedimentary materials in the wide river valleys and along the coast. Once all the ice in an area had disappeared, thus removing the cause of excessive meltwater erosion, the climatic and landform-producing conditions would have been similar to those now operative. Forests and open grasslands would have covered much of the glacial deposits and there would have been considerable growth of peat where drainage was impeded. As the ice sheets melted vast quantities of water meant rapid fluvial erosion.

Research into the nature of deposits in lowland Britain has thrown light on the nature of glacial advances and retreats. In highland areas it is more difficult to discover the chronology of anything other than the final retreat stage of the glaciers. Much of the most interesting work in unravelling the sequence of events has come from studies in East Anglia (chapter 12).

Present knowledge would suggest that the Quaternary period in Britain should be divided into three units — Upper, Middle and Lower, together with the very recent (Holocene) epoch marked by the Flandrian inundation following the last ice retreat. The preglacial stages show the gradual onset of ice conditions (Figure 8.19). The deposits of Ludhamian, Thurnian, Baventian and Pastonian age are all marine in origin showing a higher relative sea-level. In the overlying Beestonian there is evidence of Arctic freshwater plants, and these occur again in the Cromerian together with evidence of permafrost conditions.

The oldest glacial deposit is that of the Lowestoftian Drift, which contains a blue-grey, siltlike till (Cromerian) and a red, silty till (Norwich Brickearth) and is found on the coasts of Norfolk and Suffolk. From the evidence of stone orientation it would appear that the ice moved from the north-west, carrying with it material which was of Scandinavian origin. There was probably a twofold advance, with the first moving almost due east rather than south-east, and depositing the blue-grey clay-rich till containing Cretaceous and Jurassic material. Channels formed either by the advancing ice or hydrostatic action are cut in the underlying deposits.

The following interglacial period (the Hoxnian) was mild with a temperate flora and fauna. In the Birmingham area this was typified by peat deposits, whilst in Norfolk, Suffolk and Essex lacustrine deposits up to 20 m in depth covered large areas. Contemporaneous deposits are to be found as wide apart as the Gort area in County Galway, the Isle of Man and the Thames basin.

The Gipping advance, which followed (although these two advances formerly recognised in East Anglia are now thought to represent one major glaciation), appears to have carried the ice sheet to its maximum extent. In East Anglia it was southward-moving and brought much igneous material from the north of England as well as Chalk, probably from the Lincolnshire and Yorkshire Wolds. Many of the features of the retreat of the ice sheet which deposited the till, including ground moraines, outwash material and crevasse fillings, are found in Norfolk (chapter 12). These have been preserved because the succeeding advance did not reach as far as this and only modified the landscape by periglacial processes. In the English Midlands the Gipping advance was contemporary with the second Welsh glaciation, and three advancing ice sheets met, forming a large, temporary lake in the area now enclosed between Birmingham,

150

151

152

153

Plate 150 The snout of a glacier in the Columbia icefields, Jasper National Park, Alberta, Canada. The car park in front of the glacier gives scale. Ablation is taking place by melting of the exposed ice, with channels eroded in the surface by meltwater. The moraines suggest a former greater extension of this glacier. (National Film Board of Canada)

Plate 151 The Ferperclé glacier in the Valais of Switzerland, with the Matterhorn in the background. (Swiss National Tourist Office)

Plate 152 The Aletsch glacier, Switzerland. (Swiss National Tourist Office)

Plate 153 The Franz Josef glacier, South Island, New Zealand. Compare this with the Swiss and Canadian glaciers. (High Commissioner for New Zealand)

Plate 154 Moraine on the surface of the Aletsch glacier. Notice in the depth of the moraine, as shown in the crevasse, and its effects on the surface melting of the ice. (Swiss National Tourist Office)

Plate 155 Drumlins in Canada. (USGS)

Plate 156 The striated surface of rock, formerly under ice, in Alaska. (USGS)

154

155

156

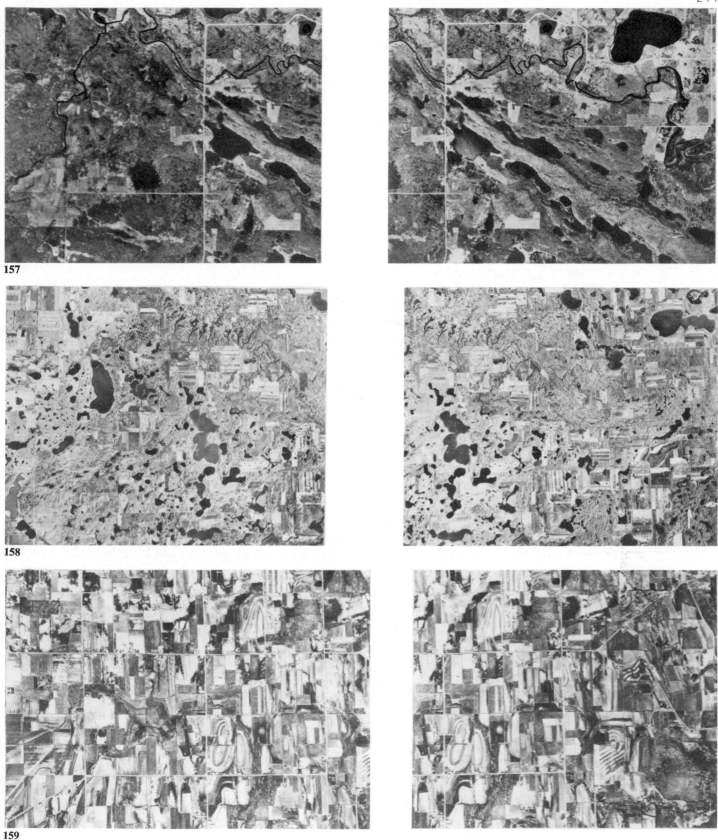

Plate 157 Stereo-pair: ice marginal deposits in Minnesota, USA, near Grand Rapids. A north-westerly trending ridge was formed as a channel filling beneath the ice; it is bordered by kettle lakes and a till plain (with more kettles).

Plate 158 Stereo pair: end moraines in Kidder County, North Dakota, showing lobe-shapes and parallel ridges, with many small lakes.

Plate 159 Stereo-pair: drumlins in Wisconsin in a compact calcareous clay till: 91 per cent of the stones originate in a local limestone. (All USGS)

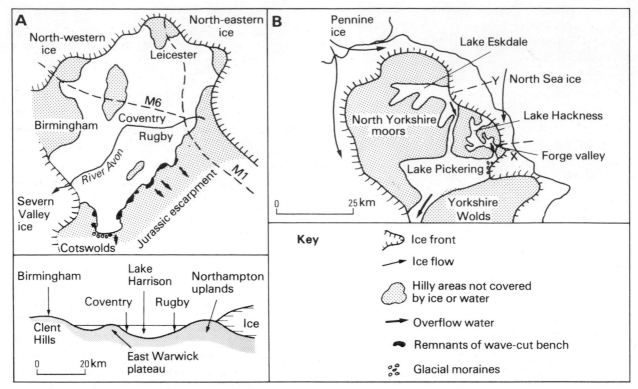

Figure 8.25 Glacial meltwater lakes in England.
(A) Lake Harrison in the Midlands, formed by an early advance of the ice southwards. Notice the extent of the lake, positions of shoreline remnants, and areas possibly covered by lake deposits. (After Brunsden and Doornkamp, 1974)
(B) The proglacial lakes of the North Yorkshire Moors area, formed during the Devensian ice advance which left much of the area uncovered. Meltwater lakes were dammed by the ice, leading to the carving of overflow channels. The river Esk subsequently reoccupied its old course at Y, entering the sea at Whitby, but the Derwent did not flow to the sea at Scarborough (X) again and instead followed the circuitous course taken by the meltwater through Forge Valley and the Vale of Pickering.

Stratford-on-Avon and Leicester — Lake Harrison, which was over 80 km long and 125 m above present sea-level (Figure 8.25). The waters are thought to have overflowed into the Thames basin by way of channels cut in the Cotswold escarpment. This was eventually destroyed by the forward movement of the eastern ice sheet which eventually terminated in the moraine at Moreton-in-the-Marsh.

The next interglacial takes its name from the deposits containing a temperate fauna and flora around Ipswich. Freshwater and estuarine deposits in the Solent area suggest that the sea-level during the Ipswichian Interglacial may have been similar to that at present.

The most recent ice advance, the Devensian Glaciation, contains a number of oscillations within it. The line of maximum advance has been well-established, and since it gave rise to the most recent, uppermost glacial deposits, it obviously left the best preserved features. For this reason its deposits were for long called the 'Newer Drift' to distinguish them from the older deposits lying beyond its maximum extent which show evidence of periglacial surface alteration which occurred during the Devensian advance. Peat and sand deposits from a number of places in the English Midlands show that the Devensian glaciation was one of considerable fluctuations. At various times ice-dammed lakes formed between the ice front and local features of higher relief. One important example of this was Lake Lapworth, which extended over the Cheshire Plain until it eventually drained to the south by way of an overflow channel at Ironbridge Gorge in the Shropshire Hills, the course of which is now followed by the river Severn (Figure 8.25). Ice moving across the Irish Sea also blocked the entrances to valleys flowing north in the Welsh hills, so that these were diverted over low cols to the east. Rapid downcutting of these new courses meant they are still followed today.

It would seem that the Pennines remained largely as isolated nunataks during this phase: the valley sides were actively eroded, whilst their summits, such as Ingleborough, were masked by frost-shattered material. During one stage the Devensian ice blocked the former mouth of the river Derwent, near where Scarborough stands today, thus damming back a lake, known as Lake Pickering (Figure 8.25).

The Devensian glaciation left relatively fresh features of deposition and erosion: large numbers of moraines were left virtually intact, including the York and Eskrick moraines in the Vale of York, or the Irish Tipperary end moraine; there are also large numbers of eskers, kames, kettle moraines and drumlin swarms (Figure 8.9). Landscapes developed on the younger drift still reflect strongly the shapes of the initial deposits. Slope retreat under humid temperate conditions has not destroyed the results of glacial activity, but stratified lacustrine deposits now fill many glacial lakes and depressions in the kettle moraine.

It is almost impossible to correlate the erosion of upland masses with particular glacial periods, but there are good reasons for thinking that cirque glaciation existed for long periods after the close of glacial periods, so sharpening the high-level relief at a time when the lower areas were subject to ameliorating conditions. Cirques are thus the product of multiple occupation by ice; many have a moraine ridge at their mouth as a result of the last cold phase. It is possible that the residual snowfield on the north side of Ben Nevis which survives through some summers is part of this cirque glaciation, most of which disappeared 8000–10 000 years ago. The positions of erratics would suggest that sometimes valley glaciers were penned back by the movement of lowland ice, while at other times lowland ice had to give way to the more powerful valley glaciers. Only highly detailed knowledge of the meteorological conditions which existed would explain the relative dominance of particular ice in an area at any one time, and such information is unlikely to be found.

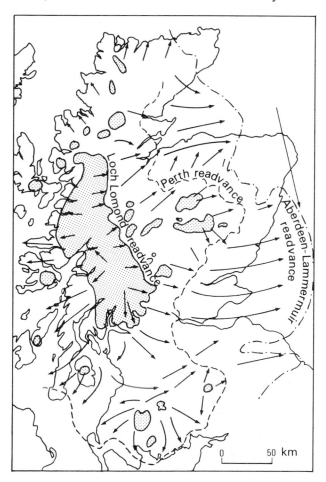

Figure 8.26 Successive stages in the retreat of the last ice sheet in Scotland. The landforms are related to the time over which ice covered them: thus the clearest evidence of glacial erosion is found in the shaded areas. (After Sissons, 1967)

Interesting work in the Lake District has shown how the major glaciers reacted during the closing stages of the last glaciation. Five different terminal moraines are to be found at the southern end of Coniston Water, each one having been cut through by meltwater emanating from a later position of the glacier front. To the south, occupying Morecambe Bay and the Fylde Peninsula, are the sands washed away from deposits of Lake District ice, giving Britain's best-studied examples of fluvioglacial materials. As the lowland ice collapsed and disappeared it would seem that minor glaciations, accompanied by minor advances, continued in the hills. Coniston Old Man contains a number of cirques above 350 m which extruded small glaciers, although they melted away before reaching lower ground.

Knowledge of the final phases of glaciation in Scotland is also reasonably complete in outline (Figure 8.26). The last phase, the Loch Lomond readvance, persisted for a long time in a generally retreating state until it occupied only the cirques on the north-east facing slopes of the mountains of the north-west highlands and islands.

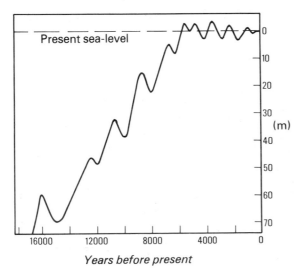

Figure 8.27 Changes in sea-level since the melting of the ice sheets. (After Fairbridge, 1961, in Sissons, 1967)

The final stage of the Quaternary is known as the Flandrian. With the gradual waning of the ice sheets during the last 15 000 years the sea-levels rose some 6000 years ago to a level slightly higher than they are at present (Figure 8.27). This rise has been neither continuous nor even. Temporary advances within the Devensian were accompanied by slight falls in sea-levels. Periods of rapid temperature increase have seen rapid rises in sea-level, and, though isostatic adjustment has done much to compensate in certain areas, the adjustment was not quick enough to prevent further large-scale flooding. Noah, from his home in the Fertile Crescent of the Tigris–Euphrates, illustrates well what must have happened, while similar stories from China show that these effects were widespread.

Eustatic changes meant the final separation of Britain from continental Europe and the flooding of the Solent river. It meant the formation of a large number of rias along the south and west coasts, fjords in the north-west, wide estuaries in the lower areas and the inundation of areas like the Fens in which peats formed later. Isostatic adjustment is still compensating for the Scandinavian ice cap, and this is causing some parts of the east coast to sink.

With the warmer and calmer conditions in the North Atlantic just before 1000 AD the Vikings sailed to Iceland and Greenland, and possibly to North America by this route, but they lost contact with their north-west Atlantic colonies soon after this when the climate deteriorated again. In the early eighteenth century Londoners roasted oxen on the frozen Thames in winter fairs so cold had the winters become. From the mid-nineteenth century onwards the climate improved, but since 1940 another deterioration seems to have set in. It is an open question as to which way the climate will trend in the future, but these fluctuations in the last few thousand years are on a small scale compared to those which were involved in the glacial advances over much longer periods of time.

9

Wind

The sculpturing effect of wind is all too often regarded as being associated uniquely with desert landforms, and its universal impact is frequently ignored. Wind is as obvious a feature of weather as is temperature or precipitation. It may become a powerful agent of transportation, leading to the development of large-scale depositional features; when armed with a load of particles it may also be a potent agent of erosion, although on a small-scale. Under humid temperate conditions, where soil moisture and plant cover normally prevent the wind picking up soil particles, it is hard to appreciate the full effect of wind action. Although the winds of deserts are no stronger than elsewhere, their effect is increased by an uninterrupted passage across a sparsely vegetated or bare landscape.

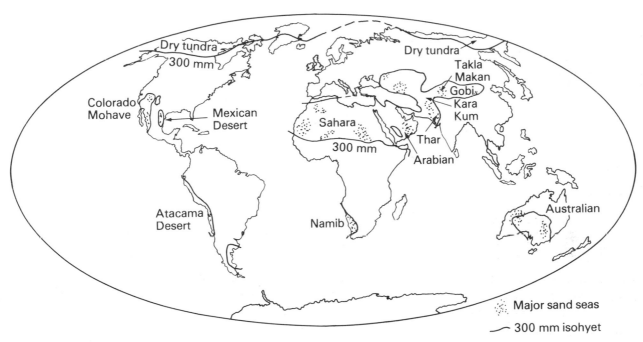

Figure 9.1 The world's major deserts, as defined by the 300 mm isohyet. These include the tropical deserts, temperate deserts and tundra areas. (After Allen, 1970)

Wind becomes more important as the amount of surface, or near surface water in the soil decreases. This is related closely to the production of an effective vegetation cover. The position of the 300 mm isohyet on the world map (Figure 9.1) shows clearly the dry areas of the tropics and subtropics. Even in temperate areas with relatively high precipitation the wind can become effective in removing fine soil particles and beach deposits if the plant cover is removed, or cannot gain a hold. Thus ploughed fields in East Anglia suffer wind erosion in dry periods, when dust clouds testify to the removal of the finest soil particles. Dunes forming at the tops of sandy beaches, where vegetation finds it difficult to develop because of excessive atmospheric salt, partially repeat the depositional landforms produced in deserts. In cold deserts, including many tundra areas, vegetation may be sparse, and the wind can assume an important role in erosion and deposition. Four environments can be identified, therefore, in which the wind acts to varying extents in the creation and modification of the landforms (Figure 9.2).

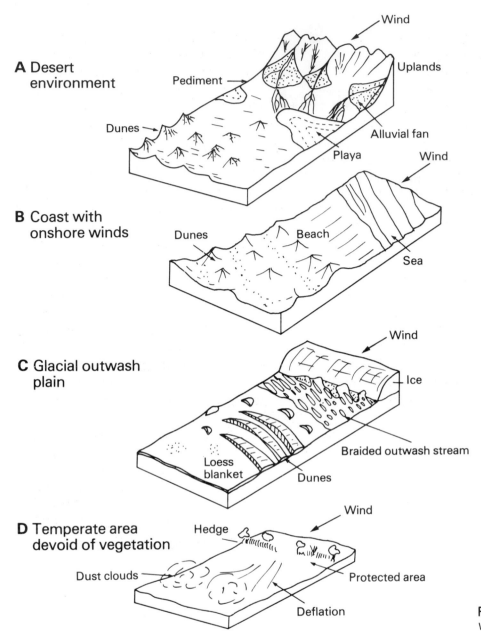

A Desert environment

Wind
Uplands
Pediment
Dunes
Alluvial fan
Playa

B Coast with onshore winds

Wind
Dunes
Beach
Sea

C Glacial outwash plain

Wind
Ice
Braided outwash stream
Loess blanket
Dunes

D Temperate area devoid of vegetation

Wind
Hedge
Dust clouds
Protected area
Deflation

Figure 9.2 Environments of wind action. (After Allen, 1970)

Wind transport and deposition

Winds are movements of the atmospheric gases, and as such their capacity for the movement or rock debris contrasts with the more buoyant medium of water in its liquid or solid phases. The rock debris is supplied from weathering (chapter 5) of muds and shales (to give dust), sandstones and granites (sand), and the fragmentation of other rocks into larger pieces: the latter take a very long time to reduce to smaller debris under dry conditions.

The mechanics of wind transport

Wind blowing over a dry and vegetation-free surface acts as a sorting agent for particles of different sizes. The strength of the wind determines the size of the particle which can be carried. Small particles (dust, under 0.05 mm diameter) are carried in suspension; medium-sized particles (sand, 0.05 – 2.0 mm diameter) are moved by saltation (page 218); and larger material (equivalent to the bed load of a stream) moves by creep, if at all. Wind currents are far more variable in strength and direction than those in water, and even when winds are blowing strongly gusts may be separated by quiet lulls. In addition, flow around obstacles leads to contrasts in the carrying capacity of the wind. Essentially the wind has to overcome the terminal velocity of a particle (i.e. its steady falling speed) before it can lift it off the ground. Since the wind is blowing horizontally over the surface, and the upward currents in its turbulent flow are one-fifth of the ground velocity, the velocity of the latter has to be five times the terminal velocity in order to lift a particle into the air. Terminal velocities of dust particles are very low (e.g. 0.01 m/s for particles 0.01 mm diameter), so they are easily moved even by light breezes and are kept aloft in suspension until they become involved in cloud droplet formation and are brought back to Earth in raindrops. As the dimension of the particle increases, the terminal velocity increases too, but even more rapidly. The terminal velocity of a particle of 1 mm diameter is 8 m/s, and a wind speed of 40 m/s, required to lift it off the ground (i.e. five times the terminal velocity), is gale force. Most sand particles moved by the wind are finer than this, in the range 0.15-0.3 mm diameter, and they are seldom raised more than 2 metres above the ground surface. Larger particles are moved slowly along the ground.

a) **Suspension.** The finest material, derived from silts and clays and having a diameter of less than 0.05 mm, is carried to great heights by the wind in the form of dust clouds. These dust clouds can move vast quantities of material. It has been demonstrated that air moving across the Atlantic Ocean in the middle troposphere (1.5-3.7 km altitude) carries 25-37 million tonnes of Saharan dust across a line north and south through Barbados every year. Much is deposited, via rain, in the journey across the ocean, and this is sufficient to account for the total non-carbonate ocean-floor accumulation of sediment (chapter 2). Such dust gives rise to hazy conditions in the atmosphere, where concentrations of up to 4000 tonnes/km^3 may be carried.

On land dust storms create havoc, particularly in farming areas. It has been calculated that 850 million tonnes of dust are carried up to 2300 km from the dry south-western USA every year; that Europe has received enough Saharan dust in the last 100 years to form a layer 15 cm thick; and that the surface of the Nile delta has lost a depth of 2.5 m of material to dust clouds since it was first cultivated under the pharoahs 3600 years ago. A dust storm in Melbourne, Australia, in 1920 was judged to have resulted in the deposition of 16 tonnes of dust per square kilometre. In tundra areas adjacent to ice sheets strong winds blowing off the ice entrain the finest particles of silt in the outwash lands.

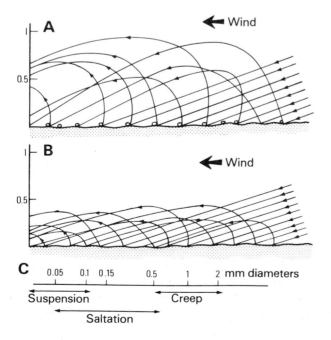

Figure 9.3 Movement of sand by saltation.
(A) On a rough gravel surface the particles jump higher: one measurement showed that 90 per cent of the sediment moved was below 87 cm.
(B) On a sandy surface saltation raises particles to lower heights: the same measurements showed that 90 per cent of particles were below 31 cm.
(C) Particle sizes involved in movement by the wind.

b) **Saltation.** When wind reaches a critical velocity over loose, dry sand, grains begin to roll and accelerate, and after a few centimetres may jump into the air, travel many times their diameter and finally return to the surface after a long parabolic path. The minimum wind velocity needed to start this saltation process is 5.37 m/s (20 km/hr) for particles up to 0.2 mm in diameter. Such a wind initially causes a particle to roll forward and, on knocking against an immovable object, it bounces up into the air. This bounce may be of the order of 2 m, while exceptional records of 6-19 m have been observed. It is rare, however, to find saltation acting above 1 metre, and 10 cm is the average height reached. This will vary with a number of factors — wind strength, particle size and the nature of the surface on which the saltation takes place (Figure 9.3). Once entrained the particle is carried by the wind, returning in a parabolic path to strike the gound at an angle of 3-10 degrees. This is low enough to persuade the observer that it is travelling parallel to the surface. When it strikes the ground it either bounces up again, or causes other particles to bounce. Saltation, therefore, involves the constant interchange of particles at the surface. Maximum efficiency in movement by saltation would require a strong and continuous wind, but most winds in desert areas are essentially erratic in their daily range of strength, direction and duration.

Particles carried by saltation in air will be far more abrasive and abraded than would similar particles carried by water. The density of air is only one eight-hundredth that of water, and the lower viscosity of air reduces any cushioning effect. A particle will strike another particle, or a boulder on the ground, with almost the same velocity as the wind. This is important in generating the bounce of saltating particles and in explaining the intense action of 'sand-

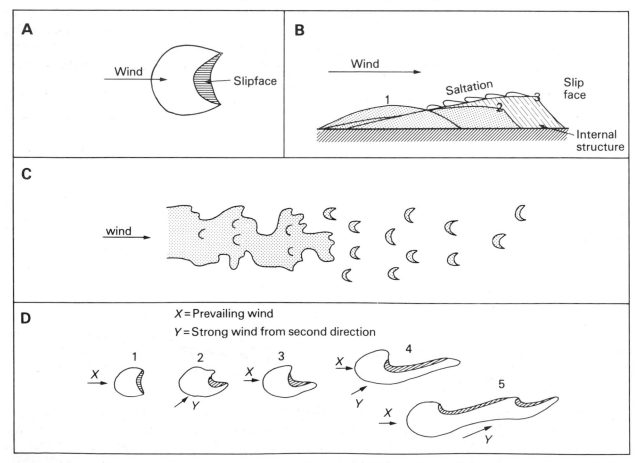

Figure 9.4 Barchan dunes. (A) Plan view. (B) Section, showing how the dune slip face develops. (C) A typical belt of barchans. (D) One suggestions as to the formation of longitudinal, or seif, dunes. (After Bagnold, 1941)

blasting' as opposed to corrasion and abrasion in water. The wearing down of particles involved in saltation proceeds at up to a thousand times the rate of particle abrasion in water.
c) **Creep.** Particles too large to be lifted off the surface by the bombardment of saltating particles may, if they are not too large, be gradually and erratically moved along the surface. At low wind velocities there is a jerky movement of a few millimetres at a time, but the distance involved increases with wind velocity until in high winds the whole surface may appear to be creeping slowly forward. Laboratory experiments have shown that between 20 and 25 per cent of the weight of sand particles moving past a particular point may be due to this creep process. Whilst the majority of the particles involved are too large to move by saltation, they receive their kinetic energy from that process. It has been shown that a high speed grain in saltation can move by impact a surface grain six times its own diameter — or more than two hundred times its own weight.

Wind deposits: their nature and associated landforms in deserts

Landforms produced by wind transport and subsequent deposition are the most significant results of wind action. Such features occur on a large scale in deserts: it has been estimated that 99.8 per cent of aeolian sands are found in ergs (sand sheets) of over 125 km²; a medium size is 188 000 km²; and the largest area, the Rub al Khali of Saudi Arabia, is 560 000 km². These areas also demonstrate the complex interrelated factors which are at work. The fascination of such deserts is referred to by one of the most famous investigators of desert landforms, R.A. Bagnold:

'Here, instead of finding chaos and disorder, the observer never fails to be amazed at a simplicity of form, an exactitude of repetition and a geometric order unknown in nature on a scale larger than that of crystalline structure. In places vast accumulations of sand weighing millions of tons move inexorably, in regular formation, over the surface of the country, growing, retaining their shape, even breeding, in a manner which, by its grotesque imitation of life, is vaguely disturbing to an imaginative mind.'

Sand is deposited in a variety of forms, which vary in shape and size.

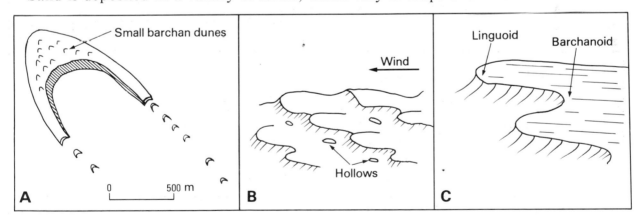

Figure 9.5 Draa and aklé dunes.
(A) A large, draa-sized barchan-type of dune system in Peru. Whereas barchans move approximately 9 m/yr, this draa moved only 45 cm/yr and the small dunes migrated over its surface.

(B) Aklé dunes. A larger quantity of sand is available than in the case of barchans. The major elements are shown in (C). (After Cooke and Warren, 1973)

a) **Large-scale features** are related in their shapes to the major prevailing wind directions. Barchans are the best-known forms related to a unidirectional pattern of winds (i.e. with a deviation of less than 20 degrees from the mean direction) (Figure 9.4). They occur in areas of limited sand supply where the rock pavement is often exposed, and are thought to originate from a random thickening of a sand-patch. The increased height causes saltation and creep on the windward slope, so leading to a steepening and increase of height in the dune. As the dune grows in height, so it moves into zones of greater wind velocity. At this stage the sand begins to be removed and further growth is inhibited. This process effectively sorts the sand particles, ensuring that only the finer particles are found at the crest. The slip slope maintains a constant angle of response so that there is a continual overbalancing of the sand accumulating at the crest. A barchan is, therefore, a mobile structure progressing downwind at a speed controlled both by its own mass and by wind strength and direction. Its crescentic shape is maintained by the wind sweeping round its flanks and constantly extending its horns, the lower side areas moving more rapidly than the high centre. Barchans are found in their most perfect form when they occur singly on a rock pavement, but are more frequently found in swarms covering large areas. Speeds of movement as great as 47 m/yr have been recorded in Peru, and 15 m/yr in Egypt.

In areas of greater sand supply the barchan type of dune is replaced by somewhat similar forms, known as aklé dunes (Figure 9.5). Aklé dunes consist of long, sinuous ridges transverse to the wind direction with a dune wavelength of up to 200 m. The ridges comprise crescentic barchan-like formations alternately facing up- and down-wind; the segments are known as linguoids and barchanoids respectively. No one theory for their formation has so far gained wide acceptance, although it is possible that they are at least partially the result of the interaction of fast- and slow-moving parallel airstreams with the sand.

Multidirectional winds give rise to elongated dune forms, known as seifs (Figure 9.6). These are long, sinuous forms, often stretching for more than 100 km, and crowned with low summits at regular intervals. They have side slopes of some 20 degrees. At times there is a

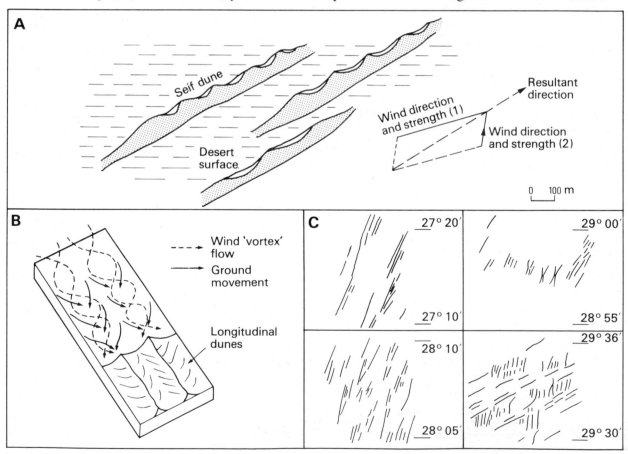

Figure 9.6 Seif dunes.
(A) A general diagram of seif dunes, relating orientation to wind direction.
(B) A second theory of their origin, related to vortex-like winds. (After Cooke and Warren, 1973)
(C) A variety of seif dune patterns in the Libyan desert. (After McKee and Tibbitts, 1964, in Pitty, 1971)

steeper slip slope nearer the summit. Seifs reach 100 m high in Egypt and over 200 m in southern Iran, their transverse width being of the order of six times the height. They often occur parallel to each other and up to 10 km apart. The upwind end of a seif is invariably rounded, whilst the downwind form is sharp. Smaller, but similar features are found in Australia, where heights of 15 m and lengths of 50 km are common. These are found far closer together — normally less than 2 km apart. Recent research has suggested that, instead of the alignment of the seif being that of the prevailing wind (with the barchans having been joined together as a result of cross-winds), its alignment is that of the resultant of the two winds. This extension of the original theory has received widespread support with one example observed in Libya showing seifs having an alignment parallel to the resultant of morning and evening winds. In Israel resultants of summer and winter winds would seem to account for the alignment. If this seems too simple it must also be noted that from areas where there is considerable sand cover a more complex relationship with the winds is found. Here seifs may be the oblique element of a pattern created by a dominant wind and emphasised by a cross-wind.

Pyramidal, or star-shaped, dunes (sometimes known as rhourds) are further distinctive forms. Their pattern is often confused, but normally contains steep, radiating ridges culminating in one, or several, peaks with a superficial similarity to a glaciated mountain. The larger forms approach 300 m in height, and are up to 2 km across at the base. They seldom stand alone, but appear to merge into other alignments. They are clearly formed by winds from many directions, but the detailed explanation for some of the sand movement is not as yet available. An attractive, if improbable, theory for their origin is that the rhourds exist at the centre of a convection cell. Strong, inward-blowing winds of equal strength would be responsible for heaping up the sand and producing the highly peaked summits. Unfortunately, however, it seems that winds of sufficient strength are not available for this type of movement to be possible.

b) **Small-scale features** include ripples, ridges and forms initiated by obstructions. A flat sand surface becomes unstable when particles are passing over it. The wind creates saltation; the saltation produces creep; and the size of the particles, the resulting surface relief and the nature of the sand movement interact to determine which form of deposition, or removal of the sand, take place.

The smallest of the features produced when a steady, if slow, rate of deposition is occurring is the ripple. Ripples are formed at right angles to the wind direction and stretch laterally for considerable distances. Their wave-length varies from 2 to 15 cm, depending on the wind strength. In cross profile they are slightly asymmetrical with a maximum height of little more than a centimetre. Invariably the largest particles are found in the troughs. It is thought that such ripples are initiated by slight irregularities in the surface, which lead to increased saltation from those surfaces facing the wind. This is then followed by the production of other ripples whose wave-lengths correspond approximately to the length of the saltation path. The larger particles are moved forward by creep and collect to the leeward of the ripple where, because their is virtually no saltation to move them, they remain.

Where sand is being removed from the surface a ridge becomes characteristic. Like a sand ripple this is transverse to the wind, and asymmetric in cross profile. Ridges are larger than ripples, and commonly have wave-lengths of up to 3 m and a maximum height of 15 cm. Exceptional wave-lengths of up to 20 m and heights up to 60 cm have been recorded in the Libyan desert. The classic explanation for their origin is related to the high percentage of coarse sand granules and small pebbles. It involves rapid saltation which, because there is little or no supply of fresh material, leads to removal of the sand. In this process the larger particles are piled by creep into the parallel ridges. As the finer saltating material which produces the creep is thought to be in a diminishing supply, it is often suggested that these ridges are a product of considerable age. The essential difference between a ripple and a ridge is to be found in the relationship between wind strength and particle size. A ripple is prevented from further upward growth by winds strong enough to carry particles away, whilst the ridge grows because the wind can remove only the smaller particles and leaves the larger to accumulate. The pattern of both ripples and ridges is changed by long-term alterations of wind direction. The full development of ripples and ridges is inhibited if their particle size range is too great.

Over vast areas of almost flat sand sheets, such as those which occur in the Kalahari and Selima deserts, no ripple-type formations occur. On closer examination it is found that the sand contains some particles whose diameters are at least ten times greater than those of the sand particles.

Obstructions to wind flow are caused by larger boulders. Conditions for deposition arise since they deflect the wind and cause local turbidity currents. Consequently an obstacle creates two wind shadow zones — one on the down-wind surface and another, much smaller, on the up-wind side. The zones are separated by a surface of discontinuity between the deflected windflow and the turbulent flow (Figure 9.7). Many examples of this process can be found. On sandy beaches old concrete caissons, oil drums, or the wrecks of boats, all develop windward-facing drifts and dunes downwind. Frequently the two merge and virtually bury the obstacle. This process has helped to preserve many sites of archaeological interest in arid and semi-arid areas. Such features may be so common that early travellers in arid areas believed all dunes to be the result of accumulation around fixed obstacles like plants, ruins and dead camels.

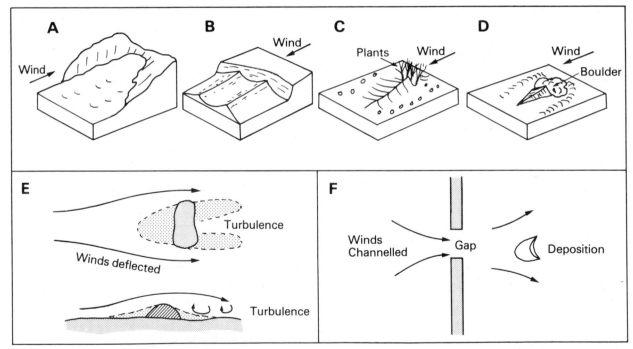

Figure 9.7 Sand deposition and obstructions. (A) Drifting sand upwind of tall cliffs: 'climbing dunes'. (B) Ridges to the lee of a low cliff: 'falling dunes'. (C) Drift in the lee of a bush: a nebkha. (D) Drifts around a boulder. (E) Pattern of sand accumulation around an obstruction. (F) Deposition in relation to a gap in an obstruction. ((A)-(D) after Allen, 1970)

In addition to the fixed dunes other depositional forms are associated with the shadow zones caused by the deflection of the wind. Drifts occur where winds blow against steep cliffs as in the Peruvian desert, where drifts have been built to heights of several hundred metres. Sand-bearing winds passing over a scarp face will deposit material rapidly in the form of sand pans, whose detailed shape is controlled by local turbidity currents. Deposition may also occur totally unattached to an obstacle where the wind pattern is disturbed by obstacles upwind. Such drift deposits are often found downwind of a gap in a wall, cliff-line or plantation.

Sand dunes along coasts with onshore winds

If the retreating tide exposes a sandy beach it dries out quickly, becoming unprotected and vulnerable to onshore winds. Due to their long, uninterrupted fetch these winds are more effective than offshore winds in moving the sand. Saltation and creep proceed and gradually the sand moves inland. If the coast is low-lying the sand will migrate until it becomes stabilised by an obstruction such as water or vegetation. Dunes will then begin to develop, although it is uncommon for these to grow beyond 20 m in height. In some areas where there are adequate supplies of sand (e.g. the Culbin Sands of Morayshire and Les Landes in Aquitaine) the dunes may penetrate far inland. The dune coast of the Netherlands has played an important part in man's occupation of the area. Frequently, such dunes become naturally, or artificially, stabilised with marram grass. Very rarely, however, do perfect forms develop, because of the frequency of what are termed 'blow-outs'. These involve the removal of large quantities of sand, producing a hollow in the dune belt. They are related to the destruction of the vegetation cover by storm wave action or excessive use by holiday-makers. A new pattern of dunes then becomes superimposed on the old and results in an overall chaotic appearance.

Loess

Dust fractions seldom fall back to the ground in the arid zone: they are commonly carried considerable distances and deposited following rain. Large areas of the USA, the North European Plain and Central USSR were covered with great thicknesses of this material during the Quaternary Ice Age, when cold winds blew off the ice sheets to the north and winnowed the fine dust called loess, or limon, from exposed regolith in front of their margins. The soil which develops on loess forms some of the world's most fertile farmlands (Figure 9.8). Similar deposits are also found in the semi-arid zones marginal to the present deserts, and it has been estimated that loess deposits cover up to 10 per cent of the world's land surface.

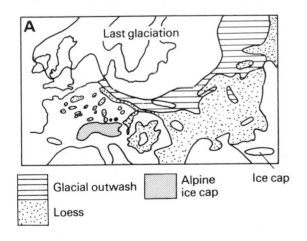

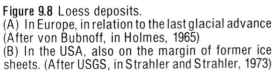

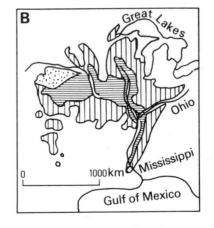

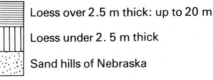

Figure 9.8 Loess deposits.
(A) In Europe, in relation to the last glacial advance (After von Bubnoff, in Holmes, 1965)
(B) In the USA, also on the margin of former ice sheets. (After USGS, in Strahler and Strahler, 1973)

Loess is composed dominantly of silt fractions (70–85 per cent in the range 0.007–0.003 mm diameter), but also includes some clay and fine sand. The clay helps to bind the deposit, and a high calcium carbonate content in some leads to heavy leaching. Loess has a buff colour due to weathering and is unstratified, since it accumulated in such fine layers that they do not show up. Root structures become the main reason for a vertical fissuring and this gives the deposits a high permeability. Loess is eroded easily by running water, when it contributes a high quantity of suspension load (cf. Yellow river in China).

160

161

162

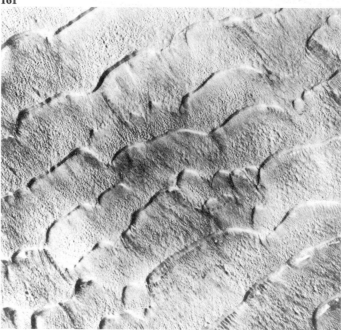

163

164

165

Plate 160 Dunes in Salado Wash, New Mexico. Note the type present, the rippled surface, and the relationships of the dunes to the surrounding relief. (USDASCS)

Plate 161 Dunes in a 'sand sea' in the northern Sahara, Algeria. (USGS)

Plate 162 A rippled accumulation of sand across a road in Kansas. In which direction was the wind blowing? (USDASCS)

Plate 163 Major transverse dunes, with crests spaced 1–2 km, south of Kheran Kalat, Pakistan. The main dune-forming wind is from the south-west, and the dunes have steep north-eastern faces. (USGS)

Plate 164 Loess-like silt being blown from an unvegetated and braided channel floor in Alaska. This process is thought to have been responsible for the formation of much of the loess in areas formerly marginal to ice sheets. (USGS)

Plate 165 Loess soil is a deep, unstructured deposit, eroded easily when vegetation is removed in a humid climate. This section is along the lower East Red river in Louisiana. (USDASCS)

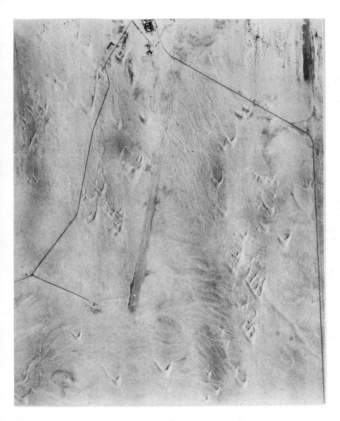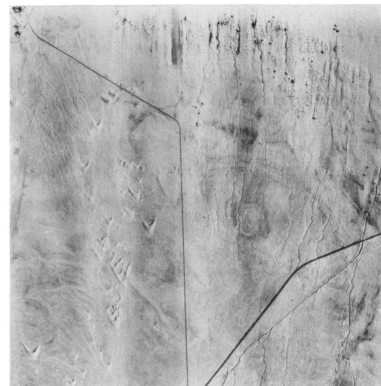

166

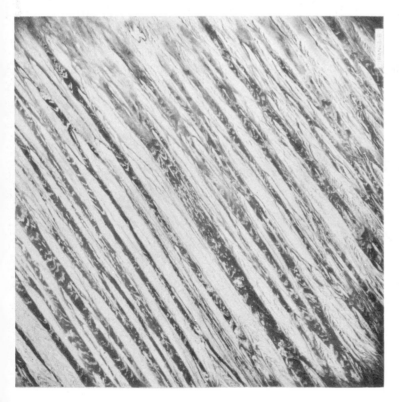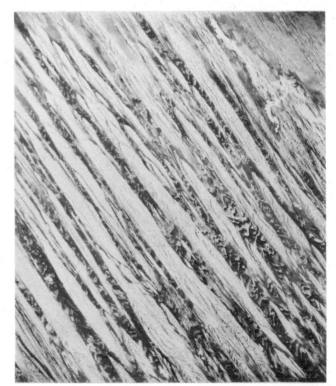

167

Plate 166 Stereo pair: Barchan dunes west of the Salton Sea, California. The dunes are open to the east with horns 50–250 m apart and heights of 3–10 m. They moved 100–300 m from 1956 to 1963, and rest on a sloping surface covered by stones. (USGS)

Plate 167 Stereo pair: wind-scoured troughs and yardangs of the Shahr-e-Lut (lit: 'city of the desert') in Iran. Residual ridges up to 500 m wide and 30 m high have blunt up-wind ends. The troughs are carved in fine lacustrine material which accumulated in a broad desert basin. (USGS)

Wind erosion

Wind is not an effective agent of erosion in terms of corrasion, although some small-scale effects may result from this process. It can, however, remove large quantities of loose, unconsolidated rock debris to leave a hollow or to lower the level of an area: this process is known as deflation, and it can act in desert areas, or in areas where the vegetation cover is removed by ploughing.

Soils consisting of fine particles are particularly susceptible to deflation. The soils of the Great Plains of the USA often consist of fine alluvial material bound together by the natural grassland, and the area receives a low rainfall. The land was developed for farming wheat in the early twentieth century, but after a period of moist years drought led to the formation of the Dust Bowl, in the Great Plains area between the Rocky Mountains and the river Mississippi, dominated by the loss of wind-removed material. During the months when the grain was germinating the dessicating winds removed the top layer of the humus-rich soil before the corn stalks could offer any protection. The wind, together with the winter rains, quickly reduced the area to a wasteland of highly dissected and barren countryside. In the Missouri area the surface was lowered by 18 cm in 24 years: if it had been left under grass it has been estimated that this would have taken 3000 years.

The same problem has become apparent recently in East Anglia, where the soils contain a high percentage of fine particles derived from glacial sand, silt and especially peat. Strong winds blowing in late autumn and early winter can have serious consequences. Locally this is becoming more serious as fields are enlarged by the removal of hedgerows, which protect the land from wind action for a distance of about five times the height of the hedge on the windward side, and up to thirty times to the leeward (Figure 9.9). Deflation ceases, however, when the underlying damp soil is revealed, since this is more coherent and will rapidly support some form of vegetation.

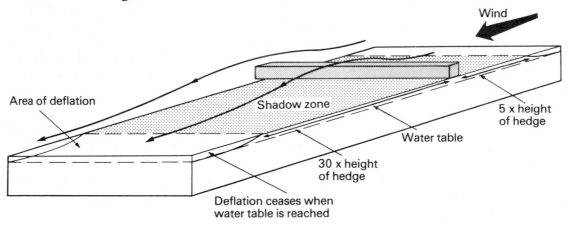

Figure 9.9 Deflation of ploughed fields.

In arid zones deflation may take place on a large scale because water is not present near the surface. It is feasible for all the fine regolith material to be removed, leaving behind either a completely bare rock surface, or a mass of moderate-sized boulders, known as a lag deposit. Expanses of rock are exposed forming the rocky desert, known as the hammada in the Sahara. The majority of this is devoid of any superficial deposits, apart from the occasional sheltered location where some sand and rubble may be found. This is not, however, as common as the stony expanses of moderate-sized boulders, with occasional outcrops of solid rock. It is characterised by an almost complete absence of sand — due to deflation — and a prepronderance of shattered, angular rock fragments giving an environment totally hostile to life. Its widespread occurrence is emphasised by the variety of names applied to it: 'reg' in Algeria; 'rig' in Iran; 'serir' in Libya; 'the gibbers' in Australia.

Depressions have been noted in many deserts, and have often been attributed to this form of wind action. While this is a likely explanation there would seem to have been little real research to date proving this, or establishing the mechanism which could be responsible. The numerous pans of South Africa, varying in size from a few hundred square metres to 300 km², are one example of such depressions, whilst the Qattara and associated depressions of Egypt are the best known (Figure 9.10).

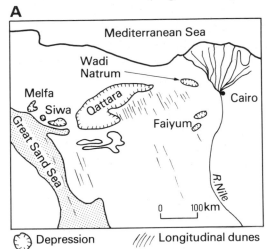

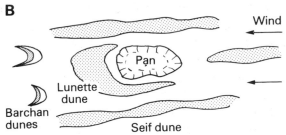

Figure 9.10 Effects of deflation.
(A) Depressions in the Egyptian desert. These may be due to deflation, but it is thought that other processes (e.g. limestone solution in past wetter periods) also helped to remove and break down rock material. (After Holmes, 1965) (B) Pans (wind eroded basins) in the Transvaal. (After Wellington 1955, in Cooke and Warren, 1973)

Why and how these hollows are initiated is not fully understood, but their continuance and expansion would seem to be more straightforward. Once the hollow has started to form, sufficient disturbance would be caused to the winds blowing across them to produce considerable turbulence. Providing that the wind is strong enough, and the particles derived from the country rock are small enough to be transported, then deflation will continue. Many of the smaller pans appear to be aligned with the prevailing wind whose involvement with their formation would seem to be strengthened by the presence of sand dunes directly downwind.

Deflation is also responsible for the smaller, but probably far better known, features of arid lands — the oases. Normally associated with sand deserts, but sometimes also found in rock and sand areas, deflation has uncovered water-bearing rocks. The position of this has been quickly stabilised by the growth of vegetation — commonly palm trees. Some oases are very small with only a few trees, whilst others are large enough to support moderate-sized townships surrounded by gardens and date palms.

Corrasion takes place where the wind uses the transported sand (dust is not effective in this way at all) to wear away rocks by a variety of sand-blasting. Sand-blasting machinery is used widely to clean up stonework in old buildings by removing the outer affected layer. Corrasion is mainly effective as part of saltation, and this limits the height at which it can oprate. For a wind to be significantly corrasive it must not only be strong, but must have a long fetch across a source area of suitably-sized particles. Sand-blasting must continue for a long time to be effective on rock, and thus necessitates a continuous supply of particles for saltation. As these particles are rapidly reduced by abrasion the most effective corrasion occurs close to the supply area. The Sphinx of Egypt have, on successive occasions, been partially covered by drifting sand and, at each of these temporary levels, have been subject to corrasion. As the levels have fluctuated, so their positions have been recorded by niches etched into the rock. Telegraph poles in the desert are completely protected from corrasion if a pile of stones surrounds them to a height of 1–2 metres.

The effect of wind corrasion on exposed surfaces is to remove the softer material and pick out every detail of structure like a delicate etching tool (Figure 9.11). Sandstone rock outcrops may become honeycombed, while conglomerates may have their cement so weakened that the larger stones fall out. The exposed rock pavements of the hammada, and the rocky wastelands of the reg, owe their origin partially to wind corrasion. Frequently angular particles, up to 10 cm long, may be found in such areas. These have, apparently, been subject to wind corrasion for a long time and have become moulded by this action into a wedge shape aligned in the direction of the prevailing wind. Two, and sometimes three, well-defined edges may be found on a single pebble representing the effect of different dominant and prevailing winds. Collectively these are known as ventifacts, and a three-sided form is called a dreikanter. When preserved in sediments of some earlier geological period they record evidence concerning winds prevailing at that time. Of similar historical interest is the growth on exposed rock surfaces of the so-called 'desert varnish', resulting, it is thought, from constant exposure to Sun and wind. Sometimes this is almost translucent, progressing through light brown to an almost jet black. Apart from the fact that it gets darker with age (2000 years for the translucent stage and up to 50 000 years for the jet black), little is understood concerning its origin.

Where resistant rock strata overlie softer material, and where the predominant wind action corresponds with a fault or gully, wind corrasion may produce formations known as zeugens as it dissects such a structure. These are long, mushroom-shaped features, often standing up to 30 m above the general level of the land. They undoubtedly owe their origin and continued existence to the scouring and abrasive nature of the wind. The exact mechanism is not yet known, however, and research has not concentrated on such features. They occur in the Sahara, and may be called gours or garas, but are not particularly common. A more frequent feature is the smaller yardang, in which, because of the vertical layering of the rock, corrasion

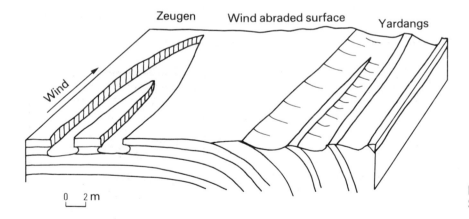

Figure 9.11 Sand-blast effects. Small scale features are produced.

has been able to remove the softer material and smooth the more resistant strata. Yardangs exist in a variety of forms, frequently reaching 6–7 m in height, and have a straight, curved or fluctuating spine depending on the original rock characteristics. They may be up to 1 km long. It seems likely that the present form of yardangs reflects the role of wind, water and chemical activity in the early stages of development.

The largest feature associated with wind corrasion is probably the groove, a feature known for some time, but the extent of which has become apparent recently from satellite photography. The most spectacular area for this is the Tibesti Massif, where an area of 90 000 km^2 is scoured with grooves up to 1.5 km wide, running approximately parallel to each other and 0.5–2.0 km apart. They are eroded in sandstone, aligned with the path of the north-easterly winds and actually cut across drainage channels. The bottoms are flat and the sides steep. The rock itself is frequently varnished, suggesting that these are features of considerable age.

Conclusion

Wind is thus the least effective of the major processes interacting with the Earth's surface features. It is able to carry large quantities of dust around the world, but sand grains can be lifted only short distances and tend to be shifted from one part of a desert to another in well-sorted piles, becoming stabilised if they are blown into more humid areas. Coarser material is seldom moved at all and forms an increasing lag deposit. Such a pattern of sediment movement has little effect on erosional landforms, and many features of deserts, where wind has its greatest effect, are due to the action of running water (chapter 7). This may be a reflection of the slow activity of the wind — the small quantity of precipitation received is more effective than the wind — or it may be due to a more rainy climate in the past. This would fit into the pattern of changing climates associated with the Quaternary Ice Age, which has been stressed in other sections of this book. The problem will be discussed again (chapter 13).

The origin of sand in deserts

Quartz is the most commonly found substance in sand particles. Derived from silica — the most common of rock-forming materials — it is normally resistant to chemical weathering, to solution and to abrasion, as well as being tough enough to withstand fracturing during movement. It is, however, not the only substance of which sand is made, and sand is essentially a grade size (0.05–2.00 mm diameter): it may be formed of quartz, feldspar, mica, mud pellets, rock fragments, crystals of gypsum, or of mixtures of these. There are a variety of ways in which such sand-sized particles may be formed (Figure 1).

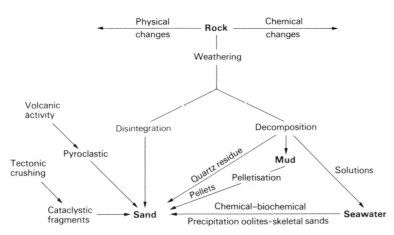

Figure 1 Processes of sand formation. Most people think of sand as being yellow or white in colour, and receive an initial shock when visiting such beaches as those in Tenerife in the Canary Islands, where the volcanic rocks have broken down to form a black sand.

Sand, whilst it is found widely over the Earth's surface, is certainly not evenly distributed. Despite the fact that quartz-bearing rocks are themselves found almost universally over the Earth's crust, sand is concentrated into certain well-defined areas: the great deserts, along coastlines, in river estuaries and in glaciated areas. Coastal sands provide a clue to the reasons for this concentration, since they are brought together by the action of water and wind. Such sand is being constantly re-worked within a narrow zone, but is also being added to by material entering the local beach system whilst further sand is being lost to deeper water (Figure 10.15). The sands of the great deserts have similarly been gathered from a variety of sources into often vast seas like the Rub-al-Khali covering 560 000 km² in Saudi Arabia. It seems that water may have been an original agent in the concentration of such large masses of sand in deserts: the wind has subsequently sifted the sand from the gravel, which it could seldom move. The wind has also affected the sands by rounding them to a greater degree than is possible in water transport. Much of the sand-grade debris concentrated into sand seas in deserts today is thus the result of greater fluvial activity in the Quaternary. This pattern of events is related closely to the occurrence of so many landforms resulting from stream action in deserts (chapter 13).

10

The sea

The sea meets the land along the shoreline in a zone, the upper and lower limits of which vary daily because of tidal movements (Figure 10.1). Over longer periods of time the vertical movements of sea-level have had a greater range, particularly during the Quaternary Ice Age, when sea-level fell over 100 m below the present level (chapters 7 and 8). Sea-level is thus not permanent (Figure 8.21), and the present level has resulted from a rise which began approximately 15 000 years ago and was completed about 3600 years ago, although slight fluctuations, up and down, have occurred since.

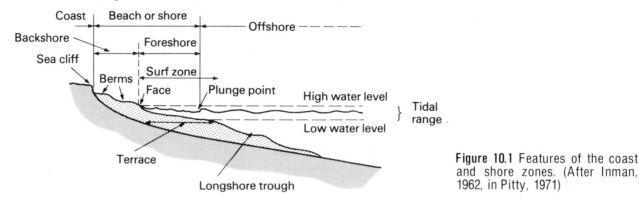

Figure 10.1 Features of the coast and shore zones. (After Inman, 1962, in Pitty, 1971)

The coast and shore are zones where atmospheric processes are also active. A consideration of coastal features thus requires a variety of factors to be borne in mind: the marine processes in relation to changing sea-levels; the nature of the rocks and materials being affected; and the activity of weathering, mass movement, running water, ice and wind above the water level.

Marine processes

The study of marine processes repeats a feature common to other processes acting at the Earth's surface. For much of the time the observer records small waves lapping at the coast, tidal changes taking place inexorably but slowly, and movement of sediment occurring with little visible result. Storm conditions result in catastrophic changes within a few hours, and when combined with high tides can lead to extensive flooding, such as occurred around the southern North Sea in 1953, or to erosional damage. The fact that such events are memorable, however, emphasises their rarity.

The sea is a mobile mass of saline water, subject to a variety of movements: waves, tides and currents all produce particular conditions which give rise to different effects, and the meeting of saline and fresh water in estuaries and deltas has further distinctive results.

Wave environments

Waves have for long been regarded as the main agents affecting coastal landforms, but it is only since World War II that considerable and systematic measurements have been made of their nature and distribution. Waves are generated mainly by the wind blowing over the surface of the ocean. Higher wind velocities give rise to larger waves. In addition other important factors affect wave size: the greater the distance of water over which the wind blows (the 'fetch') the greater the wave size up to a distance of 1500 km; the length of time during which the wind blows also leads to the gradual building up of waves, the size of which may increase over a period of 48 hours for strong winds; and the initial condition of the sea will influence the length of time needed to whip up large waves. Large waves (i.e. over 5 m high) are generated only in stormy conditions with gale force winds (Beaufort Scale Force 8 and above: over 17 m/s). These occur regularly in the temperate cyclonic storm zones in the North Atlantic and North Pacific, but especially in the 'roaring forties' zone of the southern hemisphere. The northern hemisphere zone is particularly active in winter, but that in the southern hemisphere exists all year round. The tropical hurricane and typhoon areas also generate large waves, but these are more local and irregular in occurrence, direction and velocities. Initially a complex system of waves is produced, consisting of shorter wavelength

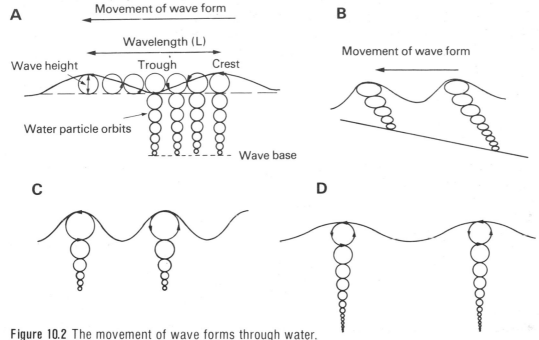

Figure 10.2 The movement of wave forms through water.
(A) The form of the wave moves, not the water: this circulates, as shown. Within the wave the circular path described has a diameter equal to the wave height. This decreases with depth.

$$\text{Wave velocity} = \frac{\text{Wave length}}{\text{Wave period (time between wave crests)}}$$

(B) As the wave enters shallow water, wave base is below the sea-floor, and the orbits of the water particles become elliptical, whilst on the bottom there is just to-and-fro motion.
(C) Water movement in storm waves.
(D) Water movement in swell waves. What will be the effect when storm and swell waves each enter shallow water?

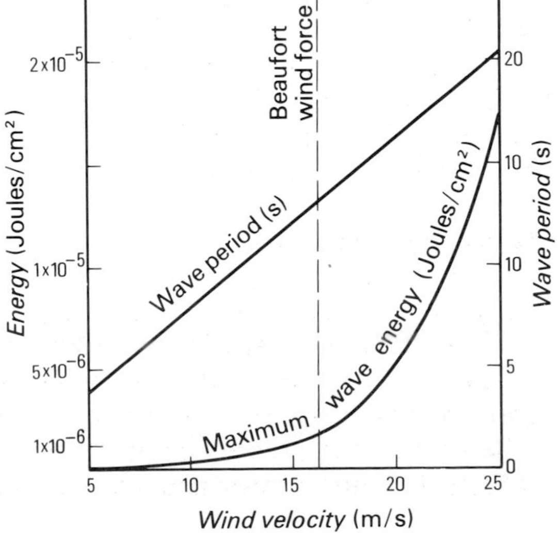

Figure 10.3 Maximum wave energy of a fully arisen sea at different wind velocities. Notice the rapid rise in energy generated at wind speeds greater than Beaufort Force 8 (gale force). Compare the maximum energy of subtropical swell wave environments and of the storm waves. (After Pierson, Neumann and Jaines, 1960, in Davies, 1972)

components of greater amplitude and longer wavelength components of smaller amplitude. The former travel more slowly and are attenuated more rapidly than the latter. Thus as the waves travel out from the area of generation the long (swell) waves eventually become completely dominant. In this movement it is the form of the wave which moves, and not the mass of water (Figure 10.2): this results in Great Circle paths for the waves, rather than deflection by the Coriolis Force.

Other waves are generated by the lower, but steady, wind velocities in the trade wind (tropical easterly) belt, and these move at 3-10 m/s, rising to 13-14 m/s in the monsoon area of the Arabian Sea during the summer monsoon. The energy generated at such velocities, however, is one-fiftieth of that in the temperate storm zones (Figure 10.3), as swell waves which have travelled from those areas may often be more active than the locally generated waves in the tropics.

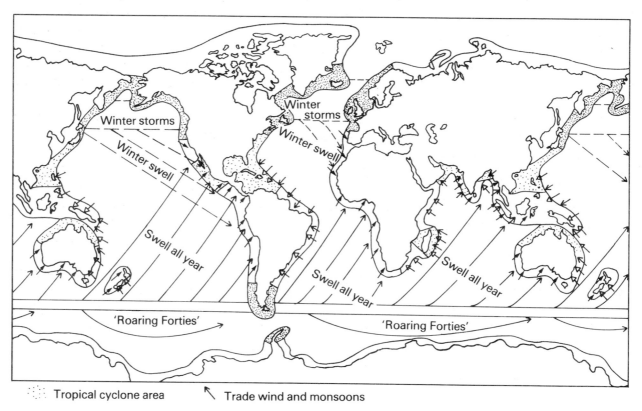

Map	Wave environment	Potential wave energy	Wave characteristics	Directions
	Storm wave (temperate storm belts)	High all year from south, less in summer from north. Swell waves too.	Height often over 5 m Short waves (period 6-12 s) with shallow wave base	Local storms cause variation. Refraction has less effect near shore. More constant swell wave background
	West coast swell	Medium-to-high, diminishing from south, Little variation.	Low waves, declining in height from centres. Occasionally (W. Mexico) subject to tropical cyclones. Long waves (period 14-16 s) from south	Very consistent, especially from the south. Refraction important due to deep wave base
	East coast swell	Medium-to-low, and less regular. High variation in tropics.	Low waves, and long. Subject to tropical cyclones	Fairly consistent, except in the tropics. Refraction important
	Protected	Very low. Little swell penetrates. Some with monsoons.	Very low waves. If subject to tropical cyclones, extreme contrast occurs	Variable

Figure 10.4 Major world wave environments. (After Davies, 1972)

Another factor which affects wave height is the temperature contrast between the water and the air. Thus where the water is at least 5.5° C warmer than the air above it (as in the Gulf Stream and other warm ocean currents off east coasts of continents in winter), waves generated at Beaufort Scale Force 6 may be 25 per cent higher than if the water and air temperatures were equal. Some east coasts can thus experience storm wave conditions, although they may not be affected by strong winds frequently.

The study of wave environments has suggested a world division of coasts (Figure 10.4), which appears generalised, but is a valuable basis for comparing the effects of wave action on coasts.

Wave action

As waves move into shallow water they are slowed by friction with the sea-floor. Wavelengths thus become shorter and the waves become taller, the upper part of the wave growing until it topples over. Wave-break takes place at the plunge line, when the depth of water and wave height are approximately equal (Figure 10.5). Above this line the water rushes shorewards, carrying sand with it. On a sloping beach there will also be a seawards movement (backwash, or undertow) following the upward rush (or swash) of the breaking wave, and this will continue until the next wave breaks. At times the seaward movement of water following wave-break may be concentrated in a rip current.

The relationship between the waves and the shore profile determines the effect of the waves on the beach materials. With small, long wavelength waves and/or a shallow shore profile the wave front steepens gradually and spills over instead of breaking forcefully. Swash is greater than backwash, and much of the material moved up a beach in the water does not return. Backwash is not sufficiently strong to impede the following wave-break and swash. Such conditions lead to beach construction. Destructive waves are those with short wavelengths and high crests occurring on a steeply sloping shore. Such waves break with an almost vertical plunge of water, and strong backwash often retards movement up the beach by the following breaker.

The two types of wave effect alternate with summer and winter on many British beaches, since the storm wave regime of the North Atlantic is important in winter, but virtually disappears in summer. Whole beaches have also been removed in a single storm, and years may pass before the reconstruction is complete. The 1953 storms destroyed beaches in Lincolnshire, which did not regain their original form until 1959.

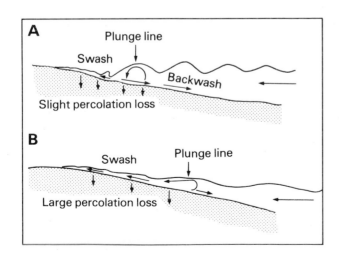

Figure 10.5 Breaking waves and the passage of water on a beach. Describe the features of (a) destructive waves, and (b) constructive waves. The position of the plunge line varies with wave height and state of the tide. (After Lewis, 1931, in Sparks, 1960)

Wave refraction

Waves approaching a coast are affected by the interaction of wave-type with bottom conditions offshore. If there is deep water, as occurs off many ocean islands, the waves will be largely reflected and little energy will be transformed into shoreline activity. Where there is

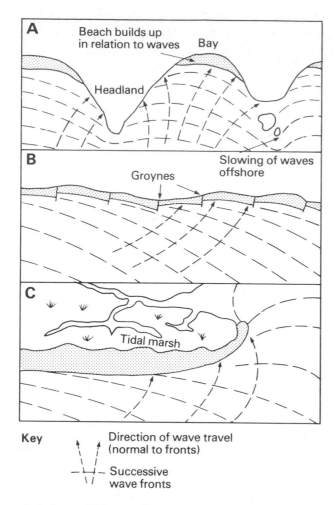

Key

↓ Direction of wave travel (normal to fronts)

—↓— Successive wave fronts

Figure 10.6 Wave refraction. How are the wave fronts in each of these situations altered by the shallowing conditions? How are they related to (a) the concentration of wave attack on headlands, and (b) movement of sediment? (After Small, 1970)

shelving offshore the wave path will be modified. The velocity of the waves is greater in deep water, and slower under shallow conditions, leading to refraction of the wave fronts (Figure 10.6). Storm waves are characteristically high and short, move rapidly, and have shallow wave bases: this means that much of the energy they contain is transferred closer into the shore zone. Swell waves are longer and the wave-base is deeper, resulting in greater slowing down and breaking farther offshore. The main effects of wave refraction are to concentrate the attack of waves on a headland, and to spread out the wave energy in an enclosed bay. Longshore drifting may be related to the direction of waves determined by offshore bottom conditions, as well as to the wind-controlled wave direction. On a larger scale material may be moved just offshore in the direction of the dominant swell movements: there is thus a northward movement along the west coast of South America. In the process of longshore drifting some sediment will be moved into bays and trapped there for long periods of time, unless removed by storm waves, and other sediment may be diverted out to sea and even to the ocean-floor via rip currents and submarine canyons (Figure 10.7).

Tsunamis and tidal surges

Very large and catastrophic waves may be generated by means other than the wind blowing across the ocean surface. Large waves caused by submarine disturbances, such as volcanic eruptions and earthquakes, have for long been known as tidal waves. Since they have nothing to do with tides the Japanese term 'tsunami' has now been adopted — it is the Japanese for 'tidal wave'! The rapid motion of the eruption or shock gives rise to a train of large, rapidly-moving waves: these have wave-lengths of several hundred kilometres and move at up to 700 km/hr (Figure 10.8). They may have a height of only a metre or so in deep ocean water, but rise to heights of over 30 m in shallow water as they approach the coast, and cause great damage. Nearly all tsunamis are generated around the margin of the northern Pacific Ocean, and the zones of generation, like Alaska and Japan, are most affected along with the large ocean islands like the Hawaiian group. There is often first a withdrawal of water along the

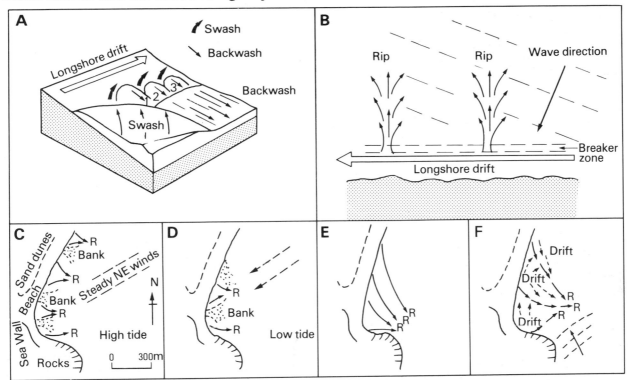

Figure 10.7 Longshore drift and rip currents.
(A) The role of swash and backwash in moving sediment along the shore. (After Strahler, 1969)
(B) Rip currents are concentrated backwash, which often moves large masses of sediment out to sea instead of along the shore. They form when waves approach the shore at an angle and water piles up until it breaks out seaward through the surf zone. It then fans out and the flow is dissipated. The shore zone may thus be divided into a series of cells of approximately equal size. (After Press and Siever, 1974)
(C)-(F) Changes in the position of rip currents at Dee Why beach, New South Wales, Australia (R = rip current). (C) and (D) are normal conditions; (E) shows the enlargement and southward movement along the coast during a north-easterly storm; (F) is related to southerly swell. (After Mackenzie, 1958, in Pitty, 1971)

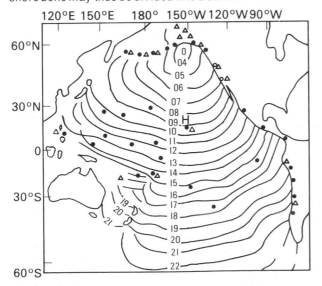

△ Seismograph stations   • Tide stations

Figure 10.8 A tsunami caused by the Anchorage, Alaska, earthquake of 1964. Figures refer to hourly positions of the wave (Greenwich Mean Time). The map also plots reporting stations of the Tsunami Warning System, centred on Honolulu (H). (After Wilson and Torum, 1968, in Strahler and Strahler, 1973)

coast, followed by its almost immediate return. The worst feature of tsunamis is that there is no warning of their generation, except for the detection of the causing earthquake: a warning system has now been established around the Pacific Ocean.

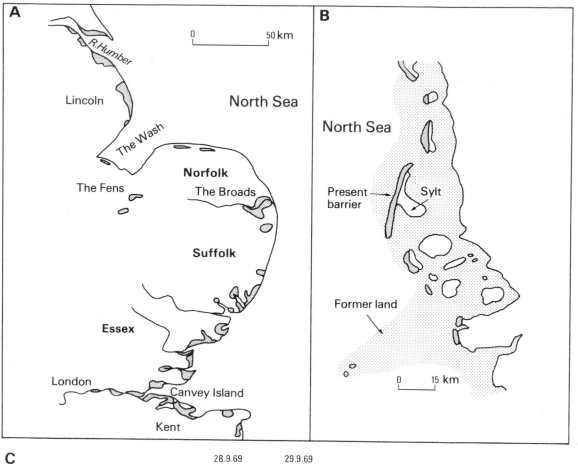

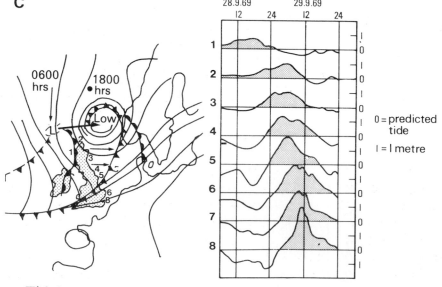

Figure 10.9 The results of storm surges.
(A) Flooded areas of the east coast of England, February 1953. (After Steers, 1953, in Steers, 1971)
(B) Land lost on the North Sea coast of Schleswig-Holstein in the storm surge of 1362. (After Strunk-Hasum, in Davies, 1972)
(C) A tidal surge in eastern Britain on 28-29 September 1969. The weather chart shows how a depression moved into the North Sea, bringing westerly winds. The table shows how the surge exaggerated the tides as they moved southwards. (After Levett, 1974)

Tidal surges occur where water piles up against a coast due to strong onshore winds. They occur most often in connections with hurricanes and typhoons, but may also affect temperate coasts. Much of the 1953 flooding in eastern England and in the Netherlands was due to the combination of a storm surge with winds blowing from the north-east and high tides. Much of the coastland zone of Schleswig-Holstein was lost in a storm surge in 1362 and has not been reclaimed (Figure 10.9). Storm surges are particularly devastating on such low-lying coastal areas.

Tidal environments

Tides are regular movements of the ocean waters due to the gravitational attraction of the Sun and Moon on the Earth. This attraction causes the waters to be gathered at two opposite sides of the Earth (Figure 10.10), so that on an Earth wholly covered with water two areas of high water would be produced. These bulges are held fixed beneath the Moon as the Earth spins through them, giving high and low tide twice a day (i.e. a semidiurnal distribution). The interruption of ocean basins by the land masses creates extra complexities, so that a variety of tidal patterns are experienced (Figure 10.11). This characteristic is important in marine processes, since it affects the length of drying between tides.

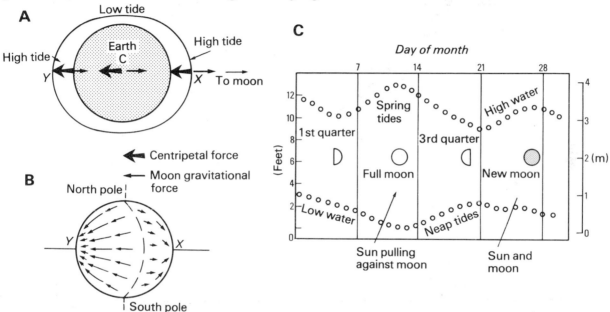

Figure 10.10 Tides.
(A) The centripetal force due to the rotation of the Earth-Moon system acts equally over the Earth, and either with or against the gravitational pull of the Moon on the Earth. The pull of the Moon is greatest at X, but slightly less at C and slightly less again at Y. The centripetal gravitational forces are balanced at C, but one is greater at X and Y, resulting in the tidal bulges. The Sun has a similar effect to that of the Moon as shown on this diagram: although it is so much farther away, its huge mass means that its gravitational pull is twice that of the Moon, and the greater distance means that there is less difference in the Sun's gravitational pull on opposite sides of the Earth.
(B) On a motionless planet covered by water there would be movements towards X and Y. Since the Earth rotates every 24 hours, the bulges stay in one place, whilst the Earth moves beneath.
(C) Neap and spring tides. When do the greatest, and smallest, tidal ranges occur (After Strahler, 1969)

Tidal range (i.e. the difference between high and low tide) varies considerably throughout the world, the highest range having been recorded in the Bay of Fundy (17 m); other high ranges also occur in funnel-shaped arms of the major oceans. On coasts facing the open oceans the range seldom exceeds 2 m, and in enclosed seas, like the Mediterranean, Baltic and Red Seas, there is hardly any range. This factor affects the strength of tidal currents, which are formed where water flows rapidly through narrow straits as high tide affects one side at a different time from the other: the tidal current between Manhatten Island and Long Island in New York is reversed with the change of tide through a narrow passage known graphically as Hell's Gate.

Tidal environments can be recognised, in which the combination of tidal type and tidal range are considered together (Figure 10.12). Macrotidal environments are those of semidiurnal type with ranges of over 4 m at springs; they have strong tidal currents, wide exposure of

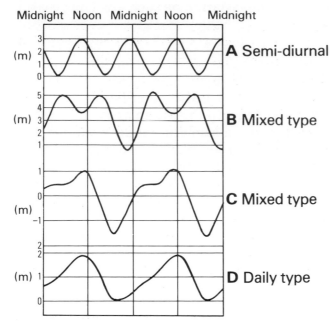

Midnight Noon Midnight Noon Midnight

A Semi-diurnal

B Mixed type

C Mixed type

D Daily type

Figure 10.11 Tidal types.
(A) Semi-diurnal type, typical of Atlantic coasts (Portland, Maine, USA)
(B) Mixed type, showing a dominantly semi-diurnal régime, together with variations related to shallow waters (Seattle, Washington, USA).
(C) Mixed type, with a dominance of diurnal type and a period of still-stand between high and low water (Port Angeles, Washington, USA)
(D) Diurnal type: the semi-diurnal constituent is so small that it makes little difference to the diurnal curve. (Manila, Philippine Islands) (After Rude and Marmer, in Strahler, 1969)

the shore zone at low tide, and extensive vertical dispersion of wave attack. Microtidal environments are those of truly tideless seas: there are a range of mesotidal environments between these two extremes.

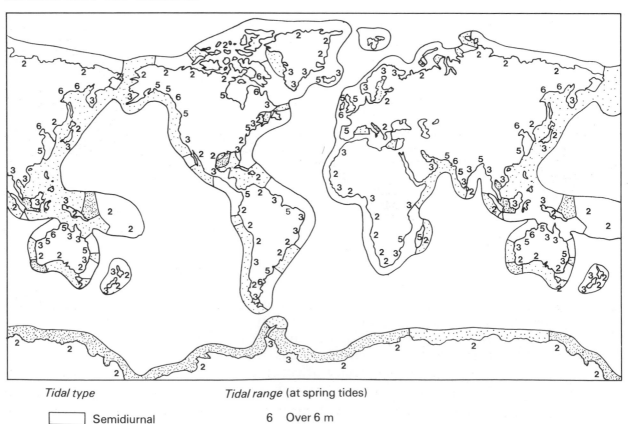

Tidal type		Tidal range (at spring tides)
☐	Semidiurnal	6 Over 6 m
▨	Mixed	5 4-6 m
▦	Diurnal	3 2-4 m
		2 Under 2 m

Figure 10.12 Tidal environments. Relate the tidal type and tidal range to a classification into macro-tidal, mesotidal and microtidal environments, and notice their distribution. (After Davies, 1972)

Currents in the sea

The major ocean currents have little or no effect on shore processes, except where a mass of warm water like the Gulf Stream may give rise to higher waves. Local rapid movements of water, however, may give rise to conditions which do affect coastal forms and sediment. River estuaries in areas of high tidal range are subjected to strong flows of fresh water at low tide; tidal changes also produce movements of water and sediment through confined channels, and effects of this type are thought to keep the floor of the Straits of Dover swept clear of sediment; winds may induce movements within storm surges; and waves produce longshore and rip currents, particularly in conditions of repetitive swell waves.

Seawater

The nature of seawater as a saline solution is important in a number of reactions with coastal rocks, leading to solution or cementation. The variations in overall salinity (Figure 10.13) do not seem to give rise to particular effects, except in the very local deposition of crystalline salt on shore platforms or cliffs. The calcium carbonate distribution is more important, since it affects the solution of coastal limestones, cementation and biological activity on coasts (pages 94 and 259). Ocean waters are supersaturated with bicarbonate ions, and this state increases with temperature — i.e. the tropical waters have a higher proportion.

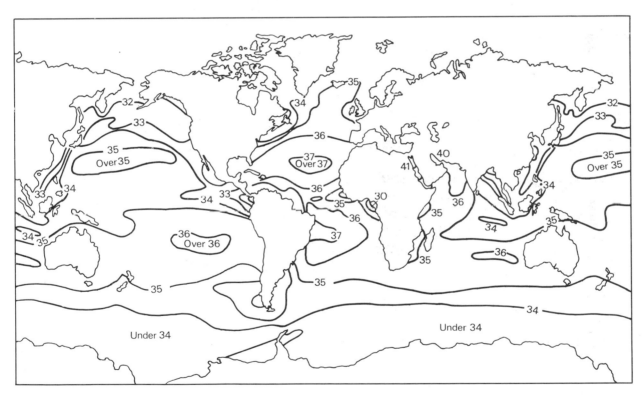

Figure 10.13 World salinity map. Figures indicate parts per thousand of dissolved salts in seawater. (After Sverdrup, 1942, in Gross, 1972)

The temperature of seawater has also been found to affect processes. Thus removal of beach material is more rapid in colder water, due to the higher density (and a similar effect occurs in cold rivers). Sea ice tends to damp wave activity and thus protects the shore, although the freezing of water on rock surfaces in polar zones may help to break up the rocks there.

Erosion, transport and deposition by the sea

The various marine processes combine to act on the coastal zone in erosion, transport and deposition of sediment. The erosional processes affect both solid rock in cliffs and the unconsolidated sediment of beaches, although the latter tend to be removed and then re-built, whereas a cliff can only be worn away by the sea. The most intense erosion takes place on coasts facing the storm waves of temperate zones: southern Chile and the west coast of Britain are amongst the areas which experience the highest rates (Figure 10.14). The sea accomplishes erosion largely by wave action. Storm waves lashing tonnes of water against a coast remove unconsolidated materials rapidly, and the trapping of air in rock crevices may lead to sudden increases and decreases of air pressure and to shattering of the rocks and loosening of large blocks. This type of hydraulic action is sometimes known as quarrying. The effect of water armed with pebbles and boulders is known as corrasion, and has been likened to the effect of a saw acting at a cliff-foot. The rock fragments themselves are broken up during this process. Solution of the rocks also leads to erosion, but the high salt content of seawater, and its calcium carbonate concentration has led to some doubt as to whether solution by seawater is

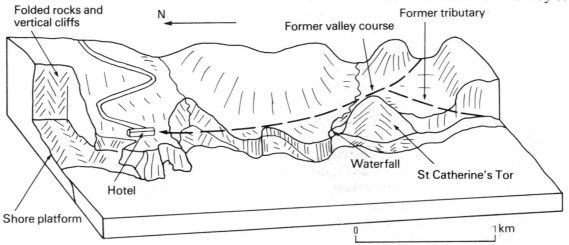

Figure 10.14 The west-facing coast at Hartland Quay, north Devon. This coast faces the storm coast advancing from the North Atlantic. The old quay, just to the north of the hotel, was swept away early in this century

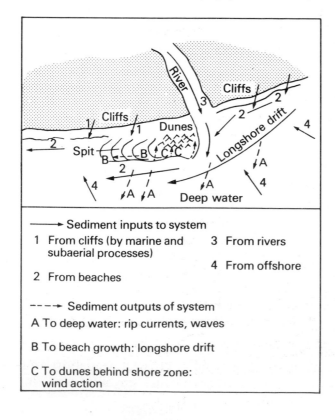

Sediment inputs to system

1 From cliffs (by marine and subaerial processes)
2 From beaches
3 From rivers
4 From offshore

Sediment outputs of system

A To deep water: rip currents, waves

B To beach growth: longshore drift

C To dunes behind shore zone: wind action

Figure 10.15 The coastal system. Notice the interacting processes involved in the supply and movement of sediment. The balance between input and output will determine the dominance of erosion or deposition, but both will occur within a few kilometres of each other. (After Press and Siever, 1973)

as effective as that by freshwater with added organic acids on the land. This problem may be resolved by the fact that many organisms remove the calcium and bicarbonate ions from solution, or add to the carbon dioxide content of the water at night when precipitation could take place. The higher solubility of carbon dioxide in cooler waters suggests that solution of coastal limestones may be more effective in temperate areas than in tropical.

Sediment is transported largely by the process of longshore drifting, and is deposited in areas where there is protection from movement. Coasts where longshore drifting is prevented by headlands are known as coasts of impeded transport. A coastal system can thus be suggested (Figure 10.15), in which inputs of sediment and energy are received and the different processes acting in the coastal environment lead to a variety of outputs. It is important to see the coastal system as one in which erosion, transport and deposition are balanced: the case of Hallsands illustrates this (Figure 10.16).

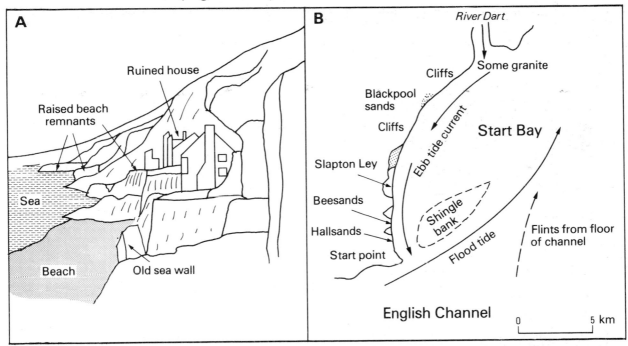

Figure 10.16 Erosion and deposition on a coast in south Devon.
(A) A sketch of the ruined settlement of Hallsands, destroyed by a gale in 1917.
(B) The movement of sediment (mainly shingle) in Start Bay. Material is carried from the shingle bank towards the north of the bay, whence it migrates southwards along the coast. Much of this coast is thus protected by beaches, and small lagoons, such as Slapton Ley have formed behind. Very small additional inputs come from the river Dart and the floor of the English Channel, but most sediment circulates within the bay, with the bank acting as a store. In 1897 650 000 tonnes of shingle were removed from the bank for use in extensions to the Devonport dockyard at Plymouth. The beach soon disappeared from in front of Hallands and within 20 years the village was destroyed. (After Robinson, 1961, in Perkins, 1971)

Changes in sea-level

If the marine processes act largely at the shoreline zone, any changes in sea-level will result in the abandoning of one level, and the resumption of activity at another. The facts that such changes have taken place around the world, especially in association with the Quaternary Ice Age, the causes of the changes, and their effects on stream valley forms, have already been discussed. Any consideration of the action of the sea on coasts must also take them into account.

A lowering of sea-level relative to the land (i.e. caused by uplift of a section of land, or a fall in sea-level) results in what is known as an emergent coast. The former beach or shore platform/cliff landforms are stranded above and out of the influence of marine processes: they will be affected only by subaerial weathering, mass movement and fluvial processes. Such raised beaches are common in western Scotland, which has recoiled isostatically since the ice sheet melted, and in California, where the land has been uplifted, but also in south-west England, where neither effect seems to have been important. Lowering of the sea-level also has the effect of a shallowing of the water offshore (Figure 10.17). Levels above 30 m, and up to at least 200 m in England and Wales have also been identified as old shore platforms.

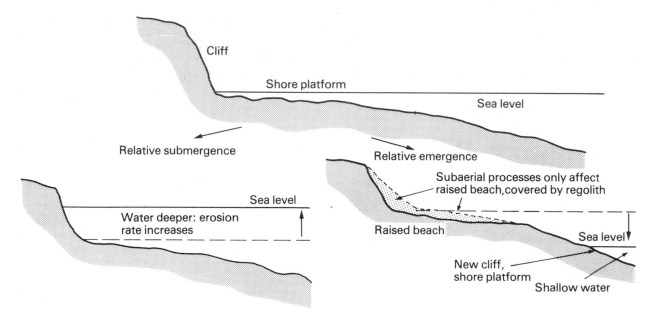

Figure 10.17 The effects of a relative rise of sea-level on coastal features and processes.

A relative rise of sea-level will deepen the water offshore, encouraging further erosion of a rocky coast, and producing features of submergence. It will also lead to the drowning of coastal features and valley mouths: the estuaries, rias and fjords of British coasts testify to a recent process of sea-level rise, along with the submerged forests and the retreat of features like pebble ridges.

It is often dangerous to say that a particular coast is one of emergence or submergence, since it will often contain effects of both. Thus Plymouth Sound is a ria (formed by submergence), but there are raised beaches around it (formed by emergence), and the feature is essentially a valley which was carved into older and higher marine platforms (further emergence). Such a complex pattern can be related to the changes of sea-level (both up and down) during the Quaternary Ice Age (Figure 8.21), and it seems that the sea could have been at, or near, the present level on several occasions. This is an important consideration in the origin of 'modern' shore platforms and the sediment in the coastal system. In addition the emergence during ice advances was accompanied by isostatic effects of ice building up on the land, whilst the submergence following the ice melting would be accompanied by isostatic effects beneath the increased load of water on the continental shelf areas.

The edge of the land

Marine processes act on the coast, which may be formed of resistant rock or unconsolidated sediment, arranged in a variety of ways with respect to the effects of marine processes. The land adjacent to the sea is also subject to different climates and emphases on what are known as subaerial processes — weathering, mass movement, running water, ice and wind. The interaction of sea, rock, relief and subaerial processes is basic to an understanding of coastal features.

Major coastal types

On the world scale coastal types can be related to the distribution of major relief features in terms of plate tectonics (chapters 1–3). Such a division emphasises a number of important distinctions (Figure 10.18), particularly in terms of the mobility of land areas, and the orientation of coastal features.

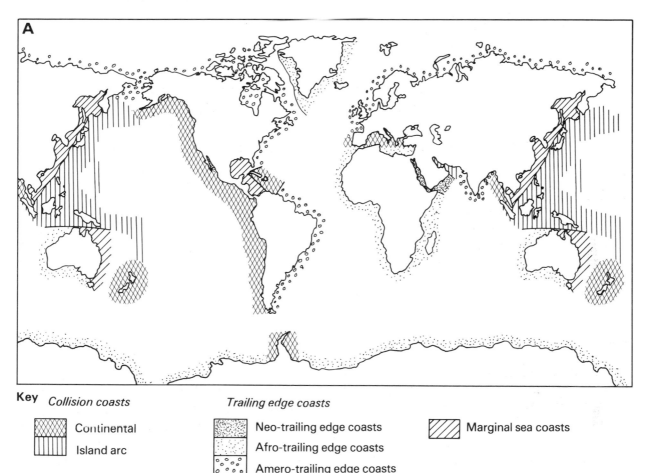

Key *Collision coasts*   *Trailing edge coasts*

▨ Continental   ▦ Neo-trailing edge coasts   ▧ Marginal sea coasts
▥ Island arc   ░ Afro-trailing edge coasts
                                 ○ Amero-trailing edge coasts

Figure 10.18 A classification of coasts in relation to plate tectonics.
(A) World distribution. (After Davies, 1972)
(B) Main features of the coastal types.

Local relief and rock-type

The study of a particular stretch of coast will need to consider the character of the inland relief, and the nature of the rocks present, together with structures affecting them.

1) Coasts can be differentiated simply into upland coasts where the relief gives rise to cliffs over 50 m high, and lowland coasts, where almost flat land lies inland: there is a transition between the two. Either type of coast may be subjected to a mixture of erosion, transport and deposition, together with movements of land or sea, but such a description gives an initial character to the coast.

2) The finer detail of coastal forms must always be related to the physical and chemical properties of the rocks which form it, including the frequency and type of joints, faults, bedding planes or other fissures which cut them. Rocks which are more compact and resistant to wave attack take longer to wear back than others, and an alternation of different types will often give rise to differentially eroded headlands and bays. Unconsolidated coastal materials may be eroded and redeposited rapidly, with many changes of coastal feature (Figure 10.19).

B

Coastal type	Plate situation	Characteristics
I *Collision coasts* (a) Continental (b) Island arc	Especially Pacific Ocean; also East and West Indies, Caspian, North Mediterranean. Thin ocean plate plunges beneath thicker continental plate: fold mountain ranges. Thin ocean plate plunges beneath another: trench plus volcanic island arc.	Structural lineations parallel coast ('Pacific type'); coasts relatively straight with tectonically active and high relief hinterlands; many movements since Quaternary Ice Age — faults, warping — so that sea-level changes and results confused with such land movements.
II *Trailing edge coasts* (a) Neo-trailing edge (b) Afro-trailing edge (c) Amero-trailing edge	Red Sea, Gulf of California: early stage of ocean formation. Trailing coasts with no high mountain ranges on opposite coast: small sediment loads supplied — Africa, W. Australia, Greenland, Antarctica. Trailing coasts with high mountains on opposite coasts of continent: higher sediment yield and wide continental shelf: eastern North America and South America, N.W. Europe, Bering Straits, India.	Little volcanic or earthquake activity May have steep relief inland (II(a)), but mostly plateau or plain. Old structures truncated at coast ('Atlantic type') — discordant grain. Local coastal flexure due to sediment loading or former ice mass loading. Other areas show Quaternary sea-level effects to better advantage.
III *Marginal sea coasts*	Coasts shielded by island arcs: East Asia, Caribbean, E. Australia.	Varied types: may have heavy sedimentation and other features of II(b), II(c).
IV *Oceanic islands*	Along, and either side of, ocean ridge (Azores); or over plume (Hawaiian Islands).	Volcanic islands with coral reefs in tropics.

Figure 10.18 continued

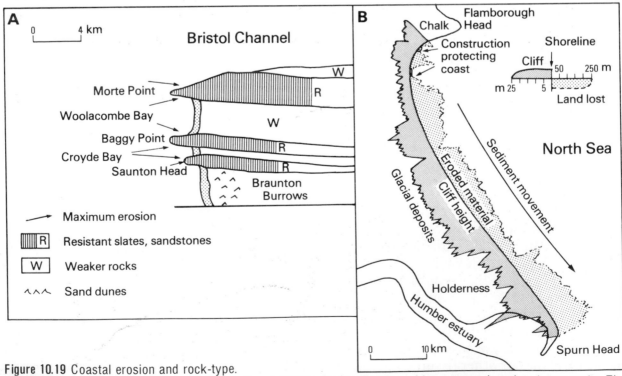

Figure 10.19 Coastal erosion and rock-type.
(A) Differential erosion on the north Devon Coast. Relate the headlands and bays to weak and resistant rocks. The resistant rocks are often slates, which form the headlands, whilst sandstones form the bays. (B) Erosion of unconsolidated glacial deposits along the Holderness coast in east Yorkshire. This shows how the erosion is related to cliff height. The annual rate of cliff recession was 1-2 m along most of this coast over the period 1852-1952. (After Valentin, 1954, in Steers, 1971)

Joints and bedding planes and faults may also regulate the speed of erosion and the shape of cliffs: Chalk is a compact rock with few marked lines of weakness and gives rise to vertical cliffs; alternations of sandstone and shale produce irregular cliff profiles.

Subaerial climates and coasts

The climate of the land area adjacent to the coast forms the third major factor interacting in this zone. Its nature is important in terms of the input of sediment to the coastal system (Figure 10.15); it affects the storminess and chemical processes at the shoreline (and thus the marine processes as a whole); and it affects the features of the upper coastal area, where subaerial processes, rather than marine, are active. The effects of climate in the shore and coast zones are clearest in the extremely dry and cold climatic zones, but insufficient comparative observation is available to make finer distinctions. In terms of sediment supply it is clear that most sand comes from desert areas to the nearby coasts, and most mud from the hot tropics, where chemical weathering is at the highest level.

It must also be realised that the Quaternary Ice Age changes in climate are responsible for many of today's coastal features and sediments. This is true not only of landforms like fjords, but of the sediment on the floor of the North Sea and the Grand Banks of Newfoundland, deposited by ice sheets which covered these areas in a period of lower sea-level. Much of the sediment along the east coast of Australia and the coast of Baja California was added during wetter periods of greater local erosion and transport on the land.

It is unsatisfactory to some extent to admit that no certain distinction can be drawn between the landforms and sediments produced in one climatic zone and those in another, or between those derived from activity in the Ice Age and those being formed today. Such an admission is important, however, since it highlights gaps in knowledge and the fact that many conclusions in the study of the Earth's surface features are still at the stage of hypothesis: they require testing and verification.

One point which does emerge from present knowledge of climatic variations is that the British Isles stand in an area of storm waves and high tidal ranges, so that marine processes must be producing more rapid changes than in other parts of the world.

Order	Approximate size (km)	Processes involved	Landforms produced
I	Over 1000	Global tectonics	Fold mountains Ocean trenches Island arcs Continental shelves
II	100	More local tectonic effects, including isostatic changes Marine processes Subaerial processes (acting over a larger area)	Deltas Fjords
III	1-10	Marine processes Subaerial processes (acting on local rock and relief characteristics)	Cliffs — with caves, arches, stacks Shore platforms Beaches, bars, spits — with smaller features such as berms, cusps
IV	0.1-0.01	As for III	Caves, arches, stacks, berms, cusps

Figure 10.20 Scale in coastal landforms related to different processes involved in their formation (cf. discussion of the scale factor in the Introduction to PART I of this book.)

Plate 168 Erosion of till on the shores of Lake Huron, Michigan, USA. The dragline in the background is building a groyne to stabilise the coast. (USDASCS)

Plate 169 Lava cliffs, 7–17 m high, being attacked by Pacific swell waves on the north-east coast of Hawaii. (USDASCS)

Plate 170 Stereo pair: a series of raised beaches, or marine terraces, at Eel Point, San Clemente Island, near Los Angeles, California. Each level rises 10–30 m above that below, and may be 300 m wide; the upper zones are mantled by weathering products. (USGS)

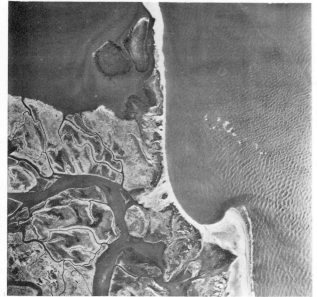

171

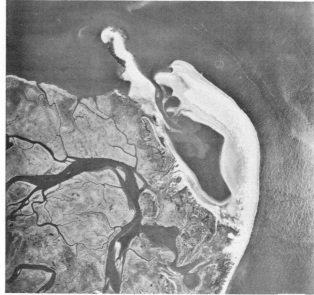

172

Plate 171/173 Changes in Island Beach, New Jersey: three aerial photographs taken in 1940, 1957 and 1963. Describe the changes taking place. It may be helpful to use the tidal creeks behind the beach as a point of reference for charting the modifications. (USGS)

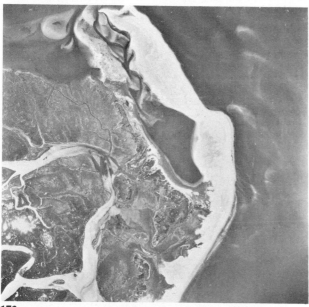

173

Plate 174 An Apollo 9 view of Cape Hatteras on the eastern coast of the USA. Notice the outer beach and the movement of sediment through this to the continental shelf. The clouds to the right have formed as cold air from the land has blown across the warm Gulf Stream current. (NASA)

Plate 175 The coasts of the North Sea and Baltic Sea give a contrast in tidal conditions. This photograph was taken by Skylab 3 in 1973. (NASA)

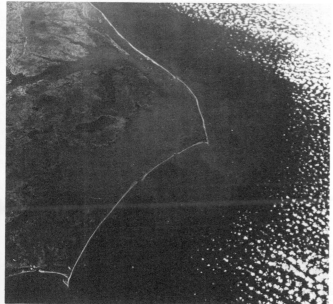

174

175

Coastal landforms

It is clear that coastal landforms are not produced by the activity of the sea alone. It is also patent that any discussion of the formation of such features must be related to a scale of size, since the processes involved change in emphasis accordingly (Figure 10.20). The scale of the first order features is too large for appreciation by individuals in the field, and these can be regarded as world-scale coastal types. The second and third orders form the basis of a study of coastal landforms.

Cliffed coasts

Cliffs and shore platforms are often associated together on coasts where erosional processes are important, but either may occur on its own. Cliffs are formed by a combination of marine undercutting and subaerial processes (Figure 10.21): the latter, and particularly slumping, may be the dominant process in areas of unconsolidated or weak clay rocks, or in areas of tundra solifluction.

Quarrying and corrasion together are largely responsible for the steep cliff and seaward-sloping platform. Both are most important in storm-wave environments, and especially on coasts facing the full force of major storm waves. Blocks of rock are broken away from the foot of a cliff, and others fall to the beach from above, having been made unstable by the activity at the cliff foot. Corrasion causes the formation of a notch at the base of some cliffs, smooths the shore platform surface, and breaks down and rounds the rock fragments. Shore platforms produced largely by such processes have a distinct slope towards the sea.

Shore platforms in lower energy environments may be formed by other processes. The effect of alternate wetting and drying of rock surfaces is known as water layer weathering, and this affects cliff faces (spray) as well as shore platforms (tidal exposure). A variety of weathering processes operate in this context, but the rock surface is reduced to a level of permanent saturation (often low tide), with the smaller details depending on the type of rock involved. Erosion proceeds more slowly than in high energy environments, but a rate of 1 cm reduction every 30 years has been estimated for southern California. The effect of solution on coastal limestones is another problem which has not been solved satisfactorily: it is difficult to isolate it as a factor from the other processes acting on the rocks, including biotic factors (chapter 11). A further problem relates to the extensive strandflat feature around the coasts of Norway (Figure 10.22).

Cliff forms vary greatly within small areas, although some general contrasts can be noticed in different climatic zones (Figure 10.23). It has been suggested that the hog-back (slope-over-

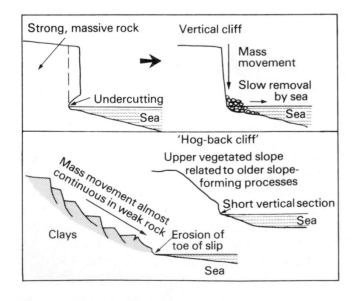

Figure 10.21 The relative dominance of marine and subaerial processes on cliffed coasts. (After Davies, 1972)

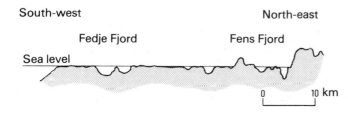

Figure 10.22 The strandflat, an exceptionally wide marine platform on the Norwegian coast. It is partly above sea-level. The origin is not clear: it is too wide and even for marine processes alone to be responsible, and some suggest a combination of frost action and glacial processes. (After Nansen, 1922, in Holmes, 1965)

wall) cliffs of temperate areas reflect an upper section lowered in gradient by periglacial conditions during the Quaternary Ice Age, but this feature occurs more widely, and may be more closely related to the occurrence of a phase of lower sea-level, removing old cliffs from the zone of wave-action (Figure 10.21). Smaller features of cliffed coasts include caves, arches, stacks and blow-holes: these are produced as a part of the cliff retreat along lines of weakness in the rocks, and are largely restricted to coasts in high energy storm wave environments. It is important to realise that they are small features of minor relevance to the overall development of a coast, and that they are not very common.

Shore platforms also vary widely in detail due to the effects of rock-type variations. Whilst quarrying and corrasion tend to produce a sloping (intertidal) form, water layer weathering results in a flat surface related to high tide, whilst biological destruction and solution produce more variable surfaces related to low tides. Coasts in temperate areas which have a development of platforms at several levels may be related to the changing pattern of sea-levels in the past. It has been suggested that the width of many shore platforms, even in the high energy storm wave environments of western Britain, cannot have been formed in the last 3600 years, during which the sea has been at its present level, and that there must have been previous periods with the sea at a similar level.

Climatic type	Processes at work	Cliff forms
Humid Tropical	Low- or medium-energy wave environments. Mass wasting and vegetative cover important.	Low-angle cliff forms due to slow retreat related to mass wasting processes; landslips often produce amphitheatre-like bays. Limestone and coral forms more like Arid areas.
Arid Tropical	Low- to medium-energy wave environments. Little plant cover or regolith.	Steeper cliff forms with flat shore platforms.
Temperature	High-energy storm environments with extensive quarrying.	Steeper and higher cliffs and extensive shore platforms.
High latitude	Low-energy wave environments. Periglacial types of mass movement.	Low angle slopes are dominant.

Figure 10.23 Cliff forms and different climatic environments.

Beaches, bars and spits

Beaches are composed of unconsolidated materials, from pebbles to sand and mud in size, and are therefore subjected to frequent changes by variable deposition and erosion. Beach forms are regulated largely by particle size, which in turn is related to the supply of rock debris and climatic conditions in the area. The geological factor is of local importance: compact sandstones give rise to boulders and pebbles nearby; flint pebbles are particularly long-lasting in the realm of marine processes; Chalk pebbles are found only in the immediate vicinity of Chalk cliffs; and many rocks disintegrate to sand or mud immediately. Sediment is also supplied, or denied, by the offshore zone. This is clear when that supply is removed

(Figure 10.16) and when the source of the sediment in the Thames estuary is considered. The climatic factor, however, is of more widespread significance than the effects of different rock-types. Most pebble beaches occur around the coasts of high latitude lands due to the storm wave environments of such regions, and to the fact that ice sheets deposited much large-size debris on the coastal and continental shelf areas during the Quaternary Ice Age. In the humid tropics mud is the most important sediment available, and sands are most important between these limits (Figure 10.24; cf. Figure 3.16). Climate also plays a part in the carbonate content of beaches — a result of broken shell material, or precipitation from solution around ooliths: beaches in the Bahamas are formed largely of such debris. The main importance of calcium carbonate in the beach material is that it can be dissolved and re-deposited as a cement to form lithified 'beachrock'. This formation is confined to the tropics, especially in areas where there is a dry season.

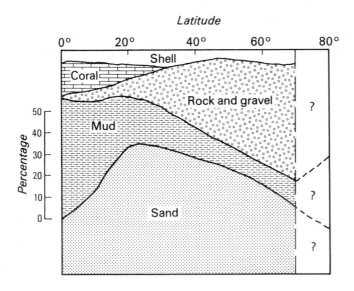

Figure 10.24 The relative frequency of sediment type occurrence by latitude on the world's continental shelves. Compare the main types of sediment in the 0-20 degrees and 50-70 degrees areas, and explain the differences. (After Hayes, 1967, in Davies, 1972)

Beaches are subject to both erosion and deposition, often known as cutting and filling, at different states of the wave environments. Thus high, steep waves ('destructive') remove material out to sea, leading to cutting, whilst low, flat waves ('constructive') move it shorewards, with filling. In temperate areas there is often a seasonal change between winter storm wave environments, which cut, and summer low energy waves which result in filling. In storm wave environments the prevalence of onshore winds leads to offshore bottom currents and intensified movement of material to and from the sea-floor offshore.

Beaches can be studied in plan view (as on a map), and in profile (cross section). The beach profile varies in height, width and gradient (Figure 10.25). The height is related to the height of the waves arriving on the beach: this determines the extent of swash on a sandy beach, and the highest point to which pebbles are thrown on a storm ridge (cf. Chesil Beach in Dorset is 13 m above the high-water mark). The width of the beach depends on the input of sediment, and an excess of input over loss will lead to extension of the beach area. The gradient of the beach (the beach face angle) is the result of interaction between the particle size and wave steepness, which governs the amount of backwash; in addition the backwash is affected by the amount of water in beach deposits — percolation is rapid through pebbles, but impossible in saturated sand or frozen beaches. Steeper beach gradients occur with swell wave environments, where material is carried up the beach, often ending in a ridge, or berm, above high-water level. Sandy beaches in storm wave environments are often very flat, since material has been carried to the line of breakers, and they may have a pebble ridge at the head of the beach, moved only in time of storm. The profile is also affected by the tidal environment, since a high tidal range allows a wider dispersion of wave attack, and the more intense activity of water draining

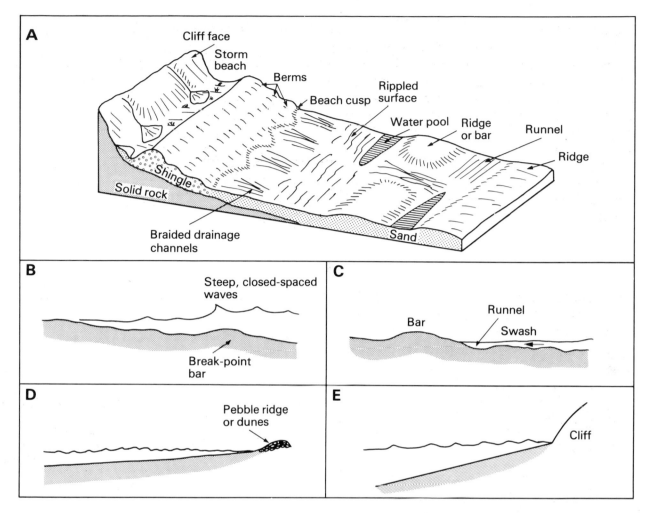

Figure 10.25 Beach profiles
(A) Idealised features occurring on beaches. (After Small, 1970)
(B) The formation of a breakpoint bar by destructive waves in association with a small tidal range (i.e. a virtually stationary breakpoint).
(C) The formation of a swash bar by constructive waves: the number of bars and runnels will be determined by the tidal range.
(D) The profile of a beach facing constant North Atlantic storm waves and swell (e.g. north Cornwall).
(E) The profile of a beach facing more variable wave conditions (e.g. south Cornwall). Explain the differences between profiles (D) and (E).

across the beach, or through an outlet point, at low tide (Figure 10.26). In high latitudes, where the sea freezes, beach features may include those caused by the pushing of ice masses and the stranding of broken sea ice: lines of boulders may accumulate at low tide on an otherwise sandy beach.

A prime consideration when studying the beach plan is its relationship to the incoming waves (Figure 10.27). Many beaches are aligned parallel to the line of maximum longshore drifting (i.e. approximately 40–50 degrees to the line of wave approach). Others, however, are aligned parallel to the crests of constructive waves — i.e. the swash. The latter group are more common on coasts of impeded transport, and where wave transport is perpendicular to the coast. Beach cusps are thought to be a feature of such coasts. These are semi-circular indentations into the berm, and may be up to 6 m from front to back (Figure 10.28). Regions of more active backwash give rise to local indentations perpendicular to the berm, and water movements are increasingly concentrated there. Larger particles and pebbles congregate in the 'horns' between the main flow of water. It seems that the 'swash alignment' of a beach is the ultimate state, since drift-aligned beaches may later revert to swash alignment and the orientation of the beach will swing round accordingly. Beach alignment is also related to

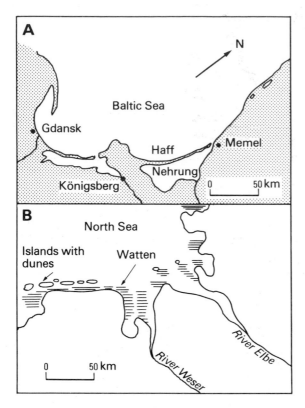

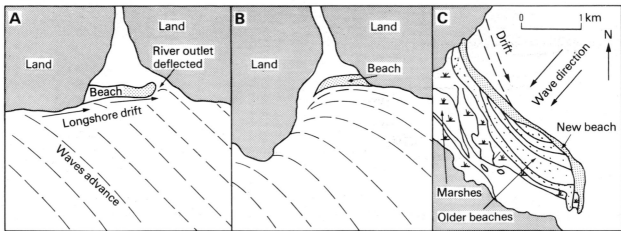

Figure 10.26 Tidal range and coastal features.
(A) The 'haff-und-nehrung' coast of the Baltic Sea, where there is a low tidal range.
(B) The 'watt-und-nehrung' coast of the southern North Sea, with a high tidal range. Watten are tidal flats.
In a similar way the coastal barriers of the eastern USA become increasingly interrupted towards the south, as the tidal range increases.

transport, and therefore to the sorting of constituent materials, within a beach. Where drift is important there may be some differentiation along the shore, but this is only clear where there is a definite source of material (Figure 10.29). Swash aligned beaches, like Chesil Beach in Dorset, show a correlation between the distribution of wave energy, particle size and berm height.

Figure 10.27 Beach alignment in relation to long-shore drift and wave approach directions.
(A) Conditions of free transport along the coast led to the formation of beaches related to the drift direction.
(B) Conditions of impeded transport along a coast of pronounced headlands. Beach formation is then related to swash. Compare the position of the river outlet with that in (A).
(C) Rheban spit, eastern Tasmania, shows how a spit may swivel from drift alignment to wash alignment. The shape thickens away from the shore. (After Davies, 1972)

The dominance of constructive or destructive waves results in the formation of one of two types of beach form.

a) **Breakpoint bars** build up beneath the point where waves break in conditions of steep,

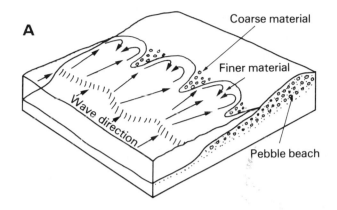

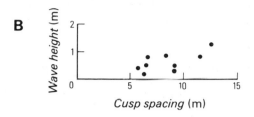

Figure 10.28 Beach cusps.
(A) A block diagram of the features, showing the distribution of beach material grade sizes and their relationship to water movement. (After Zenkovitch, 1962)
(B) The relationship of mean cusp spacing to wave height. This diagram provides a scale for (A). (After Longuet-Higgins and Parking, 1962, in Pitty, 1971)

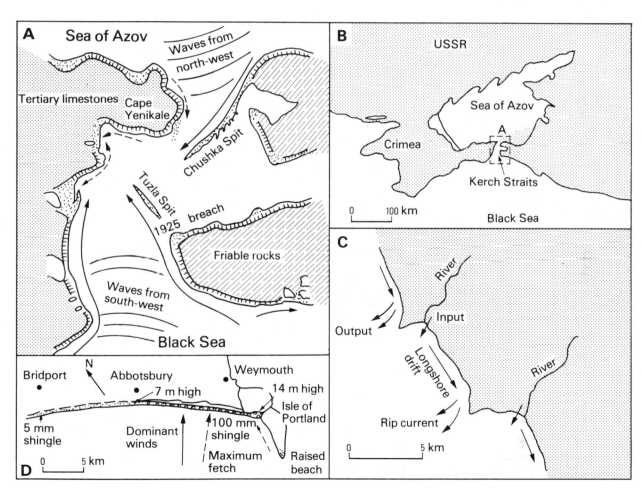

Figure 10.29 The origin and movement of coastal sediment.
(A) The Kerch Straits in the northern Black Sea (located on (B)). The rocks on the west of the Straits are resistant Tertiary limestones, whilst those to the east are poorly consolidated and easily eroded. Waves approach from the north-west and more powerfully from the south-west in the larger Black Sea. Relate the depositional features to these conditions. (After Zenkovitch, 1962)
(C) A situation where input and output of beach material can be measured.

(D) Chesil Beach, Dorset. This feature and the origin of its materials have posed a problem. The beach increases in height and width to the south-east, and the shingle grade also increases in that direction. It has been suggested that longshore drifting is responsible for this feature, but the direction of such movement is not clear. The direction of winds and longest fetch, however, suggest that this beach is a swash-aligned form, and it would seem that the grain size and beach height can be related to variations in energy along the beach. (After Holmes, 1965, and King, 1959)

destructive waves and the seaward movement of beach material. Such bars are most obvious in regions of small tidal range where the point of wave-break remains almost stationary (e.g. Mediterranean and Baltic Seas; the Gulf of Mexico), and one bar may develop in relation to storm waves with smaller bars nearer the shore. The Gulf of Mexico also experiences occasional storm surges, associated with hurricanes, during which large breakpoint bars are formed; after the storm these may appear above the normal sea-level.

b) **Swash bars** are built up by the shoreward movement of material by flat, constructive waves. These are normally submerged at high tide, and the number of such bars with intervening runnels is dependent on the tidal range. Thus in microtidal environments like the Mediterranean Sea a single berm is found; where the range of spring tides is between 2 and 4 m a berm and single ridge is common (e.g. in east Norfolk); and in macrotidal environments there may be multiple ridges and runnels (e.g. north Wales). British coasts provide a variety of wave conditions, and it is common to find a pebble ridge at the head of beaches together with rather flat ridges in sand seawards of the high tide mark, due to the dominance of destructive waves.

On a coast where the swell comes from a regular source, and where there is an alternation of bays and headlands allowing little longshore drifting, the beaches tend to have an asymmetrical plan shape. This is a common feature of many parts of the world, including the USA, Australia, West Africa and Malaya (Figure 10.30), and such bays are known as zetaform (after the Greek letter zeta).

Spits are major beach structures associated with longshore drifting, and commonly have

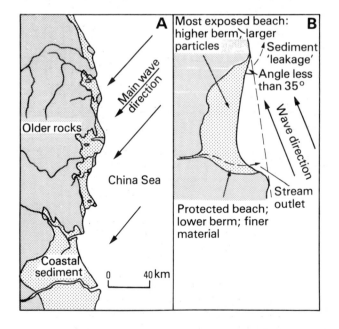

Figure 10.30 Zetaform beaches.
(A) The eastern coast of Malaya. Note the beach shapes in relation to dominant wave directions, and the positions of stream outlets.
(B) The general pattern of features on zetaform beaches. (After Davies, 1972)

curved ends in areas around the North Atlantic, but less commonly in areas like Australia (Figure 10.31). It seems that the curved end results from a combination of factors: the effect of subsidiary waves as the spit (formed in a protected situation) extends into exposed waters; the effect of refraction of the waves at the end of this constructional feature; and the effect of a large tidal range, since the end of the spit experiences deep water conditions at high tide. Once again the differences in wave and tidal environments in different parts of the world result in an emphasis on different processes, and the production of variations within a particular landform type. The student of landforms must compare conditions and forms throughout the world before jumping to conclusions that a particular feature is typical. Spit formation may begin with longshore drifting, but a variety of new environments are created as the spit is enlarged. The ponding of fresh water behind will enlarge the quiet depositional area of a river estuary, leading to mud deposition and marsh accretion, whilst the building up of pebbles and

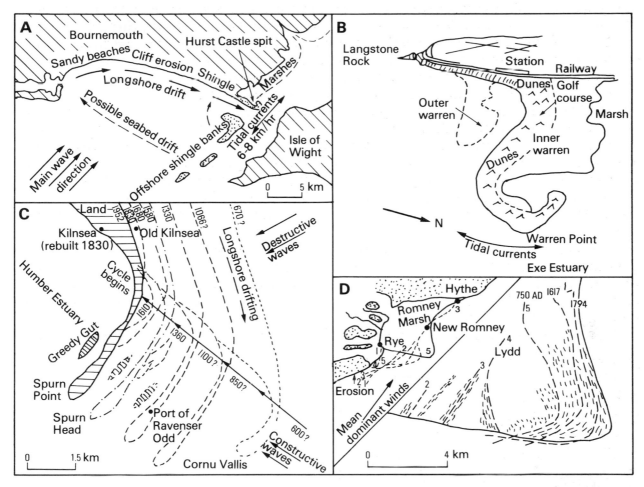

Figure 10.31 The formation of spits: a variety of British forms.
(A) Hurst Castle Spit, Hampshire. Analyse the forces acting to form this feature. (After Small, 1970)
(B) Dawlish Warren, south Devon: a sketch. The Inner Warren extends just over 2 km from the western shore of the river Exe estuary. The Outer Warren was a belt of dunes which gradually disappeared by erosion in the early 19th century, with the materials being added to Warren Point. This point is also now being eroded, and the supply of sediment has been cut by the building of the coastal railway and protection works. (After Perkins, 1971)

(C) Spurn Point, Holderness, Yorkshire. Successive positions of the spit seen in relation to the retreating shoreline (Figure 10.19B). At each stage a long spit is formed, then destroyed. This diagram has been constructed from old maps, and the arrowheads suggest dates for the time of destruction and renewed spit formation. (After de Boer, in Brunsden and Doornkamp, 1974)
(D) Dungeness Foreland, Kent. This was derived from a spit formed in Neolithic times (1–1), which then changed from a drift-aligned feature to one increasingly swash-aligned (stages 2–2 to 5–5, and then as dated by map evidence in the larger diagram). (After Lewis, 1932, in Holmes, 1965)

sand will give wind activity a chance to form dunes. Spits are thus complex environments of great interest, and they have been amongst the most closely studied landforms.

The small-scale features of beaches are related to wave type and the transport of water and sediment in the zone of wave-break and at low tide. Ripple-marks are commonly found: those with symmetrical cross-profiles, sharp crests and rounded troughs are formed by wave action; asymmetrical forms are often larger and are due to tidal currents. Small, often braided channels across the beach are formed by water draining from pools at low tide, and commonly take a meandering form if a channel is cut a few centimetres into the beach sand.

Deltas and estuaries
Deltas and estuaries are essentially areas where relatively fresh river water, bearing a variety of sediment loads, enters the sea. Their character is thus determined by the inputs of water and

sediment from the land, and the conditions of sediment availability and movement, together with the wave and tidal environments in the sea.

a) **Deltas** are formed where a river deposits a large load as it enters the sea, and are considered in chapter 7. Deposition in delta distributaries is related to the mixing of fresh and salt water, and marine beach features mark the outer margins of a delta. Often these are left as islands along with levée deposits, when a delta surface subsides isostatically under an accumulating weight of sediment once the main zone of deposition has shifted to another part of the delta (Figure 7.29).

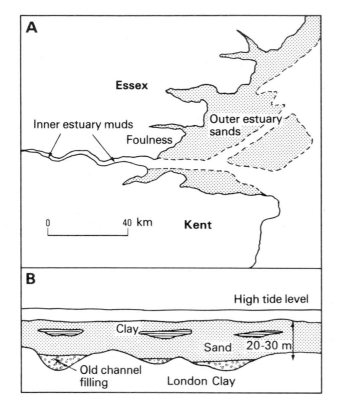

Figure 10.32 The Thames estuary.
(A) The muds of the inner estuary are brought down by the river and deposited where it reaches salt water (between Waterloo Bridge and Woolwich): the fluid, slimy mud builds up in sheltered hollows, especially in winter. The outer estuary is floored by sands, which have spread into the area from the glacial deposits on the floor of the North Sea. Tidal scour channels cut both types of sediment.
(B) A section through the outer estuary deposits at Foulness. This area is under consideration for reclamation: drying of the deposits could result in uneven subsidence. Sands compact little, but clays could be up to ten times the water-laden thickness.

b) **Estuaries** occur at the mouths of rivers bringing restricted quantities of sediment to the sea. Many river mouths have wide openings due to late glacial channel deepening, followed by postglacial drowning. The Thames estuary is one of the most studied examples (Figure 10.32), but most British river mouths have this character, as do those on the eastern coast of USA. Estuaries in general are characterised by mudflats due to settlement out of suspension of the tiny particles as they meet and react with seawater. They flocculate (i.e. cluster together) and sink. In macrotidal environments low tide channels cut as deeply as low spring tides will allow. Mudflats are exposed at low tide and soon become vegetated, leading to increased rates of accretion and providing land for reclamation by man.

The classification of coasts

The student of coastal landforms is impressed at an early stage by the variety of form, even within the restricted experience offered by a single local study. On a world scale the variety becomes daunting to many, but man appears to enjoy reducing a wealth of types to manageable proportions by means of classification into distinct groups which can then be subdivided for study.

A classification may be descriptive, or genetic (i.e. relating the forms to their origin). Descriptive classifications tend to end as a catalogue of names of coastal landforms: beaches (various types); cliffs; fjords; rias; etc. Genetic classifications tend to suffer from being related to the dominant interpretation of their time. Thus one genetic classification of coasts divided them into shorelines of submergence; shorelines of emergence; neutral shorelines (where neither submergence nor emergence can be related to the form — e.g. delta, volcano); and compound shorelines, which are mixtures. Shorelines have since been regarded as more complex than a fourfold division implied, and another genetic classification (Figure 10.33) incorporates erosion and deposition, along with submergence and emergence, in a consideration of advancing and retreating coasts.

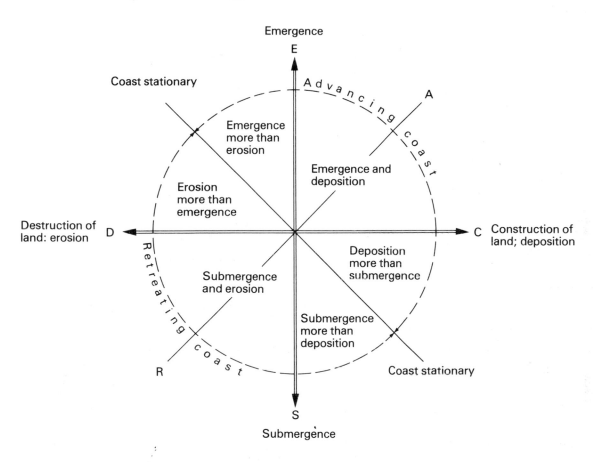

Figure 10.33 Valentin's classification of coasts: a graphical representation. (After Holmes, 1965)

Such classifications suffer from over-simplicity. Increasing investigation of coastal processes and features has emphasised their complexity. A fully genetic classification should therefore include consideration of a wider variety of factors.

a) **Geological factors.** These will include the nature of the rock-type forming cliffs and shore platforms, and the nature of the beach materials. They will also include the form of coastal relief, which is related closely to the immediate past history of the area in terms of tectonic movements, isostatic movements, subaerial processes acting on the coastal zone and of changes in sea-level.

b) **Marine processes.** The dominant marine processes are those associated with wave action, leading to erosion, transport and deposition. Tidal environments and a range of coastal currents also have effect.

c) **Climatic conditions** vary throughout the world, giving rise to different processes acting on coasts above the tidal levels, and to different types of sediment input. They are also more widely important, since they influence a number of other variables: storm environments are the main zones of wave generation; past climatic changes have resulted in changing sea-levels and have affected different parts of the world in different ways (e.g. glaciation of the humid temperate areas); and they also affect types of organism living in the coastal zone.

d) **Biological factors** have not been considered extensively in this chapter, but are important (chapter 11), especially in the evolution of smaller-scale features.

e) **Scale.** A classification must be related to features at a particular level of size (Figure 10.20), since the genetic features change in emphasis. Thus an adequate classification of coasts related to origin should have a series of levels in terms of scale.

For a classification on the world scale it would seem that a climatic basis could be the most satisfactory (Figure 10.34), but within these divisions there will be subdivisions based, in descending order, on global tectonic features, tidal environments, protection and transport conditions, and rock-type.

World climatic zone	Processes and forms
1 *Low latitude coasts*	(a) Reduced wave energy, but with consistent directions of wave approach. Few rocky coasts and weak cliff forms; horizontal shore platform development. (b) Constructional forms dominant: supply of sediment abundant — but rarely contains pebbles; massive barrier-type forms common, with lithification of beaches. Coastal dunes seldom form. (c) Biological constructional forms common: coral and algal reefs. (d) Major subdivision between coasts where it is hot and wet all year — where muds and mangroves are common — and those where there is a major dry season — where there is sandy sediment and local salt flat development.
11 *Middle latitude coasts*	(a) High wave energy in the zone of frontal storms. Hard rock coasts are common with strong cliff development and sloping shore platforms. Beaches are less common than in I, and barrier development weak, but pebbles are common in such deposits and sand dunes are formed at the back of sandy beaches. (c) Biological constructional effects are insignificant. (d) Major subdivision between coasts in areas which were glaciated during the Quaternary, and those which were not. This affects sediments and sea-level changes.
111 *High latitude coasts*	(a) Shores are frozen for a part of the year, and wave energy levels are low. Cliff forms are weak, influenced by mass movement. (b) Constructional forms are common, including barriers, and include many pebbles. There are few coastal dunes. (c) Biological constructional effects are insignificant. (d) Major subdivision between coasts where the annual freeze-up is short — so that marine processes have some effect on the coast — and those where it lasts almost the whole year. In the latter case there are few marine effects to be seen and ice action becomes dominant.

Figure 10.34 The major world coastal types. (Data from Davies, 1972)

Organic processes

The interactions of atmospheric and oceanic processes with the continental surface give rise to chemical and physical effects. This zone of interaction is also the biosphere — the narrow layer at the Earth's surface in which the immense variety of living organisms present on the planet exists. These give rise to a further interacting factor which must be taken into account in the study of landforms, although it is only in recent years that it has been given high status. The expansion of the studies of the processes affecting the Earth's surface has forced further consideration to be given to organic processes.

Organic processes affect landforms at several levels of scale. At the smallest scale there is an interaction between physical, chemical and biotic processes, in which the different strands are difficult to disentangle. Thus the development of studies in weathering began with an emphasis on physical processes; it was then realised that these were largely ineffective without accompanying chemical modification; and it is now appreciated that the part of plants and animals in these processes is highly significant as their role is considered more fully.

On the medium scale organic processes may give rise to the construction of massive reef

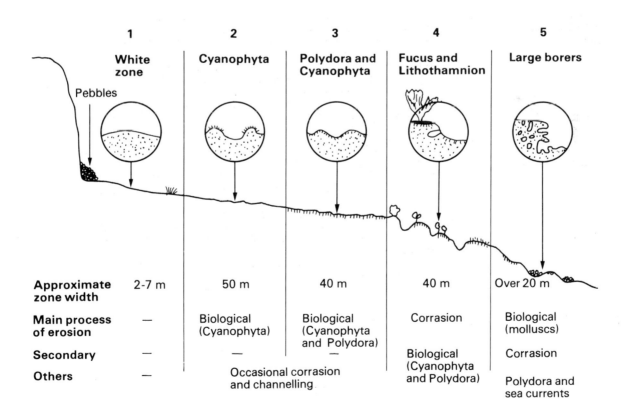

	1	2	3	4	5
	White zone	Cyanophyta	Polydora and Cyanophyta	Fucus and Lithothamnion	Large borers
Approximate zone width	2-7 m	50 m	40 m	40 m	Over 20 m
Main process of erosion	—	Biological (Cyanophyta)	Biological (Cyanophyta and Polydora)	Corrasion	Biological (molluscs)
Secondary	—	—	—	Biological (Cyanophyta and Polydora)	Corrasion
Others	—	Occasional corrasion and channelling			Polydora and sea currents

Figure 11.1 Biochemical weathering zones on a Chalk shore platform in northern France. Cyanophyta are algae; *Polydora* and *Lithothamnium* are sponges; *Fucus* is seaweed; mollusc borers are mainly bivalves. Notice how important such biological activity may become. (After Nesteroff and Mélieres, 1967, in Pitty, 1971)

structures in the oceans. The most important effect of all, however, arises from the activities of man. Whilst the interactions of physical, chemical and biotic processes can be regarded as process–response, or cascading, systems (chapter 7) man is the only process which can exert a degree of control over the natural environment on the medium scale. His involvement can be regarded as a control system.

The interaction of the biotic with the physical and chemical processes

The main effect of plants and animals is to be found in the modification of weathering and mass movement processes, but they are also effective in causing obstructions and so slowing the movement of water and wind, leading to increased rates of deposition in stream channels, coastal mudflats or aeolian dune areas.

Biotic factors, weathering and mass movement

Plants and animals work with the physical and chemical processes to either accelerate or retard weathering and mass movement. It was thought that organic acids, derived from decaying plant matter, are effective in chemical weathering, but this is now doubted: it seems that the carbon dioxide content of the soil, which is involved in the formation of carbonic acid, is more important. Where organic matter is released in an aerated soil it is oxidised to carbon dioxide and water (amongst other compounds), and the carbon dioxide content may range from 0.2 to 3.0 per cent — the higher values being in association with higher temperatures. This is particularly important in the weathering of limestones and of rocks containing iron minerals. The shore zone is also very liable to undergo weathering processes, since it often exposes bare rock, and here a variety of organisms induce or encourage weathering of the surface to take place (Figure 11.1).

Mechanical processes also help to break up and move rock material. Thus the expansion in the roots of growing plants may prise off blocks of rock (Figure 11.2). This process is thought to be particularly active in drier areas, since much of the biomass of plants is concentrated in the roots: whilst the roots of temperate forests (deciduous or coniferous) make up 15–25 per cent of the biomass, and extend down to 30–50 cm in the soil, desert plants are on average 80 per cent below the ground and tundra plants up to 90 per cent. Falling trees often lever out

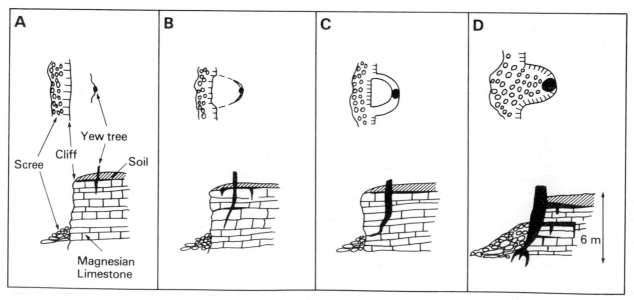

Figure 11.2 Tree roots and cliff breakdown. As the yew tree grows pressure is exerted on the rocks, forcing out a section. This situation was observed on the Magnesian Limestone ridge to the southeast of Sheffield. (After Jackson and Sheldon, 1949, in Pitty, 1971)

large boulders with their roots as they fall, especially on steep slopes. The soil fauna is important in turning over soil material, bringing oxygen into contact with more of the soil and allowing water to seep in for chemical activity to increase. This also assists the creep processes on slopes. It has been calculated that earthworms turn over all the soil in the top 10 cm within 11–80 years in temperate areas, and that termites and ants, carrying out this function in the tropics, build mounds 2–4 m high and thus expose soil particles to the processes of movement. In coastal sediments burrowing worms, like *Arenicola,* re-work the top 20–30 cm in 20 months. Thus a tremendous quantity of work is accomplished by tiny creatures.

Other processes slow down rates of weathering and movement. In particular a plant cover will intercept rain, and either absorb much of it directly or allow it to be evaporated back to the atmosphere before it reaches the ground (Figure 11.3). The combined evaporation and transpiration from trees is greater than evaporation from moist, bare soil. The regulation of water flow through a forest cover means that little, if any, flows off the surface, whilst over grassland much will be as surface flow: even with a vegetative cover there is a greater likelihood of erosion by running water with a grass cover than with a woodland cover. The build-up of decaying organic matter in the soil also restricts runoff, and therefore the erosional work it may carry out. When organic matter is completely broken down to humus it will absorb water and carry out important bonding functions with the clay particles in the soil. The high temperatures of the tropics encourage the bacterial activity which removes such humus, and leaf litter has no chance to accumulate. In cool temperate areas, however, supply of decaying plant matter often exceeds rates of decomposition, and results in the formation of a peat layer which effectively blankets the underlying rock and regolith, protecting it from

A *Evergreen rainforest of Brazil*

(a) Evaporated directly from tree crowns 20%
(b) Running down tree trunks   46%
     9.2% evaporated from trunk surface
     9.2% absorbed by bark
   27.6% reaching base of trees: 20.7% absorbed by roots
                                6.9% reaches water table
(c) Penetrates to rain gauge at 1.5 m 33%

B *Grassland:* rainfall at 25.9 mm per hour
Little bluestem intercepted 50-60%
Big bluestem intercepted 57%
Buffalo grass (partial cover) intercepted 31%

C *Temperate woodland*			
Tree type	*Gross interception* (%)	*Stemflow* (%)	*Net interception* (%)
Northern beechwood	20 (17 unleafed)	5 (10)	15 (7)
Aspen-birch	15 (12)	5 (8)	10 (4)
Spruce/spruce-fir	35	3	32
White pine	30	4	26
Hemlock	30	2	28
Red pine	32	3	29

D *Crops*		
Crop	*Interception whilst growing* (%)	*Interception before main growth* (%)
Corn (maize)	15.5	3.4
Oats	6.9	3.1
Clover	40	—

Figure 11.3 The interception of rain by vegetation of a variety of types. Compare the effects of tropical and temperate forests, deciduous and coniferous trees, and the differences between forest cover and crops. (After various authors quoted in Gregory and Walling, 1973)

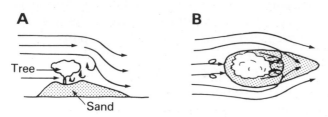

Figure 11.4 The effect of a tree on wind flow and sand deposition: a nebkha. (After Coque, 1962, in Pitty, 1971)

weathering and erosion. It has also been suggested that the uptake of minerals in the soil by plants tends to lock them into the slope system for years, but over a period of time the rate of uptake will balance the rate of return from decaying plants.

Plants and sedimentation

Another important effect is the braking of water and wind flow by plants in their path. This occurs in a number of environments from stream channels to slopes and deserts (Figure 11.4), but is most important on coasts. Seaweed on shore platforms restricts mechanical erosion by protecting the rocks and slowing the water flow. *Salicornia* and *Spartina* grasses grow on mudflats in estuaries and increase rates of mud accretion to 10 cm/year. On tropical coasts and along stream watercourses mangroves with their complex root systems encourage high rates of sedimentation by slowing the water flow so that the suspension load settles out (Figure 11.5).

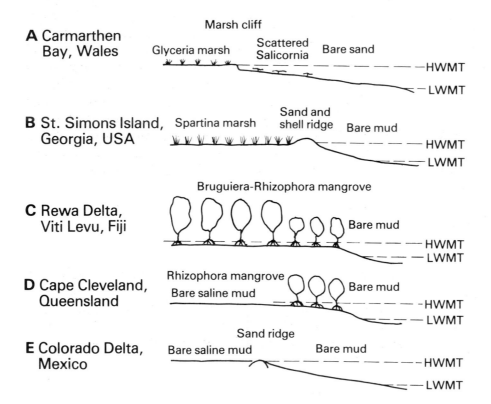

Figure 11.5 Vegetation zones in varying coastal conditions. These profiles are diagrammatic and not to scale. (After Davies, 1972)

Organic landforms: reefs

In the strictest sense reefs are built by marine organisms below sea-level, but they are often associated with island formation where broken debris from the reef is washed into piles which are built above sea-level, and are occasionally (e.g. Barbados) raised by earth movements.

Reefs have been divided into fringing, barrier and atoll varieties (Figure 11.6): this division suggested to Charles Darwin that their origin was linked to the subsidence of volcanic island features. Such an origin has been confirmed for a number of atolls by deep drilling, since volcanic rock has been encountered at over a thousand metres below the surface. At the same time the Quaternary Ice Age lowering of sea-level must have planed off and killed many reefs, followed by re-growth when the sea-level rose again. This is confirmed by the way in which older reefs have been riddled with karstic features due to exposure above sea-level.

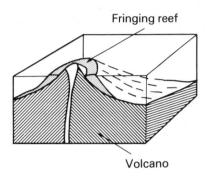 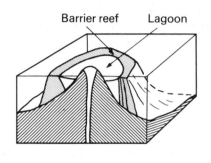 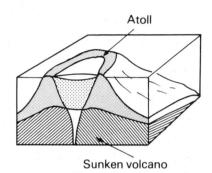

Figure 11.6 Coral reefs. Charles Darwin's interpretation of coral reef development, in which the reef developed around a volcanic island. When the volcano ceased to be active it began to sink. The reefs would build up to sea-level, or cease to exist.

Reefs are formed by a combination of processes, but the fundamental one is the growth of a framework secreted by corals, calcareous algae and carbonate-secreting worms. Other sediment and animal shells accumulate within this framework, forming a massive platform (Figure 11.7). At the present time these are confined to a zone between 30 degrees North and 25 degrees South, since few reef corals survive below 18.5° C, and the optimum temperatures for growth are 25–29° C. Growth is greatest close to the surface due to the importance of algae (plants needing sunlight for photosynthesis) in secreting calcium carbonate, and these algae have a symbiotic relationship with the reef corals. Most rapid growth is around the reef margin on its windward side, since the breaking of the waves allows the organisms to obtain more oxygen, but smaller patch reefs also grow in lagoons away from wave action. The Caribbean reefs have fewer calcareous algae than the Indian or Pacific reefs, and this is related partly to the Quaternary lowering of sea-level, which wreaked greater havoc in the West Indies, and partly due to the raising of the Central American land barrier which cut off the Caribbean from the Indo-Pacific faunal province.

Man: a control in the system

Whilst it is realised increasingly that the biotic factors are important in the origin and modification of landforms, and whilst reefs are organic constructions on a par with medium-scale landforms, man has become by far the greatest agent of change in the development of Earth surface forms. Indeed in an area of slow-working natural processes and of advanced industrial society with pressures on space, like Britain, man is now the most important of all the forces at work in modifying the landforms.

Landforms produced by man

Man's activities have been responsible for the production of highly distinctive landforms of both erosional (excavational) and depositional (constructional) origin. Some would include buildings amongst the man-generated landforms, since they are built essentially from Earth materials which have been processed by man. The large blocks of flats and offices characteristic of city centres certainly have effects on the local climate. But even if a limit is set at the movement of rock and regolith without extensive processing a large number of features are attributable to man.

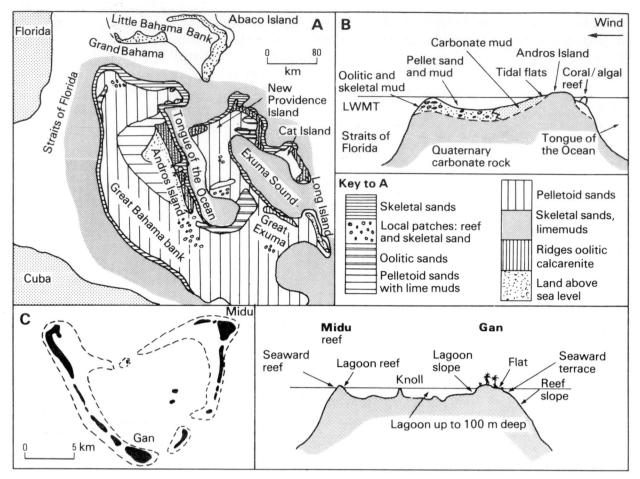

Figure 11.7 Reef features.
(A), (B) The Bahama platform. Notice the position of living reefs; ooliths occur where there is strong current activity in shallow water; finer material accumulates to the west of Andros Island. (After

Blatt, Middleton and Murray, 1972)
(C), (D) Addu atoll, the southernmost of the Maldive Islands, in the Indian Ocean. (After Spencer Davies, Stoddart and Sigee, in Stoddart and Yonge, 1971)

    The forms produced are described as terraces, embankments, mounds (all constructional), cuttings and pits (excavational). These can be related to the range of man's activities.

a) **Farming** in hilly areas has led to the construction of terraces by the rice-growers of south and east Asia; by the ancient Indian civilisations of South America; and by the medieval farmers of Britain (lynchets). The ridge-and-furrow features, so common in the fields of the English Midlands, are a result of the ploughing of heavy soils in the Middle Ages and subsequent 'fossilisation' under permanent pasture.

b) **Building** activity has been responsible for producing many mound forms. These include the ancient burial mounds, tumuli and Iron Age forts found commonly on the Chalklands of southern England, and the 'tells' of certain towns in the Middle East. More recent building activity may result in considerable levelling of a site before houses or factories are erected.

c) **Transport** by road or rail since the mid-nineteenth century has resulted in an increasing number of cuttings and embankments. Until recently these have been confined to narrow zones along the road or railway, but the building of motorways or freeways has led to the involvement of wider swathes of land. Dock construction at ports and along canals has led to the excavation of large basins and the filling of bays.

d) **Mineral extraction** has caused the most extensive alterations, producing both pits and spoil heaps, which may dominate the landscape in particular areas: the china clay workings of St. Austell and the coal tips of every coal-mining area are well-known, but the landscape is

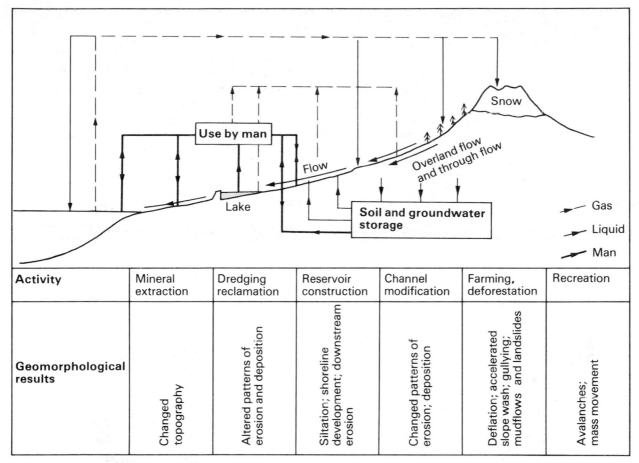

Figure 11.8 Man's effect on sectors of the hydrological cycle. (After Jones, in Brunsden and Doornkamp, 1974)

Activity	Mineral extraction	Dredging reclamation	Reservoir construction	Channel modification	Farming, deforestation	Recreation
Geomorphological results	Changed topography	Altered patterns of erosion and deposition	Siltation; shoreline development; downstream erosion	Changed patterns of erosion; deposition	Deflation; accelerated slope wash; gullying; mudflows and landslides	Avalanches; mass movement

also excavated for huge brick-clay pits around Bedford and Peterborough and for large limestone quarries in Derbyshire around Buxton and Matlock in England, and the gravel pits along most major streams in Britain have replaced land by water-filled depressions. Some of the surface excavations are being filled and the spoil heaps levelled and returned to farming or recreational use, but in either event the shape of the land has been greatly altered. It has been calculated that 300 million tonnes of materials were extracted from the rocks of Great Britain in 1972, and that this affected 2000 hectares of land; in the USA 65 000 hectares were affected in that year and the tonnage extracted was many times greater.

e) Another modern result of man's activities has been due to the increasing scale of **warfare.** The big guns of World War I left the artificially-drained plain of Flanders a morass of shell holes and trenches. Recent bombing of Vietnam has left 22 million bomb craters. Nuclear explosions have removed the whole of an atoll which extended above sea-level.

These features are distinctive because they are not controlled by the processes of the Earth's physical and chemical environment, but by economic and social processes which are man's contribution. Man can carry out a certain degree of erosion or deposition at a predetermined place. The scale of such operations, and of the landforms produced, varies with the stage of technological development: as man became a settled farmer (the last 10 000 years), civilised (the last 5000 years) and industralised (since 1800–1850), so the rate and scale of his influence on landform generation has increased. Before 1800 man was responsible for a few small landforms, but today he can contemplate Operation Ploughshare — the nuclear engineering proposal to blast a new sea-level canal through Central America to parallel the Panama Canal.

176

177

178

179

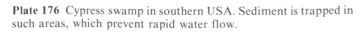

Plate 176 Cypress swamp in southern USA. Sediment is trapped in such areas, which prevent rapid water flow.

Plate 177 A beaver dam in Washington, USA. These also interfere with the flow of water and lead to deposition of sediment.

Plate 178/179 Strip mining of coal in the Appalachians of West Virginia. Notice the method of mining, and its effects on slope processes. (USDASCS)

180

181

182

183

184

Plate 180 Vineyards in California. How has man's use of the land led to such a disastrous effect? (USDASCS)

Plate 181 Gully erosion on a steep slope in California grazed by cattle. (USDASCS)

Plate 182 Eroded pasture in Iowa. (USDASCS)

Plate 183/4 Gullies in South Carolina and the remedy for stabilisation: planting with kudzu. (USDASCS)

Natural processes modified by man's activities

In addition to providing an increasing number of his own landforms, man's activities have affected the location and rate of operation of natural processes. Since man has been confined largely to the humid parts of the world, such effects are most marked in relation to the land–water interactions occurring in fluvial environments (Figure 11.8) and along coasts. As in the case of man-produced landforms, however, the scale of activity has grown from minor, small-scale effects to those of the medium scale. Man is still unable to affect the largest scale of activity, based on the energy generated in the Earth's interior and the global atmospheric circulation. He is attempting to control hurricanes and earthquakes, but such are still termed 'natural hazards' (or 'acts of God', by the insurance companies) and defy his powers of understanding and control. Perhaps, however, it is just as well that man has not yet placed his hands on the opportunity to cause so much alteration of the face of the Earth.

The interactions of man's activities with the natural processes can be grouped into two major categories: engineering and agricultural.

Engineering activities are relatively confined in the area they occupy, and have been particularly intense over the last 100 years. Along the coast dock-building and dredging of shipping channels have often upset the movement of sediment. This has piled up on one side of a harbour entrance (unless removed out to sea by dredgers), whilst the loss of supply on the other side has led to erosion of the coast. Similar effects have been experienced where groynes have been built to protect a seaside resort promenade by trapping sediment between them: the loss of the sediment down-drift has led to erosion. The reclamation of coastal lands has also altered the effects of natural processes, leaving old cliffs miles inland. The Fens of eastern England include over 500 000 hectares reclaimed since 1640, and a large proportion of the Netherlands has been reclaimed from the sea. The reclamation of the Foulness area in the Thames estuary (Figure 10.32) would not only have added more land but would have altered the pattern of sediments in the estuary, transferring the line between mud and sand sedimentation farther out to sea.

Man's need to store water for use in farming, homes, manufacturing and the generation of hydro-electricity has resulted in complete alteration of many drainage basins (Figure 11.9). The building of dams and the impounding of large reservoir lakes has led to changes in water and sediment flow, and also in channel conditions. The river Colorado now has eight major reservoirs along its length, and its waters no longer reach the sea, since a canal channels the last remaining water into irrigation projects. This river used to be the most silt-laden river in existence, but now most of the silt is trapped by the lakes. A number of unfortunate effects have resulted from this wholesale alteration of natural processes: in many cases the reservoir wall rocks are permeable and soft, leading to loss of water and crumbling of the rocks to fill the lake more rapidly; alluviation of the land and widening of stream channels by braiding above the reservoirs has destroyed farming land; and the load added to the Earth's crust by the masses of lake water has resulted in increasing numbers of small earthquakes in their vicinity.

Other effects on drainage basins include the modification of stream channels, which have been deepened, widened or straightened for navigation or to avoid flooding. Irrigation and water supply to homes and factories leads to the abstraction of water from a stream system, whilst field drainage and urbanisation increase the surface runoff after precipitation. These all have the most immediate effects on the nature of the stream channel, which responds to decreases or increases of discharge (Figure 11.10). Many streams are used for waste disposal, and the solute concentration is also increased by the draining of highly soluble fertilisers from fields. When these chemicals become concentrated in a relatively small body of water, like Lake Erie, they can lead to a complete upset of the delicate balance between the supply and utilisation of materials in natural systems. In the case of Lake Erie the phosphorus from fertilisers was increased, releasing its normal restricting effect on the production of plant

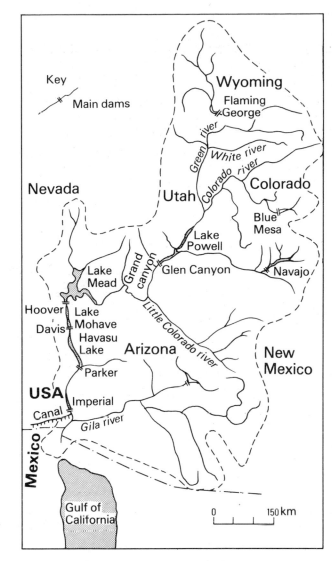

Figure 11.9 The Colorado river system: one of the most used and modified by man. The stream no longer reaches the sea.

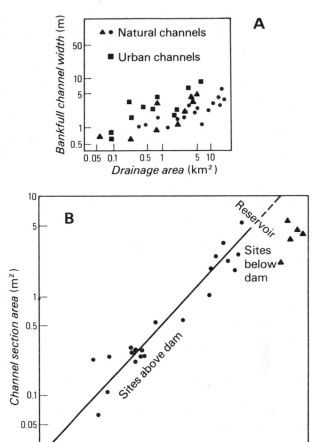

Figure 11.10 Changes in stream channels running through natural and man-modified situations.
(A) Urbanisation and channel form: the channels have increased in width to carry extra runoff. (After Wolman, 1967, in Gregory and Walling, 1973)
(B) Channel changes above and below Burrator reservoir, Dartmoor, Devon. Water is taken from the reservoir for the Plymouth water supply. What is the effect? The present dam was built just before 1900 and raised in 1930, but water has been taken from this section since Sir Francis Drake built a leat to Plymouth in the late 16th century. (After Gregory)

matter. Nitrogen-fixing blue-green algae were able to multiply greatly and the lake became foul because of the quantities of dead, decaying algae: all other life in the lake was killed.

Slope processes also involve water movement, and the concept of slope stability is important for engineers in a variety of roles. In the building of roads and railways, embankments and cuttings are graded according to the nature of the underlying materials: in solid, massive rock a cliff-like form may be allowed, but with clays a gradient of less than 10 degrees is necessary to avoid earth-flows on to the road. There is always a temptation to cut costs by making the gradient as steep as possible, but this may lead to disaster: a section of the Sevenoaks By-pass in Kent required considerable extra expense following the collapse of a cutting, but this was fortunately before it was opened to traffic. A steeper slope may be maintained by rapid grassing of the cutting sides. Similar problems are encountered in the dumping of mine wastes, and the case of the collapse of the spoil heap at Aberfan has led to increased urgency in the understanding of the processes affecting the stability of such masses.

The surface relief is altered by subsidence when materials are removed underground. Extensive areas of coal-mining and salt-extracting areas in the English Midlands and North have subsided by several metres to form basin-shaped areas in which water has accumulated. Subsidence is also caused by liquid extraction: water taken from the lake deposits on which Mexico City is built has caused 9 m subsidence since 1891, when pumping became intensive; and the extraction of oil from the Wilmington field, California, led to subsidence of 60 cm or more over an area of 65 km², reaching a maximum of 9 m in localised areas.

Intense erosion takes place wherever vegetation is removed and bare rock or soil is exposed, a response particularly common on building sites, resulting in the rapid removal of quantities of sediment which choke and pollute local stream channels and destroy land downstream (Figure 11.11). In Montgomery Country, Maryland, USA, more than 3800 tonnes of soil were eroded from a 80 000 m² site on which 89 houses were built (i.e. 3 cm soil from the whole area). Lake Barcroft, a water-supply reservoir in northern Virginia, was eventually abandoned as urban development encroached on its drainage basin. During the peak of development (1961–62) local residents attempted to maintain the lake by dredging, at a cost of £80 000, but had to abandon these efforts as too costly, and 235 000 tonnes of sediment entered the lake (over 10 tonnes for every 1000 m²). Road construction in a tributary basin of the river Potomac contributed 98 per cent of the total sediment supplied by that tributary over the three year period of construction. Measures have been designed by the US Department of Agriculture Soil Conservation Service to reduce the damage caused (Figure 11.12), and these are related closely to a knowledge of slope and stream processes.

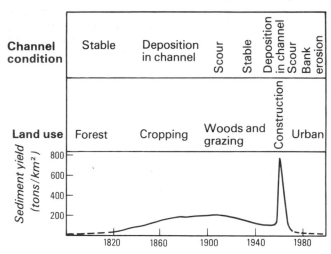

Figure 11.11 Variation of sediment yield over time, based on the experience of the Piedmont region of the eastern USA. Notice the effects of initial farming, settled farming and urbanisation. (After Wolman, 1967, in Gregory and Walling, 1973)

Agricultural activities have affected processes over longer periods of time, and are more widespread in their results. These usually begin with the removal of natural vegetation, especially forest, and replace it with a different type of cover, which may not provide an all-year protection for the ground.

Forest clearance and logging operations cause disruption of the natural systems, often completely clearing the ground. It has been shown that surface hauling of logs leads to particularly rapid erosion (Figure 11.13), and the clearance of timber results in increased runoff (up to nearly 50 per cent) and sediment transport. This effect must have occurred whenever forest has been removed from relatively steep slopes: there is an upper zone of accumulation in many Dartmoor valleys which can be attributed to the clearance of woodland from the moor since the Bronze Age over 4000 years ago. In south-west Wales coarse fluvial terraces in the valley floors have been attributed to the destruction of forest in the early Iron Age, combined with heavy runoff during the wetter climate of that time. Lower valley courses were clogged with sediment, which was incised by subsequent streams. Many parts of the

Effective practices	Significance, comments
1 Select land with favourable drainage, relief and soils.	Part of a whole-community approach to land use, based on geology, hydrology, relief and soil characteristics.
2 Fit the development to the site: make provision for erosion control.	Possible problems evaluated during site examination: how will runoff be disposed? What offsite measures will be necessary? How do roads and plots fit slopes? What will the cost of the measures be?
3 Areas not suited to building may be used for open space, recreation.	Work with local authorities: grants may be available.
4 Large tracts of land are to be developed in small units.	Soil exposure will then be reduced to the minimum.
5 Keep grading to a minimum and retain trees where possible.	Grading — cutting and filling — may be needed to increase usable land. Should always be carried out with protection measures: bench terraces, subsurface drains, runoff diversion.
6 Control runoff in storm sewers.	Runoff also controlled in grassed waterways or netting-lined channels to protect soils.
7 Protect critical areas with mulch or temporary cover crops.	Straw or hay mulch secured by netting and pegged to ground is a quick and effective protection.
8 Construct sediment basins to trap sediment.	Dams built, or basin excavated: such sediment could ruin land downstream. Temporary structure, or permanent for industrial use — graded into landscape.
9 Safe off-site disposal for runoff which increases during construction.	Need to work with local authorities: grants may be available.
10 Establish permanent vegetation to secure erosion control in long run.	Plants to fit surroundings and availability of maintenance: natural grasses, shrubs, trees.

Figure 11.12 The practice of erosion and sediment control on construction sites, as suggested by the US Department of Agriculture Soil Conservation Service. These practices are based on the principles of using land suited to development, of leaving soil bare for the shortest possible time, of reducing the volume of runoff, of retaining sediment on the site, and of releasing runoff slowly.

Y–Clearcut
N
0   800 m
610 m
X–Control
763 m
Z–Patch cut and roads
915 m
1068 m
⌇⌇⌇ Slides, scoured channels
▒ Logged areas

	Suspended load	Bedload	Total load
X	1.0	1.0	1.0
Y	4.2	2.4	3.3
Z	39.0	178.0	109.0

Figure 11.13 The effects of two types of logging practice on Douglas fir forest catchments in Oregon, USA. The patch-cut area, with roads, gave the greatest erosion and sediment yield. (After Fredrikson, 1970, in Gregory and Walling, 1973)

world have been settled more recently, and damage to these may be greater due to the greater speed of clearing and the accompanying effects of deep ploughing. In 1911 it was calculated that in Fairfield County, South Carolina, 36 000 hectares of cultivated land had been turned into rough, gullied land, and another 19 000 hectares of rich valley floor had become swampy meadow: increased runoff had incised the gullies and had led to streams exceeding the capacity of their channels and flooding the lower lands.

When forest cover gives way to farming a variety of different cropping systems, techniques and organisational systems are possible. There may be an emphasis on arable crops, which

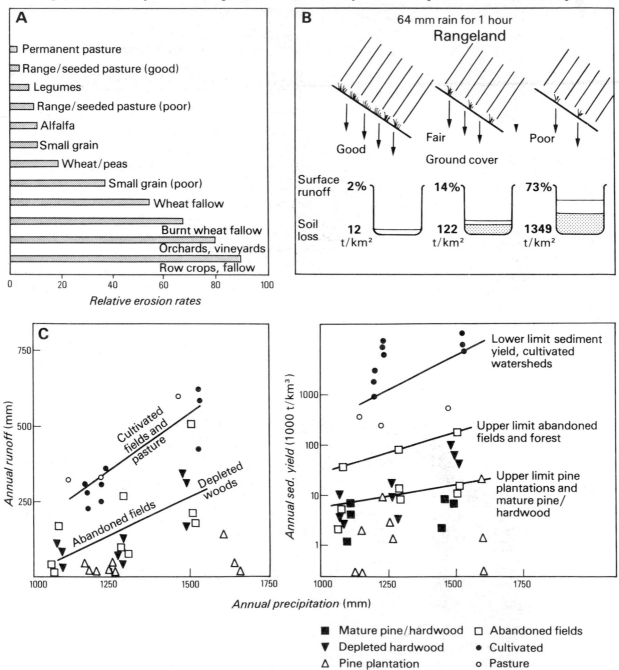

Figure 11.14 Farming and sediment yield.
(A) Relative erosion rates with different crops in the Pacific north-west of the USA. (After Brown, 1950)
(B) The effect of different grass covers in Utah rangelands. (After Noble, 1965)
(C) Runoff and sediment yield in northern Mississippi, USA. (After Ursis and Dendy, 1965) (All in Gregory and Walling, 1973)

Area	Sediment yields (m³/km² yr)	
	Forested	Cultivated
Mbeya Range, Tanzania	6.9	29.5
Cameron Hills, Malaysia	21.1	103.1
Tjiloetoeng, Java	900.0	1900.0
Barron, Queensland	5.7	13.6
Millstream, Queensland	6.2	12.3
Northern Range, Trinidad	1.8	16.0
Apiodoume, Ivory Coast	97.0	1700.0

Figure 11.15 The effects of farming on erosion in the tropics: a comparison of sediment yield from forested and cultivated area. (After Douglas, 1969, in Gregory and Walling, 1973)

may cover the ground closely or leave bare earth between; tree crops may provide a cover similar to the previous forest, or may have wide, bare zones between; or there may be a concentration on pastoral farming with a virtually permanent grass cover. Arable farming also leads to the removal of the crop at harvest, and thus to the removal of the mineral nutrients instead of their being re-cycled naturally — although fertilisers may be used to return some of these materials. There is then a period when the soil is bare and ploughed into easily-transported particles: it becomes susceptible to erosion by water and wind. Farmers with family holdings which are to be handed on to successive generations will take care to crop the land intensively and to avoid erosion, whilst a tenant farmer will attempt to get quick returns in order to make a profit above the rental costs. Farming always leads to increased surface runoff and erosion of slopes, plus the alluviation of lower lands, compared with a forest cover (Figure 11.14), being most drastic in the tropics (Figure 11.15).

Pastoral farming results in different changes, but these can also have unfortunate effects. In both temperate grasslands (steppes, prairies) and in the tropical savannas, the combination of grazing and fires (many of which have been set off purposely by man) has led to the retreat of woodland, increased surface runoff and erosion. In cases of overgrazing the grass cover has been degraded until only bunch grasses remain, so that erosion of the bare ground between occurs and the underlying rock may be exposed. Many of the Mediterranean lands have suffered from the removal of woodland, followed by careless farming methods (Figure 11.16).

Implications

The study of man's effect on the natural environment in this way has important implications. implications.

1) It once again emphasises the need to abandon the cycle concept of W.M. Davis (chapters 6 and 7), since that treats a particular process as an entity, closed to outside interference and the changing emphases of external variables.

Culture	Farming activity	Vegetation cover
Pre-Greek	Grains, herd animals in balanced farming	Woodland
Roman	Intensive grain farming; less balanced	Farming
Berber	Merino sheep; land deteriorates	Scrub
Modern	Sheep and goats; scattered food growing; deterioration continues	Herbs and grasses

Figure 11.16 The effects of man's activities in removing woodland and adversely altering the whole biosphere. This is not something new! It is the rate of change which has accelerated the problem and forced man to recognise it in the last few years. (After Tivy, 1971)

2) It brings together different aspects of geographical study in seeing man-environment relationships as a two-way interaction. This leads to a more realistic study of problems of soil conservation, slope stability, natural hazard control and damage to the environment caused by excessive use of a location for leisure activities.

3) This also provides a better opportunity for the future prediction of the effects of man's activities on his delicately balanced environment with its finite resources.

Stream processes in urban areas

One of the most significant results of the extension of urban areas is that the increase of houses, factories and roads with well-organised systems of surface drainage means that a greater proportion of rainfall input to a basin runs off directly through storm sewers to rivers. Such systems have to be designed to relate to the possibility of flooding, and most authorities attempt to estimate the level of a 30-year flood and design their prevention works accordingly. Unfortunately there is also a tendency for urban rubbish to accumulate in these channels leading to floods.

A large proportion of south-eastern London is drained by a series of small streams to the river Ravensbourne and thence to the river Thames (Figure 1). Serious flooding was common, especially in the lower reaches, and an Act of Parliament was passed in 1961 so that engineers could work on particular bottlenecks to ease the flow. The aim was to provide a channel which would take 30-year floods, and work began near the mouth. Twelve cars were dredged from Deptford Creek, and new concrete channelling was installed there, as well as in places like Ladywell Recreation Ground, where the original channel would take only 11 cumecs (as against a 30-year flood discharge at this point estimated to be over 50 cumecs). In spite of these works extensive flooding occurred again in 1968 due to exceptional rains: this was estimated to be a flood of 300 years magnitude. The river banks were covered to 2 m in places, but the lowest (improved) section at Deptford did not experience flooding. On the other hand there have been complaints from local ratepayers, who see a trickle of water passing through a wide concrete channel for much of the time: they feel that too much leeway has been provided for:

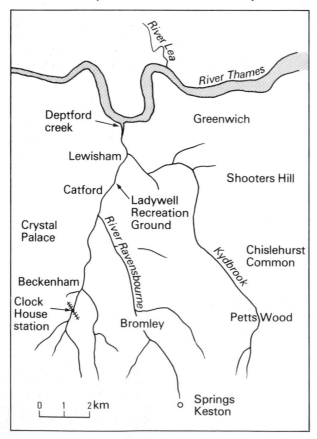

Figure 1 The Ravensbourne basin in south-east London. It is approximately 180 km² in area, and there are 75 km of streams. Points where flooding was common are indicated.

Part III

A Complexity of Landscapes

Introduction

Part III brings together the various complex elements responsible for landform origins — processes, geological conditions, past changes — in two groups of studies. In chapter 12, a variety of landscapes in the British Isles are analysed in terms of their general character, underlying geology, landforms and recent denudational history. An attempt is made to bring together these various factors which have worked to produce such a range of landscapes in a humid temperate region of the world.

Chapter 13 analyses desert landforms, studies of which have resulted in much discussion. The debates now take on a further aspect, since these regions are critical for a deeper understanding of man's environment and his use of it. The final chapter of the book considers the changing interests of geomorphologists, those scientists who study landforms, and brings together a number of strands which have emerged during the course of the book.

Climatic change and relict landforms

Relict landforms is a major concept which has to be considered in the study of any specific area. This is related closely to the view that virtually every part of the world has been affected by major changes of climate during the Quaternary Ice Ages. The existence of formerly glaciated valleys in the uplands of humid temperate regions today; the abandoned shorelines and deposits of once extensive lakes surrounding the present shrunken remnants in arid areas; the unoccupied drainage systems of deserts throughout the world; the areas of sand dunes now stabilised beneath the vegetation of northern Nigeria and Nebraska; and the presence of vast spreads of mechanically weathered rock debris at low elevations in the humid tropics can all be regarded as the relics of former climatic conditions in these areas.

As the Quaternary climatic belts moved to north and south, most areas would experience changes of some sort, but some more than others. Zones situated centrally in present climatic belts might be affected little, and remain as refuges of the same conditions for much longer periods; zones on the boundaries between belts would be subject to more frequent and greater changes (Figure 1). The central parts of the Amazon Basin, for instance, probably experienced high

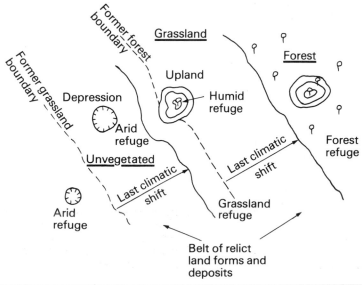

Figure 1 The displacement of climatic zones, relict landforms and refuges. Upland areas tend to be more humid; depressions to be more arid. (After Garner, 1974)

levels of all-year rainfall throughout this period, and continuing high temperatures, whilst the margins of that basin have angular rock deposits which suggest a more arid environment in the recent past. The sand dunes of northern Nigeria, and of the Sahel zone generally, which are now being exposed again to wind action, are in a boundary zone where there is reaction to only slight fluctuations. Even in the British Isles there are gradations in the degree of glacial effects from lowland to highland and from south to north.

A further aspect of the significance of relict features relates to the slow uplift of mountain ranges. It has been suggested (Garner, 1974) that ranges emerging within 35 to 40 degrees of the Equator are subjected initially to arid conditions, giving rise to typical landforms showing extensive horizontal planation. Eventually uplift takes the ranges into a humid realm, and this leads to the development of drainage nets on the flanks and also cutting across the now relict arid features. As uplift pushes the ranges higher they enter a zone where the moisture diminishes, giving a further arid zone with internal drainage and sediment accumulation in the depressions (e.g. the former river valleys). Finally the ranges may be pushed upwards into the frost zone above the snowline, where glacial forms may be generated on the same rocks which have experienced several changes of climate during uplift (Figure 2).

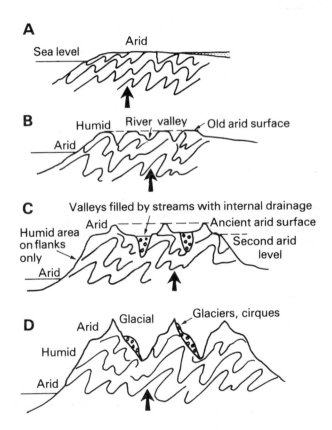

Figure 2 A sequence of landforms developed in different climatic zones as a mountain range is raised. The Peruvian Andes are taken as a basis for this model. (After Garner, 1974)

Once mountain ranges are established, they are perpetuated by isostatic effects and by the subduction of ocean floor beneath them. Isostatic compensation results in a loss of height amounting to only 20 m when 220 m of rock is removed from the peaks. Streams and glacial activity produce intensified relief: the peaks are scarcely lowered, but valleys are gouged out. Eventually, however, mountains will be reduced in height, with a consequent change of climatic influence in reverse order to that suggested for the rising mountain range. This would finish, over a major part of the world, with the extensive horizontal surfaces of aridity. It may be these which are reflected in geological unconformities, rather than any low-lying landscape (i.e. peneplain) resulting from other forms of erosive activity.

12

Landscapes of the British Isles

Although the British Isles constitute but a tiny part of the world's land surface, historical and other factors have combined to make them the home of much early growth in the geological and geomorphological sciences, as well as in several major developments during more recent years. These factors include developments in the wider fields of science, related at first to a desire to understand God's creation, and later to the changes of the industrial revolution — which began in these islands. They also include the fact that so many varieties of landforms and geological phenomena occur within such a small area, that there has always been a source of accessible examples which could readily be compared and contrasted with each other.

To choose five examples of regions with distinctive landform assemblages is not easy, even within such a small compass. Those included here could be substituted by others: it is hoped that they will serve to some extent as a basis for the analysis of other regions. The five chosen (Figure 12.1) represent as wide a range of conditions as occur in these islands, since there are no high mountains, glaciers or deserts.

Figure 12.1 Five areas of the British Isles chosen as a basis for short regional studies.

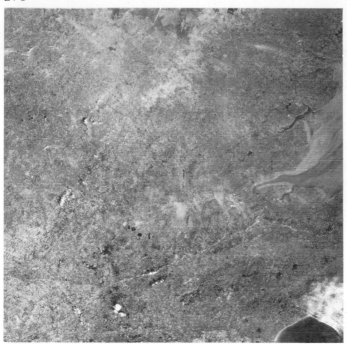

185

186

187

188

Plate 185 LANDSAT view of south-eastern England taken on 8 March 1973. Notice the way in which geological features (e.g. North Downs, central Weald), the Thames estuary, and the reservoirs and smoke of London show up. (USGS)

Plate 186 The South Downs and Arun Gap, Sussex.

Plate 187 The sandstone ridge in the north-west corner of the Weald: Leith Hill (320 m) is the highest point on the skyline. The lowland is on Weald clay.

Plate 188 Chalk cliffs and shore platform near the mouth of the river Cuckmere in Sussex. The cliff top has a deep zone of chalk fractured by frost action in the past.

(186–188 Institute of Geological Sciences, Crown Copyright Reserved)

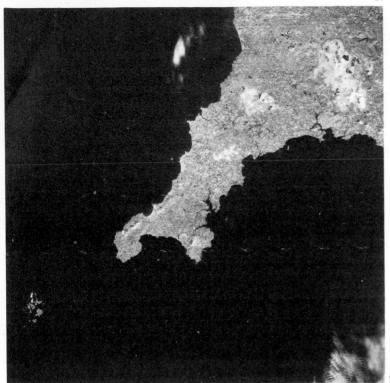

189

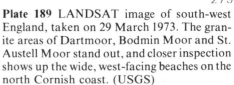

Plate 189 LANDSAT image of south-west England, taken on 29 March 1973. The granite areas of Dartmoor, Bodmin Moor and St. Austell Moor stand out, and closer inspection shows up the wide, west-facing beaches on the north Cornish coast. (USGS)

Plate 190 The north Devon coast at Westward Ho. The rocky shore platform has faults crossing it and immediately to the south is a raised beach, on which most of the settlement is built. Old cliffs are farther inland. (Air Ministry, Crown Copyright Reserved)

Plate 191 Duckpool Mouth, on the north Cornish coast. Notice the steep-sided valley, underfit stream and flat plateau tops typical of the area. (Air Ministry, Crown Copyright Reserved)

Plate 192 Rapid erosion of the west-facing coast north of Bude, north Cornwall, has resulted in this hanging valley feature. Notice the folding of the rocks shown in the shore platform. (Air Ministry, Crown Copyright Reserved)

190

191

192

South-east England

This is a lowland region (no part higher than 310 m) of relatively young rocks (Cretaceous and Tertiary age, formed 130–30 million years ago), enclosed within the Chalk ridge framework from the Chiltern Hills southwards. It includes Salisbury Plain, the Hampshire and London basins, and the Weald. The dominant landforms are escarpments and vales broken by wide stream valley floors, and the coastal features include moderately high cliffs (30–100 m) where the escarpment-forming rocks are attacked by the sea, and extensive low-lying stretches between.

Rock-landform relationships

The escarpment and vale relief of this area reflects a close relationship between the landforms and the underlying geology. The Cretaceous rocks are alternations of sandstones, clays and

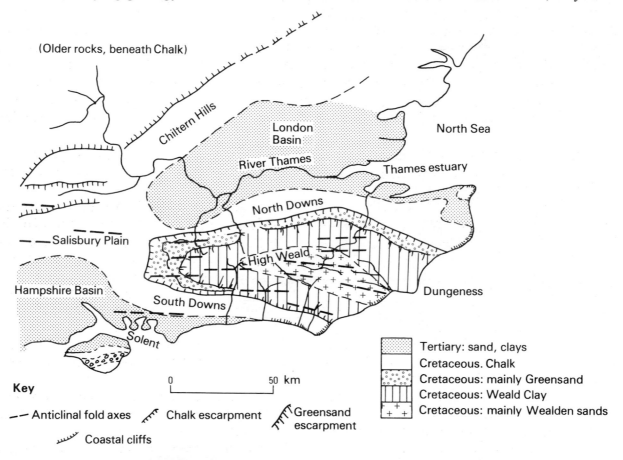

Figure 12.2 South-eastern England: the major features of the relief related to the underlying geology.

Chalk, whilst the overlying Tertiary rocks are mainly clays with some partly consolidated sands (Figure 12.2). South of London these rocks are deformed in open folds, increasing in amplitude to the south as the Cretaceous rocks become thicker. The folds are of periclinal form (i.e. upturned boat shape, dying out at each end), and are arranged 'en echelon' in east-to-west orientations. Erosion of the fold crests has resulted in the formation of escarpments on the dipping Chalk and sandstones, as vales have been eroded on the clays. This is particularly evident in the Weald area, but also occurs on the western margins of the Salisbury Plain (e.g. Vales of Wylye, Wardour and Pewsey). The London and Hampshire basins form areas of lowland on Tertiary rocks, where there is little internal differentiation of relief on the basis of rock-type.

River valleys are floored with thick alluvium, and often have paired terraces along their sides (Figures 7.39 and 7.40). Some of the escarpments are capped with gravels, and clay-with-flints is common on the Chalk, being a residual product from the solution of Chalk over several million years.

Processes at work and the evolution of landforms

South-east England also demonstrates how an area may preserve evidence of a constantly revived landscape related to changing processes and base-levels. The nature of the contrasting rock-types helps to show this aspect to best advantage.

At present running water and soil creep are the main processes affecting the landforms, but their efficacy is in doubt. Many of the major streams are manifestly underfit (Figure 7.42), few flow across bedrock, and some are canalised by man. All these points suggest that little erosion or transport can be accomplished. The load of the streams is largely in solute form or as mud, and they flow through wide valleys with gentle slopes and low gradient long profiles. The Chalk escarpments and Salisbury Plain seldom experience any surface flow of water today, although they are crossed by wide-spaced networks of dry valleys. Dry valleys are also found on the sandstone ridges. The clay vales have higher densities of surface drainage, but these streams are often small in an area where annual precipitation inputs are 600–900 mm. In the central Weald the drainage network is related closely to the fold patterns in the rocks (Figure 12.2), but around this zone the streams flow almost radially, cutting across the folds and escarpment ridges.

The London and Hampshire basins are drained by rivers into the Thames and Solent estuaries, which are both wide with low-lying shores. Mud brought by the rivers is deposited as the fresh water mingles with the sea water.

The coasts of south-eastern England include some cliffed areas, such as the section of the Sussex coast between Brighton and Eastbourne, and of the east Kent coast at Dover and in the Isle of Thanet, in all of which the Chalk forms vertical cliffs 30 – 100 m high, rising from wide shore platforms. At Hastings the Lower Cretaceous sandstones also form cliffs, but elsewhere the coasts are low-lying along the estuaries and where clay horizons have a coastal outcrop. Reclaimed marsh (e.g. Pevensey and Romney) and Dungeness Foreland (Figure 10.31) are features of these lowland coasts. The Thames estuary is so low-lying that flooding is a real danger: a barrage is to be built across the mouth with high banks below to contain this hazard. The present coasts thus exhibit a combination of erosion and deposition, although it is not certain as to how far the coastal landforms can all be attributed to the present phase of sea-level at this height (Figure 8.21).

The study of coastal and fluvial features in south-eastern England leads one to conclude that the streams and the sea have not had a very great influence on the origin of landforms in the area. A large proportion of the landforms must therefore be relics of past conditions, whether of a different climate or changed base-level nature. If the present streams are ineffective, there is also little evidence for the existence of ice in the area during the Quaternary Ice Age. Although some would suggest that ice covered parts of this region and occupied the English Channel floor as far south as the French coast, the well-established southernmost margin is in the north of the London basin at Watford and Finchley, where the path of the river Thames was deflected to a more southerly course (Figure 12.3). Farther south there is evidence that a tundra climate prevailed, rounding the valley slopes and giving rise to solifluction deposits including 'head' composed of angular, frost-shattered rock debris. In addition the Coombe Rock deposits flowed from hollows along the north-facing scarp slope of the South Downs under such conditions. Some valley floors have a thin, silty deposit, known as brickearth, and this has been identified as loess, which may also have been associated with tundra conditions (Figures 9.2 and 9.9).

Two features of south-eastern England present a particular challenge to the interpreter of landforms. One is the evidence for higher sea-levels and the drainage of streams to different

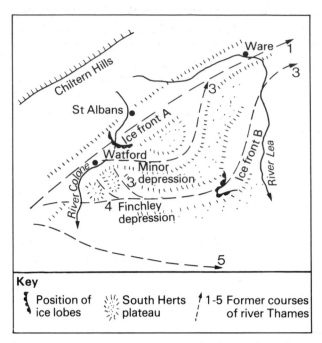

Figure 12.3 Successive courses of the river Thames. Relate these to the positions of ice lobes. (After Wooldridge and Linton, 1955)

levels than at present, and the other is the fact that many of the major streams flow discordantly to the structure — in short, that a river will cut straight across a fold in the rocks (Figure 12.2).

Deposits on the North Downs at Netley Heath show that the sea-level was 300 m above the present at the beginning of the Quaternary: the flints in the clay-with-flints deposit have been rounded like cobbles on a modern beach, and pockets of sand contain fossils of early Quaternary date. River terraces along the valley sides (Figure 7.39) are relict valley floors related to higher base-levels, and have been left high and dry as the streams have cut down their valley floors in accord with the falling sea-level. Such terraces along the Thames valley have provided the site on which central London has been built: the main levels are at 25 m and 10 m, but there are others of smaller extent, and remnants of higher levels at 60 m and 130 m are found capping hills like Wimbledon Common and Highgate. The valley floors are not even simple food plains, since they are filled deeply with alluvium. Thus the Sussex rivers draining to the English channel have 30 m of alluvium in their valley floors — a result of deposition by the river after it had cut a deep valley in relation to a sea-level which was over 30 m below the present! Such changes in sea-level can be explained by the glacio-eustatic theory (page 150), together with the more debatable idea that sea-level has been falling generally with the glacio-eustatic changes superimposed on this pattern (Figure 8.21).

The discordance between drainage and structure has resulted in two major explanations. For some years the idea that the rivers were superimposed across structures held sway, following earlier suggestions that the rivers were initiated before folding and uplift and maintained their courses during the upheaval — i.e. were antecedent. Both of these views were related to a fairly rapid, once-for-all concept of uplift, which was seen to have taken place in the middle Tertiary period, followed by erosion, interrupted only by the changes in sea-level. It seemed more reasonable to accept the superimposition hypotheses at this stage, because the supposed rapidity of uplift would not allow the streams to maintain their courses. The last few years have seen the emergence of a new approach to the evolution of fold structures in the rocks, which lays stress on the long time required to produce them. In the case of south-eastern England it is now suggested that the Weald began to be uplifted as a broad dome in the late Cretaceous, restricting early Tertiary sediments to the London and Hampshire basins, and determining the main drainage lines which flowed north and south towards the proto-Thames and proto-Solent rivers. Later in the Tertiary further stresses gave rise to the smaller amplitude, periclinal folds in the Salisbury Plain and Wealden areas, and such movements continued to have effect until at least the end of the Tertiary (Pliocene). The streams draining north to the Thames and south to the Solent maintained their courses during this slow process of uplift, and in this view should be regarded as antecedent.

With either hypothesis the result has been to produce a sequence of landforms relating to incision by the streams in response to uplift of the land and/or the lowering of the sea-level. The complex interactions of geological structure, earth movements, varied processes and their changes through time are all illustrated well from studies in this region.

South-west England

Unless the suggestions that the ice sheets once covered the entire British Isles are correct, south-western England is an area, like the south-east, which has been continuously exposed to fluvial and periglacial processes over the last few million years. It provides a contrast, however, since the underlying rocks are older and do not contain such clearly-marked contrasts in resistance to erosion as those in the south-east. Whilst some similarities can be detected in the events leading to the formation of the present landforms in both these areas, there are also many distinctive features in both. The south-west peninsula is an area of moderate relief with dominant plateau surfaces at 100–250 m broken only by the areas of moorland rising above to 300–700 m and steeply incised valleys below.

The rocks and their relation to the landforms
The most marked contrast in the rock-relief relationships is between the New Red Sandstone and younger rocks occurring cast of a line from Torbay northwards through Exeter, and the older rocks to the west (Figure 12.4). The younger rocks include sandstones, clays and Chalk,

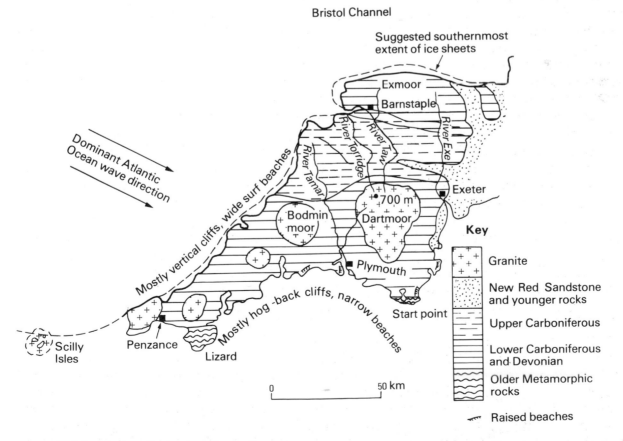

Figure 12.4 South-west England: major features.

similar in many ways to the rocks of south-eastern England, and are likewise associated with a modified form of scarp-and-vale topography. The Devonian and Lower Carboniferous rocks are mostly slates, with some masses of limestone (e.g. at Plymouth and Torbay), volcanic rocks and small intrusions of dark igneous rock (dolerite) known locally as 'elvan'. In central Devon the Upper Carboniferous rocks are mixtures of greywacke sandstone and shale (the 'flysch' association). In the north Exmoor is formed by Devonian sandstones and slates. Dartmoor, Bodmin Moor, and a series of other granite masses along the length of Cornwall, are the offshoots of a large granite intrusion into these older rocks, which took place some 280 million years ago (radiometric date). The combination of these sedimentary rock types, which have been tightly folded, and the granite intrusions, suggests that the rocks of this area once formed part of fold mountain ranges (chapter 3). The compressional stresses which gave rise to the folding also metamorphosed the rocks: the former muds and clays especially show signs of this and are now hardened and cleaved slates. The uplift of the area followed and was accompanied by granite intrusion, which heated the surrounding rocks locally and gave rise to further, but more local metamorphism. Uplift was followed by erosion, and deposition of the rock fragments in spreads of alluvial fan type features around the upland margins (seen today in the cliffs of Torbay and Dawlish). These events occurred over 200 million years ago, however, and were followed by long periods of erosion during which the granite became exposed at the surface. This is thought to have occurred at least by the Cretaceous (but could have happened earlier), since rocks of that age contain distinctive minerals derived from the granite. During the Tertiary phase of earth movements, which gave rise to the folding in south-east England, this rigid block (Figure 3.4) in the south-west reacted to the stresses by fracturing and the faulting produced a series of blocks bounded by north-west to south-east trending faults. The best known of these is the Sticklepath fault, which offsets the eastern part of Dartmoor (Figure 3.13), and is involved in the deep rift basin of Bovey Tracey at its south-eastern end. It has been suggested that this fault continued to move during the Quaternary, offsetting gravel deposits of that age.

The most spectacular effects of the rock characteristics are seen on the cliffed coasts of Devon and Cornwall, where small contrasts in resistance to erosion may differentiate between headland and bay (Figure 10.19 A). In Torbay the two headlands are formed of Devonian limestone, whilst the head of the bay has outcrops of less resistant Devonian slates and New Red Sandstone, crossed by faults which introduce lines of weakness. Inland the relief is dominated by the plateau surfaces, which cut across rock structures and any differences in rock type. Many areas are underlain by monotonous slates and sandstones. The granite moors, such as Dartmoor and Bodmin Moor, stand above the general plateau level. This may be because the granite is more resistant than the surrounding rocks, or because the low density of granite causes these areas to retain a high place in the relief. It has also been suggested that Exmoor, another high area, but formed of sandstone and slate, may owe its higher relief to an underthrust mass of low density sedimentary rocks. Certainly the investigation of the geological background to the origin of landforms in south-western England suggests that earth movements have not left such a static foundation as once was envisaged. Even in an area where the major mountains were raised and eroded away millions of years ago there is evidence of quite recent fault movements.

The processes at work and the evolution of landforms
The streams flowing across the older rock areas of the south-west occur mostly in valleys deeply incised into the plateaus. The granite moors are characterised by wider and more open valleys except around some of their margins where the incision has advanced, but Exmoor has deep valleys into its interior. The incised valleys commonly have a meandering pattern, particularly in their lower sections, but the streams are clearly underfit, as in south-east England, although the mean annual precipitation totals reach 1000–2000 mm in the south-

west. Runoff is generally swift, and, whilst some water may be stored in moorland peat, the soils and rocks are impermeable and the water from a heavy storm works its way through the stream systems within hours of falling.

The south-west peninsula has a long coastline. The west-facing coasts are amongst the highest energy coasts in the world, with vertical cliffs and wide surfing beaches (Figures 10.14, 10.21 and 10.25). These contrast with the 'hog-back' cliffs and narrower, shelving beaches of the more sheltered southern coasts. Spits and extensive sand dune accumulations occur at the mouths of the wider estuaries, as at Dawlish Warren (Figure 10.31) and Braunton Burrows, but many river mouths are narrower ria forms. The occurrence of such rias around Devon and Cornwall has tempted the classifiers of coasts to designate the whole area as having a submerged coastline, but this is too simple a view unless one resorts to a very narrow time-scale in the explanation of landforms. The most recent, postglacial, event has resulted in melting ice and rising sea-levels which have drowned valley mouths. But many rias have raised beaches around their margins (signifying higher former sea-levels), and the incised valleys were carved in response to a falling sea-level. The rias must be seen as a part of the whole landscape.

If glaciation did not affect this area, periglacial processes certainly left their mark. On the higher areas, like Dartmoor, the slopes are littered with huge boulders of granite up to 2 m in length, and the hills are often capped with tors, the origin of which has been linked to such conditions (Figure 5.12). Freeze-thaw action and solifluction must have affected all slopes, and coastal sections reveal varying thicknesses of 'head'. The tin gravels of the moorland valleys were also produced at this time. Some of the raised beach deposits (Figure 12.5) preserve a series of deposits which enable the climatic fluctuations of the later phases of the Quaternary Ice Age to be reconstructed. Apart from these coastal sections, however, few widely accepted tills or glacial landforms have been recognised in the south-west. Perhaps the ice came nearest as sea ice, floating up to the coasts of Devon and Cornwall and dropping the erratics on Saunton beach.

There are three main elements to the inland landscapes (Figure 12.6).

a) The **moorland areas** have been seen as the oldest relict landscapes, with their summits related to ancient erosion levels (e.g. Davisian peneplains), and their wide valleys regarded as the subsequent stage in erosion. This may be the case, but another view would suggest that the relief of, say, Dartmoor is very close to the original roof of the granite mass, and that little erosion of the granite has taken place. There are places, like Leusdon Common, where altered country rocks rest on the granite upper surface with a very low angle contact, and where the

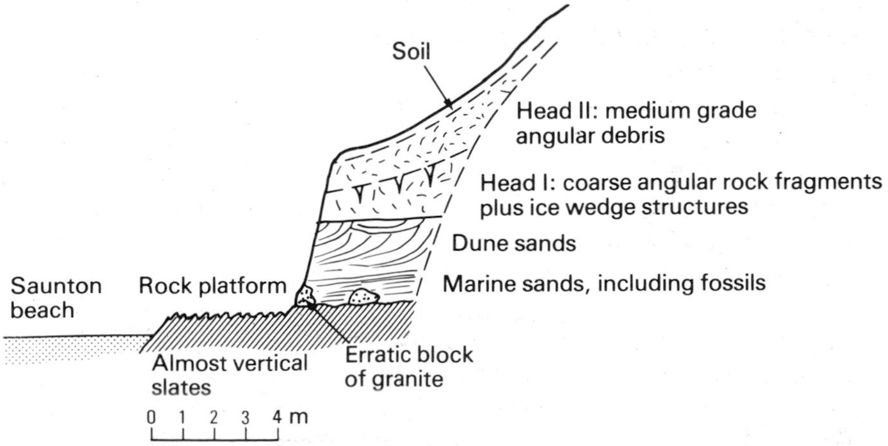

Figure 12.5 A series of events recorded in a cliff section at Saunton in north Devon. Notice the succession of different processes which have been in action.
1 Formation of the rock platform by marine erosion.
2 Deposition of erratic blocks, possibly by sea ice.
3. Formation of the marine sands during a warmer (interglacial) phase of higher sea-level.
4 Dunes formed by the marine sands as the sea-level fell in a succeeding cold phase.
5 Head formed by periglacial processes in a cold phase, sludging down over the dune sands.
6 Second layer of head, possibly formed after a further warm phase had intervened.
7 The present slope, soil and cliff profile formed.

Soil
Head II: medium grade angular debris
Head I: coarse angular rock fragments plus ice wedge structures
Dune sands
Marine sands, including fossils
Saunton beach
Rock platform
Almost vertical slates
Erratic block of granite
0 1 2 3 4 m

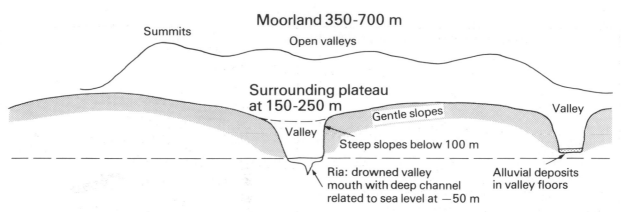

Figure 12.6 The elements of relief in the South-west peninsula. They were probably formed in the following order.

1 The moorland areas stand above the surrounding plateaus. Valleys in the moors are open and relatively shallow, except at the margins.

2 The surrounding plateaus have accordant summit levels over extensive areas, and gentle slopes below. The landscape suggested by the pecked lines may have been the immediately preglacial state.

3 The valleys have steep slopes and are incised in narrow gorges. Their floors are alluviated or drowned. Beneath sea-level there is a deep channel related to sea-levels lower than that of today. Much of this pattern has been formed during the Quaternary Ice Age.

granite roof must have been close to the present relief.

b) The **plateau areas** surrounding the moors, and stretching to the cliffed coasts, may have been formed over a long period of time, and the Quaternary movements along the major faults may have been largely horizontal as tear faults, although some vertical differences in height can be detected as well on either side of the Sticklepath Fault. These plateau levels may have been formed by a combination of subaerial and submarine processes acting on the faulted blocks uplifted by earth movements.

c) The **steeply incised river valleys** reflect a phase of rapid erosion, possibly related to the downcutting experienced in south-east England. Whether this was all due to the falling Quaternary sea-level, or whether there was some local upward movement of the land as well, is difficult to determine, but the overall effect was to carve deep valleys to over 50 m below the present sea-level. This is shown by the deep channels used by shipping in the rias and estuaries. The valleys of the south-west give the appearance of being consistently incised more than those of the south-east due to the smaller degree of differences in rock resistance in the former area.

South-west England is thus another region where a complex interplay of processes has operated. A full understanding of all the forces at work is not yet available, but the overall picture is gradually becoming clearer.

Norfolk

The northern part of East Anglia is a low-lying (below 130 m) part of eastern England (Figure 12.7). Precipitation is low, evaporation rates are high in summer, and present fluvial processes are not very active in an area of low slopes and low runoff. Much of the land area is a relict of the last stages of the Quaternary Ice Age, since its deposits mantle the solid rocks and scarcely permit them to have a surface outcrop, apart from an occasional coastal section. This is a good area for studying a formerly glaciated lowland area on the margin of the last major advance of the ice, for investigating coastal processes, and for evaluating the effects of man's activities on the landscape.

The underlying rocks

The rocks beneath the glacial deposits are those typical of south-eastern England. The oldest are in the west, where Upper Jurassic clays underlie the Fens. Lower Cretaceous sands around

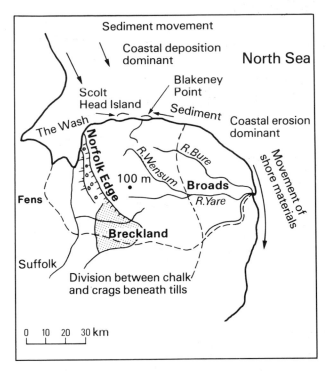

Figure 12.7 Norfolk: the main relief features of the county.

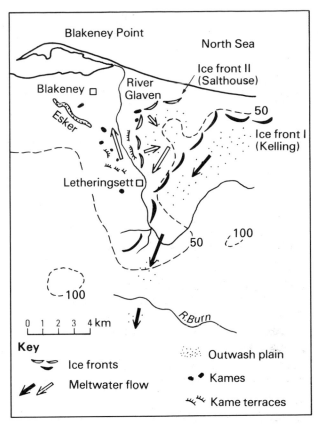

Figure 12.8 Ice marginal features in north Norfolk, related to two fronts of the Devensian advance. The kames and kame terraces represent only a few of those present in the Glaven valley. (After Sparks and West, 1964, in Sparks and West, 1972)

Sandringham, and the Upper Cretaceous Chalk, form the highest parts of the county, which slopes towards the east coast. Here the youngest rocks have occasional outcrops in the low cliffs: they are poorly consolidated sands and shell-bank sediments, known as 'crags', having been deposited just before the Quaternary Ice Age. The crags provide evidence that this part of Britain was part of the subsiding North Sea trough at this stage, since their base is over 70 m below sea-level in the extreme east, and also that the climate was becoming cooler (Figure 8.19). Apart from such contributions to the scenery and to an understanding of the preglacial events in the area, the solid rocks of Norfolk play little part in determining the detailed relief of the area.

The glaciation of a lowland area

The Quaternary Ice Age witnessed a series of advances and retreats of the ice sheets (chapter 8). In England they advanced at least to the Thames valley, and thus covered East Anglia at this maximum stage. It has become conventional to suggest that two separate phases of ice advance (Anglian and Wolstonian) had such an extent, but recent mapping by the Institute of Geological Sciences in the southern part of East Anglia suggests that all the drift in the area has resulted from deposition in the Wolstonian advance. It is never easy to interpret the differences between the deposits left by separate ice advances, since later events would probably rework the unconsolidated debris left by earlier ice, and studies of the most recent, Devensian, ice advance in Britain show that there were fluctuations within that event. Certainly the older tills cover most of Norfolk, incorporating much local material to give them a local sandy or chalky character.

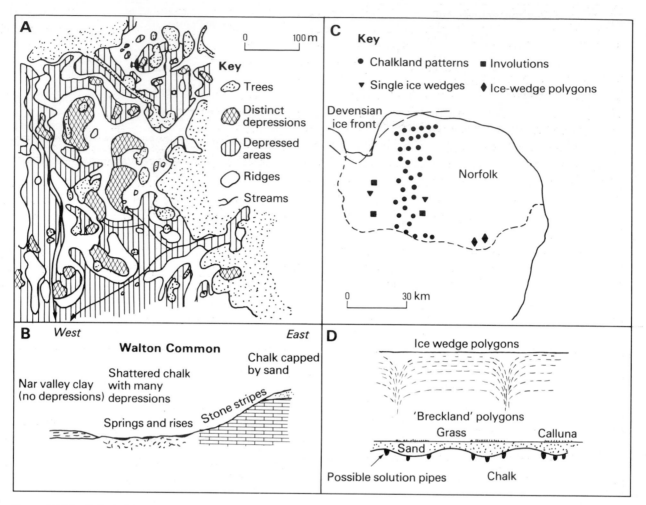

Figure 12.9 Periglacial features in Norfolk.
(A), (B) Ice mound features on Walton Common, near Kings Lynn (After Sparks, Williams and Davies)
(C), (D) Distribution of patterned ground features.

The Chalkland patterns are unsorted polygons of the 'Breckland' type, resulting in different soil colours or vegetation associations. (After West, 1968) (All in Sparks and West, 1972)

The Devensian ice sheet advanced down the east coast of England, bringing erratic blocks from the north, and only just reached the northernmost coasts of Norfolk, resulting in the Hunstanton brown boulder clay, and a series of ice marginal features in north and north-west Norfolk. Outwash fans and dead-ice outwash are found immediately south of the former ice front (Figure 12.8), and ice mounds, patterned ground and thermokarst hollows are characteristic of the Breckland sand area (Figure 12.9), itself probably a product of periglacial aeolian activity (polished stones and ventifact pebbles testify to the aeolian origin of the sands). Many of the dry valleys may also be traced back to an origin during periglacial conditions, when the ground was frozen.

The river valleys of Norfolk contain broad, low terraces, and former lake deposits on these have been dated as being Ipswichian in age (i.e. formed during the last interglacial period). During this phase the glacial deposits were re-distributed by fluvial processes, and plants and animals returned from the south. Most of the present river valleys were thus in existence by this time, and their deposits were affected by the later periglacial activity.

Postglacial effects: coasts, fens and man

The postglacial rise in sea-level has resulted in the accumulation of over 10 m of alluvium in most river valley floors, and wide, low-lying areas like the Fens in the west and the Broads in the east have become the sites for thick accumulations of peat in marshy conditions. Man has affected both of these, by drainage in the Fens, and by medieval cutting of the Broads peats. This peat-cutting became a prosperous industry from 900 to 1300 AD, at which stage severe storms led to flooding and decline: the worst disaster was in 1287, and this also resulted in the drowning of extensive areas in the Netherlands. In spite of the fact that the coast near Yarmouth was 4 m higher than today many pits were flooded, and further inundations led to a change from peat-cutting to fisheries in the area.

The coasts exhibit a variety of features of recent origin. Erosion is important in the north-east and east of the county, providing material for southward drift and incorporation in the long spits which divert the mouths of streams like the Yare and Alde (Figure 12.7) but much of the sediment also comes from the floor of the North Sea. On the north coast offshore bars have formed at Blakeney Point and Scolt Head Island, extending the coast out to sea, and the shallow water muds of the Wash provide a further coastal environment of deposition (Figure 12.10).

Norfolk thus provides evidence of action by ice, running water, wind and alternating freeze-thaw in tundra conditions, together with the effects of changes in sea-level and the action of the sea on the coasts with different orientations. As such it provides a contrast with the other two areas studied so far, but is nevertheless yet another area where complexity of interpretation is necessary for a full understanding of the features observed.

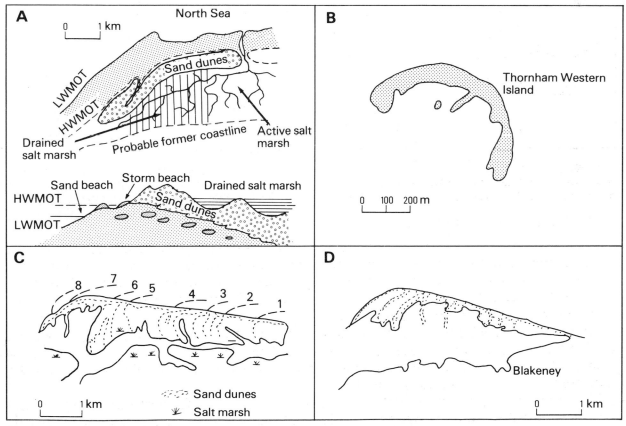

Figure 12.10 Features of coastal deposition in north Norfolk.
(A) At Holme-next-the-sea, near Hunstanton, the advancing coastline can be examined. (Open University, 1971)
(B) Thornham Western Island, 1939; the initial stage in offshore accumulation, having a crescentic shape.
(C) Scolt Head Island is a fully developed offshore bar: initial crescentic forms are eroded at their eastern ends by waves from the north-east, moving the debris produced to form a spit at the western end and giving the whole front a swash alignment. 1-8 are successive stages in the formation of the island.
(D) Blakeney Point, a spit with swash alignment.
((B)—(D) after Steers, in Sparks, 1960)

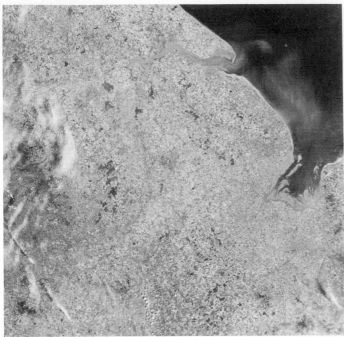

193

194

Plate 193 Part of Norfolk shown in the extreme eastern area of the LANDSAT image of eastern England, taken on 23 August 1972. The sediment in The Wash, the straightened Fenland draingage and the forests on the Sandringham Sands (dark areas) show up well. (USGS)

Plate 194 LANDSAT image of north-eastern England. The Pennine ridge is on the left (with Morecambe Bay in the south-west corner). The far northern point is just south of the river Coquet in north Northumberland. (USGS)

Plate 195 A June 1974 LANDSAT image of western Scotland, which brings out the barren nature of the ice-scoured highland zone and the radial pattern of drainage superimposed on the area by ice movement in the recent past. (USGS)

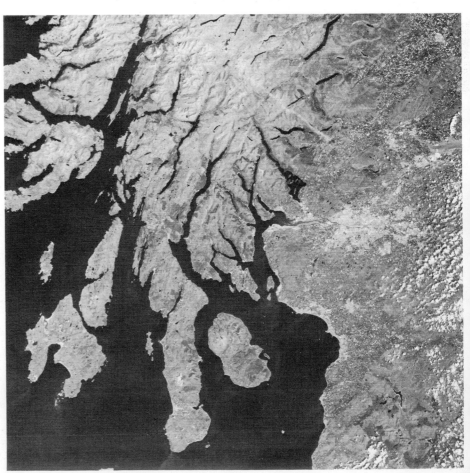

195

196

Plate 196 The view south from Ben Nevis. Notice the bare, glaciated valley troughs; the cirque near summit level, and the general accordance of summits.

Plate 197 Glaciated features near the summit of Goat Fell, Isle of Arran. (196 and 197 Institute of Geological Sciences, Crown Copyright Reserved)

Plate 198/199 A gravel pit exposure in the Great Glen of two distinct tills. The upper is more compact and clayey. (Forestry Commission)

198

197

199

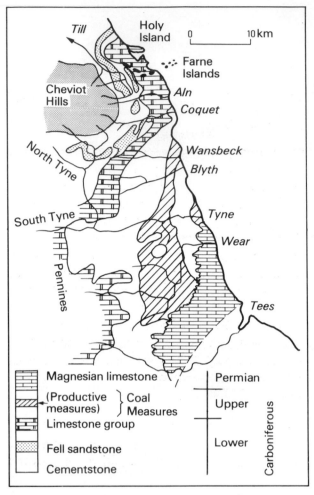

Figure 12.11 Geology and relief in north-east England. The main escarpment-forming horizons are shown, together with the outcrop of productive Coal Measures. There are also escarpments associated with sandstones in the upper Carboniferous rocks. (After Smailes, 1960)

North-east England

Northumberland and Durham provide another study of the effects of glaciation on a lowland region of Britain. In this case, however, the Devensian ice also covered the region, and the underlying geological features have a more definite effect (Figure 12.11): the scarp-and-vale landscapes of southern England recur in different but definite forms in this area.

The rocks and relief

North-eastern England consists of a coastal plain with hills rising to over 300 m, but it slopes generally to the east from the Pennine 'backbone' of northern England. The rocks of the area are of Carboniferous and Permian age: the former had been gently folded and eroded before the deposition of the latter. The Carboniferous rocks consist of alternations and repetitions of limestones, shales and sandstones, together with the coal seams which have provided the basis for industrial development. There is not a simple division into Carboniferous Limestone, Millstone Grit and Coal Measures here, as in the southern Pennines, and the lowest Carboniferous horizons include thick shales (Cementstones) and a massive sandstone (Fell Sandstone). Most of the coal seams occur in the Coal Measures, but some also occur lower in the succession. The limestones and sandstones form the higher relief features in a series of escarpments, and the Permian area is dominated by another scarp-forming horizon, the Magnesian Limestone. All of these have steep slopes facing west and north-west (Figure 12.12). In the Tertiary, 200 million years after these rocks were formed, a phase of igneous intrusion, mainly concerning the west coast of Scotland, led to the intrusion of the Whin Sill in the rocks in the north of this area.

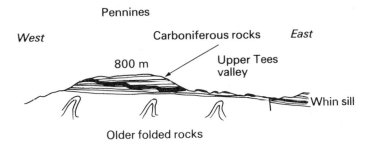

Figure 12.12 Section across the northern Pennines showing variations in thickness and horizon of the Whin Sill. (After Homes, 1965)

Stream processes had moulded much of the present relief before glaciation, although the sea-level changes during the Quaternary Ice Age led to further valley incision. The earliest drainage may have been directly eastwards, but considerable diversion had occurred in association with the downcutting of tributary valleys along the less resistant shales (e.g. Cementstone Group in the north, the North Tyne drainage system, and the south-north section of the Wear valley in Durham). Faulting in the Carboniferous rocks affected the production of a more complex series of escarpment alignments than in south-eastern England. The escarpments are also more broken in character than in the south, due to higher water tables in the rocks, which has promoted the formation of scarp slope streams which eroded their steep valleys.

The coasts of the north-east are fairly straight and often cliffed. This is particularly true of the Magnesian Limestone coast in Durham, where there are few interruptions to the pattern of cliff, shore platform and narrow beach. North of the river Tyne there are more signs of variation, since the alternations of Carboniferous limestone, sandstone and shale, and the Whin Sill intrusions, provide a more varied geological basis.

The glacial diversion of drainage
Ice reached the area mainly from Scotland and the Lake District, and erosion of the least resistant shales led to further emphasis of the main features, like the Whin Sill ridge, formed

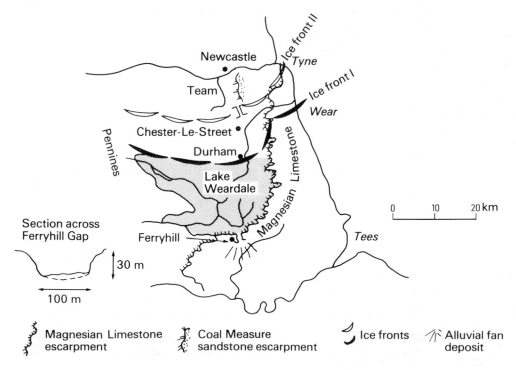

Figure 12.13 Glacial diversion of drainage in Durham; some of the features resulting from the melting of ice in this district.

on the more resistant rocks. The till deposited as the ice melted often filled valleys, and is also found capping low cliffs, but lies thinly on the escarpment summits. End moraine ridges occur in the upper Tyne at Hexham, and across the North Tyne valley at Acomb.

Removal of the ice cover led to some isostatic recovery, and rejuvenation of the rivers near their mouths, cutting into the till deposits in valleys like the lower Tyne. There are also postglacial gravel terraces in the major valleys: two steps can be seen at Hexham.

One of the most distinctive effects of the Ice Age in this area was the twofold diversion of the river Wear: one was not permanent, but the other was, and both have implications for the understanding of man's use of the area (Figure 12.13). As the Devensian ice front retreated meltwater built up just south of Durham City to form a lake trapped by the ice and the Magnesian Limestone escarpment. Water entered the lake via channels from the north-west. Discharge eventually took place to the south through the Ferryhill Gap, giving rise to extensive, fan-like spreads of debris to the south. Further retreat of the ice front revealed a lower col in the escarpment east of Chester-le-Street, and the lower Wear gorge to Sunderland was cut in this way. Most of the spillways, including the Ferryhill Gap were abandoned as fluvial conditions returned to the area, but they now provide important routeways in places (e.g. the Ferryhill Gap is used by the main railway line). The lower Wear diversion, however, was permanent, providing that river with a narrow, gorge-like valley through the Magnesian Limestone to the sea and contrasting markedly with the features of the nearby lower Tyne estuary. Whilst the shipbuilding industry has been concentrated at the mouth of the narrower Wear, it has stretched inland along the wider Tyne valley, which is easier to excavate from the soft glacial deposits. The former course of the Wear joined the Tyne via the Team valley, now a wide, drift-filled feature at the foot of an escarpment in a Coal Measures sandstone and drained by an underfit stream making its way across the glacial deposits.

North-west Scotland

No study of regional geomorphology in the British Isles would be complete without the consideration of an area in which the glacial erosion of highlands may be demonstrated. Whilst north Wales, or the English Lake District, would have served well, the north-west of Scotland is the highest part of Britain, provides the largest area of such country and has the widest range of associated features. The landforms of this area, however, cannot be understood without the consideration of a wider variety of processes than glacial erosion, and the underlying geology has also played an important part.

The rocks and preglacial relief
The rocks of the study area (Figure 12.14) are mostly ancient and resistant rocks formed in the Lower Palaeozoic era (570 – 400 million years ago), and involved in the Caledonian mountain-building events (Figure 3.4). These imposed north-east to south-west trends on the folding, and led to the metamorphism of the original sedimentary rocks. In addition there were granite intrusion of these mountains and volcanic activity. Thus Ben Nevis, the highest point in Britain (1420 m), is on the site of volcanic caldera rocks formed at this time. Faulting, including the Great Glen tear fault, broke up the rocks at the end of the Caledonian movements and formed lines of weakness which were emphasised by later erosional processes.

Much later, and in association with the most recent opening of the Atlantic Ocean (Figure 3.25), volcanic activity in the early Tertiary (50 million years ago) resulted in lava flows, volcanic cones and intrusions being formed on Skye, Mull and Arran along the west coast, and in a swarm of dykes cutting through the rocks inland. There is thus a variety of igneous, sedimentary and metamorphic rocks in this area.

Another Tertiary event resulted in the uplift and tilting to the east of this old and worn-down rigid block. Accordant summit levels in the highlands have been related to this event and/or later planations. Streams probably flowed towards the east, with some elements in the

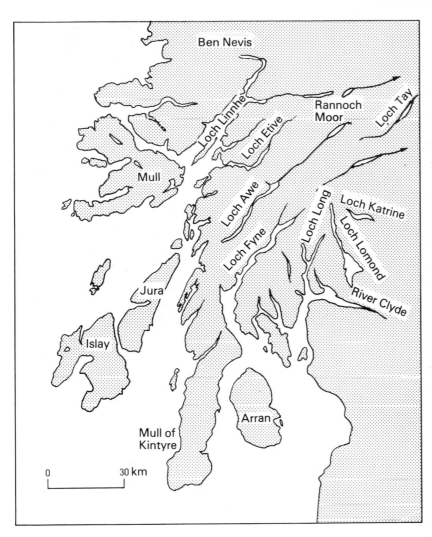

Figure 12.14 North-west Scotland. The lochs include fjords and inland ribbon lakes. Confused drainage patterns are particularly common in the Rannock Moor area in the north of this map. Raised beaches are found along Loch Linnhe, and around the islands and coast.

drainage networks eroding valleys along lines of weakness which reflect the old north-east to south-west folding and faulting structures. Such erosion by streams was well-advanced before the coming of the ice: deeply incised valleys with a weathered regolith form an important point of entry to an understanding of the effects of glaciation in this region.

The effects of glaciation

This region was probably the centre of ice accumulation and of ice movement in west Scotland (Figure 8.16). The ice flowing from cirques, and then the blanketing ice sheet, moved along the valleys, removing soil, steepening the sides and trimming away spurs to produce the typical features of glacial troughs. Such valley forms, together with the bare rock outcrops and high-level cirques, are the main results of glaciation. The drainage of the area was changed considerably by the ice, giving rise to a more radial pattern with many breached watersheds. Some valleys were deepened in rock basins to below the present sea-level.

As the ice retreated (Figure 8.26) less and less of the area was covered, although the overall retreat was punctuated by short phases of re-advance. Moraine ridges in the valley floors provided dams behind which lakes formed, but other ice depositional features are not very prominent. The last ice remained in the cirques, which thus show the clearest effects of glaciation. The cirques mostly have their openings orientated to the north and east (Figure 8.12), and rise in altitude inland from the west coast — suggesting lighter snowfalls to the east. The cirques are associated with hanging valleys, where the outlet stream from a cirque feature falls down the steep side of a glacial trough. There are, however, few knife-edge ridges (arêtes) in this part of Scotland (cf. Snowdonia and Striding Edge in the Lake District), and most cirque headwalls preserve a zone of the preglacial land surface at their tops.

Postglacial landforms

The last ten thousand years have seen small, but important changes in the pattern of relief features left as the ice melted. The first effect of decreasing ice cover was a rise in sea-level, drowning the mouths of the glacial troughs to form fjords. Heavy rains washed away many of the finer fractions of glacial deposits into these fjords and the ribbon lakes in the uneven floors of the glacial troughs farther inland, to form extensive area of flat, reclaimable land. Some of the former lakes have been completely filled and all show some signs of the extension of delta-type deposits at stream mouths.

The weight of the ice had caused the land to subside isostatically (chapter 4), and it was some time before the readjustment to the removal of ice took place. The coastal zone, and most sea lochs are now marked by a series of raised beaches up to 30 m above the present sea-level, and these have become one of the main sites for settlement and communications in the area.

The re-establishment of drainage in an area of breached watersheds has resulted in many strange network patterns: major streams may have very low divides between them, and there are many boggy depressions filled with peat. Man's influence in the area has been to heighten the effect of bareness which is supposed to be a feature of glaciated regions. There was once more forest at least up to the 300 m contour, and cutting of this timber, together with overgrazing in the last 200 years has removed much of the soil from the valley sides. Man is thus responsible for accelerated rates of erosion and deposition, and has emphasised the stark nature of the formerly glaciated landscape.

Summary

The examples chosen in this chapter have served to emphasise the complex interdependence of natural processes working to produce landforms in an area over time. The shortness of each treatment has merely provided an introduction to a varied group of landform-process associations, and in each case a whole book could have been written. It is hoped, however, that the reader will wish to delve deeper, and to apply the ideas and principles to other parts of Britain and the world.

13

Arid landscapes

More than one-third of the Earth's land surface is desert (lit. 'empty of life') or semi-desert, and Europe remains the only continent of which some part is not desert. An appreciation of thirst and aridity is naturally limited in the humid temperate regions, where population growth and increasing domestic and industrial water consumption have led to a consideration of the problems of water shortage only in the last few years. Whilst there is no single word in the English language expressing 'to die of thirst', Arabic possesses some eight degrees of expression for thirst ranging from al-'atash, the lowest level, through al-Huyam, a vehement thirst (also meaning 'passionate love'), to al-Juwad meaning 'the thirst which kills'.

Climatic changes taking place at the present day are causing such arid areas to be extended, particularly along the southern margins of the Sahara in Africa. Soils are drying out for want of moisture, anchoring vegetation is dying and exposing old, stabilised dunes, and the wind is drifting sand and dust to bury cultivated lands. The relative lack of water in deserts has always posed immense problems for man. Students of desert landforms and processes can make a contribution to the understanding of the arid environment which may help the lot of peoples inhabiting these regions and their margins.

Deserts are caused by the inter-relationship of climatic factors, including rainfall, temperature and evaporation rates. The hot tropical deserts have high potential evaporation rates, which may exceed annual rainfall totals by over 500 mm. Cold parts of the world may also be deserts where the annual precipitation is less than 50 mm. Areas with a low total rainfall and winter maximum are less likely to be deserts than areas of similar rainfall totals, but a summer maximum, owing to the higher rates of evaporation in the summer.

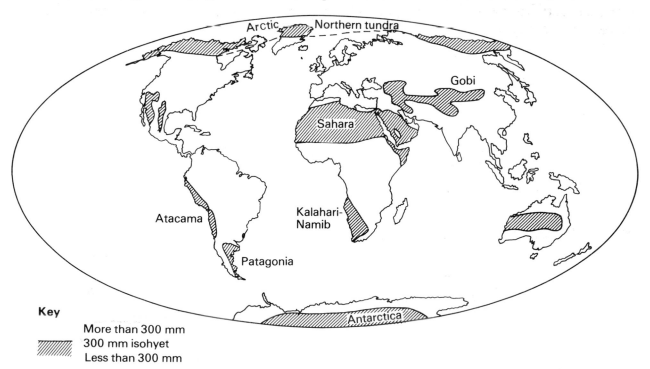

Key
More than 300 mm
300 mm isohyet
Less than 300 mm

Figure 13.1 The arid areas of the world, as defined by the 300 mm isohyet. Notice the worldwide distribution of these areas: they are not restricted to the tropics.

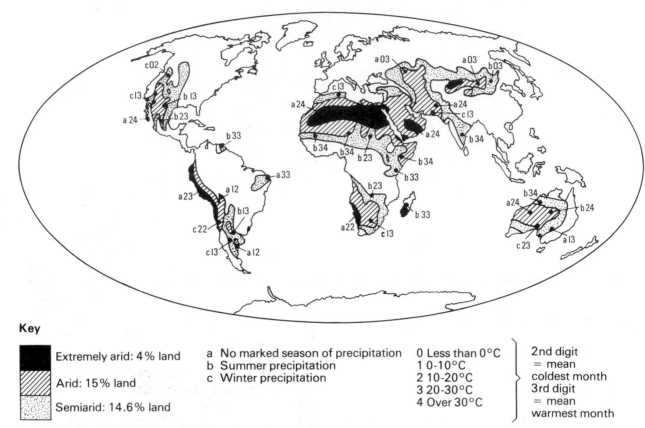

Key

Extremely arid: 4% land

Arid: 15% land

Semiarid: 14.6% land

a No marked season of precipitation
b Summer precipitation
c Winter precipitation

0 Less than 0°C
1 0–10°C
2 10–20°C
3 20–30°C
4 Over 30°C

2nd digit
= mean
coldest month
3rd digit
= mean
warmest month

Figure 13.2 World distribution of arid lands: a more refined definition. (After Meigs, 1953, in Cooke and Warren, 1973)

Definitions of aridity based on precipitation totals alone are thus unsatisfactory. The 300 mm isohyet has often been adopted for this purpose (Figure 13.1), but areas within this boundary may have periods of years without rainfall, followed by a few weeks in which the rain total for three or four years may fall. More people in arid areas die from the unpredictable floods caused in this way than from thirst. Some of the driest deserts (e.g. the coastal Atacama desert of northern Chile) have high humidity levels in the atmosphere. Various indices of aridity have been put forward, the most widely used being those of Thornthwaite (see p. 315), and these give a more refined analysis (Figure 13.2). The vegetational response to these conditions also suggests that a total of 33 per cent of the world's land surface is arid: the proportions, however, are different, since semi-arid vegetation (brushland, thorn forest, short grasses) occupy 5 per cent, arid vegetation (desert grass savanna) composes 24 per cent, and extremely arid lands (no vegetation) make up 4 per cent. Another approach to defining aridity is based on the distribution of surface drainage systems which do not reach the sea (i.e. endoreic), and those which do not have any surface drainage (i.e. areic). Once again approximately one-third of the world's land surface is affected (Figure 13.3): this approach can be seen to have a relation to the basic precipitation-temperature-evaporation factors.

Deserts are thus not simply 'hot, dry regions' of the world, and their margins fluctuate with small changes of climate, such as are affecting the southern edges of the Sahara in the 1970s.

Processes of arid erosion

The study of landforms in deserts has been affected very much by the approach of those who attempted to construct a general model of landform development in the late nineteenth

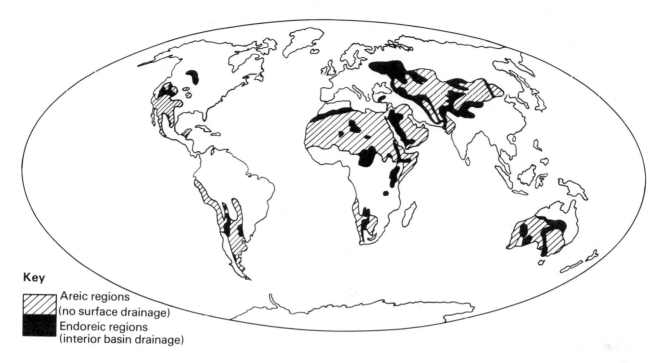

Key

///// Areic regions
(no surface drainage)

■■■ Endoreic regions
(interior basin drainage)

Figure 13.3 Areas of endoreic and areic drainage. Compare these with the maps of aridity. (After de Martonne and Aufrère, 1928, in Cooke and Warren, 1973)

century. At this stage it was assumed that the wind and mechanical weathering processes were dominant in the origin of desert landforms, and this impression is still abroad today. Just as advances have been made in the understanding of weathering, slope and hydrological processes by increased measurements, so the more intensive study of desert landforms has changed this view concerning the relative importance of processes. In particular it is now seen that fluvial processes and chemical weathering are probably more significant than aeolian processes in deserts and also that deserts contain many relict features of previous periods, which have not been destroyed by presently active processes.

Weathering and mass movement

The presence of angular broken rock material and the low rainfall totals suggested that rocks disintegrated and split under the influence of temperature changes. Laboratory experiments, involving speeded up alternations of heating and cooling, suggested that little could happen unless water was present. Observation in the field has not produced conclusive results, but has so far shown that rock expansion by heating does occur; that chemical changes in the presence of water — and particularly in salt deposits — are more effective still; and that lichens may also contribute to these weathering effects. In short, the processes of weathering in deserts are more complex than at first envisaged (see further discussion of weathering processes chapter 5).

Once pieces of rock have been broken from the main mass they will move downslope, often without resistance from vegetation. One frequently finds a pile of broken fragments at the foot of a slope. Mudflows are also common in arid areas, since the water from an occasional rainstorm may mobilise a dry mass of rock material: mudflows can be regarded as an extension of stream activity having a very large suspension load (chapter 6). Smaller-scale features of the desert surface include patterned ground (Figure 13.4).

Erosion by the wind

Deflation hollows (chapter 9) are common features of arid landscapes with the breakdown of softer rock and eddying of the wind which further deepens the hollows. Downward deflation is halted by the presence of a water table, and such hollows may develop to form oases. Removal of finer dust and sand also leaves behind the stone and boulder lag deposits common on the surface of many deserts.

Corrasion, or erosion by sand using the sand load, has been very much over-emphasised in the past. Sand grains, transported by saltation (chapter 9), rarely have effect at heights of over 2 m above the ground. This severely restricts the zone in which any corrasion may be effective, although there is evidence of some polishing of bare rock surfaces at greater heights. Rock pedestals (or zeugens, page 228) have been attributed to wind corrasion, but it is felt increasingly that water at and about the surface containing salt concentrates, is here the dominant agent of rock destruction by chemical weathering. Further, it has been held that wind corrasion may lead eventually to the formation of flat, plane surfaces. This is now strongly refuted as the process is so inconsistent with the fact that eddying leads to the accentuation of uneven zones rather than to subdued relief.

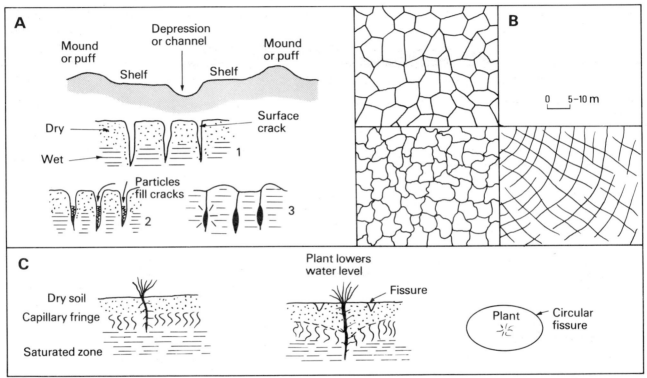

Figure 13.4 Patterned ground in arid regions.
(A) Gilgai: small-scale features in the Australian desert due to alternate drying and wetting of the ground. (After Verger, 1964)

(B) Some examples of desiccation crack patterns.
(C) The formation of ring fissures as a plant lowers the water table. (After Neal, 1965) (All in Cooke and Warren, 1973)

Erosion by running water

Most of the major erosional features of arid landscapes have been carved by running water. Sheetwash may occur for a short while after storms. Streams either rise outside the desert and flow into it, or may gather sufficient water from springs or occasional storms, but few maintain their flow right across a desert region. Water is lost by evaporation and percolation with little or no fresh input to the system. Thus areas of interior drainage characterise arid landscapes today, and fluvial material is deposited within the boundaries of such areas. Base levels are not controlled by sea-levels, but instead are strictly local and independent of each other, tending to rise through time with increased deposition. Where the deposition of debris occurs in the wide, shallow stream courses subject to fluctuations of discharge, braided channels result.

The nature of the stream channel depends much on its general location within the arid region. Over plains intermittent streams carve wide (up to 100 m), but shallow (up to 10 m), wadis which have steep sides and flat floors masked by large quantities of alluvial debris. Many of these channels, separated by extensive plateau tracts, are dry for periods of several years. In upland areas of resistant rocks the stream channels have steep sides and sharp interfluves, whilst in areas of less resistant, unconsolidated rock, similar erosion produces badland topography with steep, close-spaced gullies. Run-off rates are high, helped by the steep slopes and lack of vegetation cover. Hence downward erosion is especially active. In semi-arid regions, where rainfall occurs more often than in the most arid sectors but there is still a sparse covering of vegetation, this results in the most rapid rates of erosion experienced in the world.

Depositional processes in deserts

Areas of bare rocky plains, mountains, hills and plateaus dominate the world's deserts (Figure 13.5), but there are also extensive areas of rock debris accumulated by the activity of wind, running water — and ice in the cold deserts. Much of this debris accumulates in deserts due to the inability of the processes to remove it. Thus whilst the wind removes dust particles beyond the bounds of desert areas, it can move the sand grains only within the confines of the desert since they become stabilised by vegetation on the margins. Larger sized pebbles and boulders are not moved at all. In a similar way the debris moved by streams within a desert is limited in its distribution by the fact that most desert streams are short-lived and drain to internal basins.

Surface feature	South-west USA	Sahara	Libyan Desert	Arabia
Playas	1.1	1.0	1.0	1.0
Bedrock fields and hammadas	0.7	10.0	6.0	1.0
Desert flats	20.5	10.0	18.0	16.0
Riverine desert (bordering through-flowing rivers)	1.2	1.0	3.0	1.0
Fans and bajadas	31.4	1.0	1.0	4.0
Dunes	0.6	28.0	22.0	26.0
Dry washes	3.6	1.0	1.0	1.0
Badlands and subdued badlands	2.6	2.0	8.0	1.0
Volcanic cones and fields	0.2	3.0	1.0	2.0
Desert mountains	38.1	34.0	39.0	47.0

Figure 13.5 The proportions (%) of various relief features in the major world deserts. (After Clements et al., 1957, in Cooke and Warren, 1973)

Deposition by the wind

Sand seas cover varying areas of deserts, ranging from 12 per cent of the Sahara to over 30 per cent in Australia. The forms of these deposits and of other sand and boulder deposits are described in chapter 9.

Deposition by streams

The lack of vegetation cover and the often unconsolidated nature of sedimentary deposits in arid environments have been noted. High rates of run-off under such conditions in upland areas mean that vast quantities of debris are carried as stream load. The sheer volume of load, together with rapid loss of gradient, evaporation and seepage into unconsolidated sediment leads to a rapid reduction in stream discharge as it reaches lower ground or an area less confined by valley sides, and deposition occurs. Alluvial fans are common features along mountain fronts in deserts, and often have a decreasing size of particle from apex to base (Figure 7.28). These fans may coalesce to produce an alluvial apron (Figure 13.6), and their

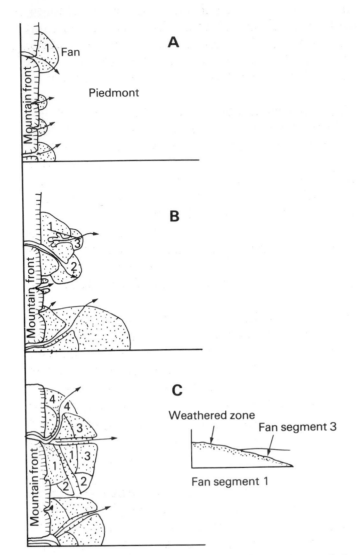

Figure 13.6 The development of an alluvial fan system, including several segments trenched by streams. (After Denny 1967, in Cooke and Warren, 1973)

gravelly surfaces result from the deflation of the residual lag deposit (page 226) rather than a true depositional feature. The extension of these fans into open plains or wide basins leads to the formation of a bahada, which is a complex mass of coalescing fan deposits up to 100 km wide and includes sediments formed by the evaporation of stream water. Direct gravity falls from mountain slopes may also be significant in contributing material to conical landforms at the foot of slopes, or to stream load.

Farther beyond the mountain front the finer debris is carried to the lowest part of the basin to form playa deposits. These are essentially dried lake beds floored by salt-and-clay plains. Most of the world's major deserts (e.g. Iran, Gobi, Australia, western USA) show evidence that the former shorelines of these lakes were very much more extensive: the indications of well-ramified drainage systems and shoreline features like beaches, cliffs and spits demonstrate that rainfall was formerly more important, and this is attributed to pluvial periods during the Quaternary Ice Age. The rain belts moved away from the poles as the ice advanced. The salts accumulating in playas often have high concentrations of gypsum and sodium carbonate, as well as sodium chloride. Wind erosion of these lake bed deposits may form pans or 'lunettes'.

Fluvial and aeolian processes and desert landforms

The core problem in the analysis of the origin, evolution and form of arid landscape features is one of the relative interaction of surface processes. An area dominated by the climatic characteristics of temperature extremes and limited precipitation initially suggested a landscape moulded little by water. Weathering dominated by temperature variation and the resultant differential expansion and contraction, with erosion primarily executed by strong winds set up by steep pressure gradients, seemed to be indicated. Such conclusions are now seen to have been false.

Aeolian action is most evident over a dry, unconsolidated land surface. Though not necessarily any stronger than in other parts of the world, desert winds are particularly effective in gathering the finer surface material as they blow across open, sparsely vegetated or non-vegetated areas. They arm the lowest 1–2 m of the atmosphere with a load which may be active in bombarding, and so modifying, bare rock structures and outcrops. It is important, however, to assess the scale and time duration of wind activity in the land forming processes. Aeolian action is most effective over dune and dried lacustrine saltflat areas, where it is the dominant process. The zone of deflation is limited by the depth of the water table, and even above the desert surface is it increasingly clear that moisture in the rocks, or condensing on the surface, is probably as important as sand bombardment in erosion. Capillary action draws salts to the surface and these are responsible for flaking on the lower parts of features like rock pedestals.

In many deserts there is abundant evidence of climatic change with more humid conditions prevailing over what are now hot desert areas during the Quaternary period. Data from some of the more recently dried areas like the Central Desert of Iran shows that aeolian activity increased as the Kavirs (internal drainage basins) became increasingly arid . It seems that the modification of desert landscapes by wind is very much a recent and present process.

In the case of the cold desert, aeolian action is not a land-shaping agent, but may operate to give special ice surface forms, as in the Antarctic. Though strong winds occur frequently over these deserts they lack the sand or dust load which characterises the hot and temperate regions. Without such material to assist erosion, wind is an ineffective agent of denudation. It simply blows around the surface snow.

The role of water in the denudation of arid landscapes similarly needs viewing by reference to both scale and time. Whilst in cold deserts precipitation is at a very low level, hot arid areas experience not only limited amounts of rainfall, but highly sporadic and localised falls in the form of relatively heavy downpours, thunderstorms or cloudbursts. These are not great inputs when compared with heavy downpours in humid regions (the 100 mm maximum for 24 hours received in a desert contrasts with the totals of up to 1000 mm received from thunderstorms in the south-east USA), but it is their impact on the dry, unvegetated rock debris which is most important. Rain has been recorded in nearly all arid climates, and it must be realised that marginal arid and semi-arid regions, especially in the tropics, receive more rain in a year than does south-eastern England.

The majority of fluvial activity in temperate regions occurs during periods of flooding, and this rule is followed in arid areas. The impact of raindrops is important in eroding bare soil or regolith, and runoff is enhanced by the presence of impermeable surface encrustations to produce large volumes of surface water flowing at high speeds. Where the surface materials are unconsolidated large quantities are carried away by the water, leaving the landscape scarred and gullied. Intermittent storms and the associated flooding are the most important single agents of landscape denudation in arid lands. They are also responsible for most of the large-scale denudational forms and for depositional features. Whilst aeolian activity is responsible for the extensive dune areas, it must be remembered that these cover a maximum area of 15–30 per cent of the total surface, and that they are related to sandstone source rocks.

Such a summary considers only present processes and some would suggest that these have been over-emphasised. Arid landscapes show evidence of both present and past activity. Relict features (page 275) are therefore present in desert areas, many of which owe their origin to past climatic environments prevailing in these regions. In particular the Quaternary period saw a narrowing of climatic zones over the Earth's surface due to the ice sheets advancing, particularly from the north. Arid landscapes thus contain evidence of chemical weathering, running water and even frost action dating from these periods of higher precipitation and lower evaporation in the pluvial periods. Furthermore, tectonic activity has clearly influenced the climatic environments of arid areas, leading to changes in the rainfall and temperature regimes. Areas within and adjacent to the deserts of North America, Africa and Asia were subjected to considerable uplift during the Tertiary and Quaternary periods. This often led to increased precipitation and increased fluvial activity.

The influence of water in arid lands is thus evident both at the present and in the recent past, when it may have been dominant. Few desert landscapes are dominated totally by wind-produced forms. Zones of dune formation, and others where the reworking of fine surface regolith is evident, are characteristic of wind action. Truly arid landscapes (covering only 4 per cent of the world's land surfaces) show little present erosional activity with slow rates of change and a tendency to landform preservation rather than destruction. Fluvial processes, either past or present, therefore may be dominant in producing the major erosional features of the deserts. If a greater emphasis is placed upon past processes, deserts are seen essentially as collections of relict features in what is virtually a fossil landscape.

The significance of such climatic changes makes it apposite to summarise the effects which may occur during the changes from humid to arid and arid to humid conditions. As a humid region, with a complete coverage of vegetation and drainage nets, becomes drier, so the rainfall becomes more localised in the upland areas. In these sections the vegetation cover has been lost and any rainfall rapidly sweeps rock debris into the stream channel. The short-lived occurrence of streamflow is not supported by baseflow, and intense evaporation together with permeable valley floor materials soon cause the stream to peter out, depositing its load. The middle and lower portions of the old valley system become filled with this debris. Streamflow eventually ceases beyond the upland elements of the old networks. Processes of slope retreat cause the valley divides to be worn back, and after a long time horizontal surfaces, or pediments, formed by a combination of erosion and deposition, will become dominant, whilst the hilly source areas for precipitation and runoff will be reduced further in their size and effectiveness.

Such an arid region, when subjected to an increasing rainfall again, begins without a cohesive drainage network — any former system will have been deranged. New streams may flow as sheets of water across flat surfaces, or may divide and reform in anastomosing patterns. Such patterns are common in southern Venezuela, which may have been arid during the glacial periods of the Quaternary. The old drainage lines, filled with debris in the arid phase, would be flushed out and the deposits reworked as the precipitation and runoff increased.

This is a simplified picture of the sort of changes which could occur in the processes, landforms and deposits of an area.

The analysis of arid landscapes

Any examination of arid landscapes faces the major issues of the relative roles of past and present processes, and of the fluvial and aeolian processes. The arid lands can be seen as a morphogenetic system where landforms are evolving or being modified at present through both of these denudational agents. What must be stressed, however, is that in the past the role of wind has been overemphasised and that when water has been considered much attention has been given to the present water balance of arid lands, and little to the evolution of what are now relict features under a climatic regime of higher precipitation and lower evaporation. It is this emphasis which leads to the present view of arid landscapes as being dominantly fossil landscapes.

Arid upland landscapes

It is in upland areas that arid landscapes show most clearly the influence of structure. The role of structure is thus a major constraint. Horizontal plateaus, folded and block features, volcanic necks and plugs are typical landforms in such areas which are structurally controlled. Such forms have often dominated American writings on arid landscapes, for they characterise the deserts of south-western USA and western South America.

In terms of modification by erosion processes, there is strong evidence of the importance of water. Deep canyons, wadis and badland topography all testify to the importance of stream drainage. The basic pattern of drainage in such areas is closely textured, crowded streams showing strong integration into and around the wadis. This is due possibly to the high rates of runoff leading to the development of large numbers of drainage gullies. High evaporation and percolation rates mean that the flow of water soon ceases after a storm, and hence no major channel can develop to exert influence over the whole system. Additionally, sharp ridges separate the gullies as vertical erosion is often great. Steep channel sides and generally steep slopes are indicative of the inefficiency of streams to move quantities of debris, and this is not helped by the high rates of evaporation.

Present rainfall regimes seem ineffective in upland arid regions and they cannot be responsible for the deep gullies and wadis which characterise the landscape. In the High Atlas Mountains of Morocco features such as cirques, glacial troughs and a whole range of periglacial forms are evident. It thus seems that these upland arid zones show a major degree of structural control with mainly fluvial landforms, but that many of these are to be attributed to former pluvial conditions and can therefore be seen as 'fossilised' under the present climate.

Pediment landscapes

The origin of pediments and the associated upland masses, known as inselbergs, which they fringe, offer some of the most controversial problems associated with arid landscapes. Studies of pediments and inselbergs have frequently gone together, drawing evidence particularly from North America and Africa. North American cases, however, show few or no inselberg features and so analysis concentrates on the pediment, whilst African studies are dominated by these residual uplands. Broadly the problems arise in a consideration of the modes of formation of pediments, in assessing their morphological significance in arid areas, and in evaluating their relationships with other features of arid landscapes.

Pediments are low angle slopes, where the rock is near the surface: they may be mantled by a lag deposit, but no more than this (Figure 13.7). They occur in areas of low annual average rainfall and low humidity and are usually located in the piedmont area, fringing an upland massif, but sometimes in wider lowland zones. They vary considerably in extent away from the mountain front from benches of five or ten kilometres to pediplains up to 100 km wide, or more. Structure may occasionally influence the extent of these features. In plan pediments are fan-shaped, a marked variation existing in the profile gradient. Arizona has slopes varying 10-60 m/km; New Mexican examples average 80 m/km; whilst in Western Australia and in Africa gentle slopes at less than 40 m/km are recorded. None seem to be absolutely flat, and it is suggested that the slope corresponds in some way to a hydraulic curve, indicating a fluvial origin. In cross profile, however, over considerable distances, the pediments are often level, though both convex and concave forms have been observed. Pediments often remain free from depositional features, but may have a thin cover of stones or sand which increases in thickness downslope. Gullying and simple drainage patterns are sometimes evident. The two most noticeable features are the sharp niche, or break of slope, at the upper end of the pediment at the base of the upland front (the piedmont angle), and the residual upland masses (inselbergs) which frequently interrupt the otherwise smooth, extensive surface of the pediment.

Explanations of formation and assessments of the significance of these landforms vary. Some regard pediments as the only true desert forms, and the only features which can be attributed directly to arid conditions operative at present. Others see the pediment slope,

200

201

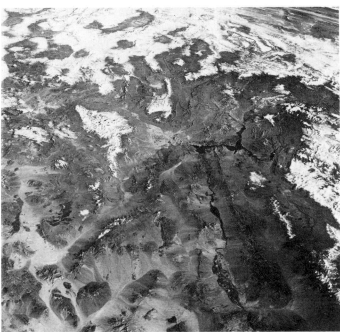

202

203

204

Plate 200 'Channel country' in Queensland, Australia, a semi-arid region where annual floods fill the complex network of channels, billabongs and swamps. (Australian News and Information Bureau)

Plate 201 A Gemini IV photograph of the Hadramaut plateau in southern Arabia. An old stream system is now quite dry and choked with sand. (NASA)

Plate 202 An Apollo 9 (1969) photograph of the mountain-and-basin landscape of Arizona in the foreground, with Lake Mead on the river Colorado just west of centre. (NASA)

Plate 203/204 Antarctic: two views of the ice sheet at its edge, where outlet glaciers form. The effects of wind on the surface snow can be seen. (USGS)

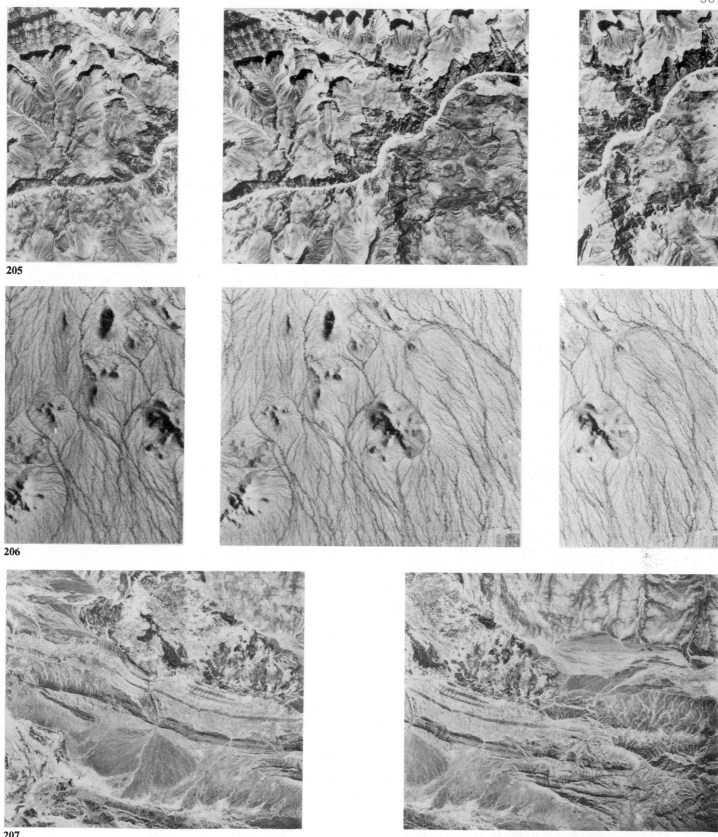

205

206

207

Plate 205 Stereo pair: the Grand Canyon of the Colorado river at the confluence with Bright Angel Creek. Notice the nature of the valley slopes in the arid environment. (USGS)

Plate 206 Stereo pair: a pediment is crossed by an intricate drainage network on the north-west side of the Sacaton Mountains, Arizona. Most of the area is mantled with regolith, through which the inselberg hills protrude. (USGS)

Plate 207 Stereo group: folded rocks and a salt plug in southern Iran, an area which shows landforms typical of upland arid regions. (USGS)

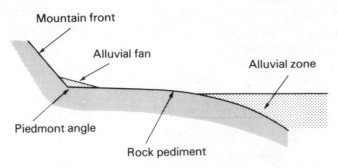

Figure 13.7 The general features of the pediment area.

drainage patterns and the requirement of some agent of debris transport for large particles as all suggesting a fluvial origin. Bahadas are depositional forms which have similar shapes to pediments, and the material in many bahadas has been transported across the pediment. The pediment can thus be seen as a transitional zone between the erosion of the upland desert and the deposition of the bahada. The piedmont angle then becomes a point of delicate balance in a state of dynamic equilibrium. There have been a number of stages in the development of ideas concerning the origin of pediments:

1) At first they were attributed to corrasion and deflation by the wind, but this is now regarded as an ineffective agent, and it is seen that pediments occur all around upland areas, even where a prevailing wind direction is dominant.

2) Water has thus become associated with the formation of pediments. One suggestion is that there has been spring sapping at wadi heads, but this is unlikely to have occurred all around an upland area. Another idea suggests that the streams issuing from a mountain area swing from side to side and erode a wide area at the foot of the mountains. Much of the evidence for this idea is based on the occurrence of rock fans, but these are not common in pediment areas. There is also no evidence of indentation or scalloping which would be expected with such a lateral planation process. It is also difficult to envisage the ultimate removal of complete upland massifs by this process, since the streams issuing from the remnant uplands would possess so little energy in the later stages. Sheet-flood erosion is another suggestion which has been made, but it seems that the pediment form is responsible for this process, which is largely one of transport, rather than the other way round.

3) Penck's ideas on parallel slope retreat (chapter 6) have been brought into the debate. In this view the pediment slope reflects the balance between the energy expended on the slope, the volume of material to be removed, and the efficiency of the transporting agent. This can account for the piedmont angle and emphasises the speed of retreat as reflecting different climatic environments. It can also be joined with ideas of sheet flooding and lateral planation as a composite theory for the origin of pediments. Yet if any of these ideas are correct, it would seem that pediments are features resulting from fairly rapid activity in the midst of many fossil landforms. This is where it becomes important to view the landscape as a whole, and to take a wide range of similar features into account. The possibility of the pediments themselves being relict features has to be considered, and there is evidence to suggest that this is the case. Quaternary laterites have been found on pediments in Australia, and some in Africa show evidence of depositional features of Tertiary age.

4) In recent years more attention has been devoted to the link between inselbergs and pediments. Investigations in Africa have stressed the importance of deep weathering and rotting in such inselberg areas through the humid environments of the late Tertiary and Quaternary periods (Figure 13.8 A). There is a concentration of weathering in the rock areas of marked jointing, and the rotted areas are soon removed by fluvial activity. Pediments are thus developed by streams in the areas of weathered material, whilst the preserved rock remains as the residual inselbergs. It seems likely, therefore, that pediments represent the lines of former weathering fronts, and a composite hypothesis as to their origin may be put

forward. This begins with deep rotting by weathering in the Tertiary period, followed by the removal of the debris largely by fluvial activity, but in some cases including aeolian action. The least affected areas of rock emerged as residual inselbergs and established slope forms, which retreated in parallel fashion. The modification of the former weathering front by fluvial and aeolian processes resulted finally in the present extensive pediment areas.

5) This idea is not, however, satisfactory in regard to North American examples of pediments. Studies of these suggest that fluvial processes, related to the stable, or rising local base-level of inland drainage, are responsible for the formation of pediments (Figure 13.8 B).

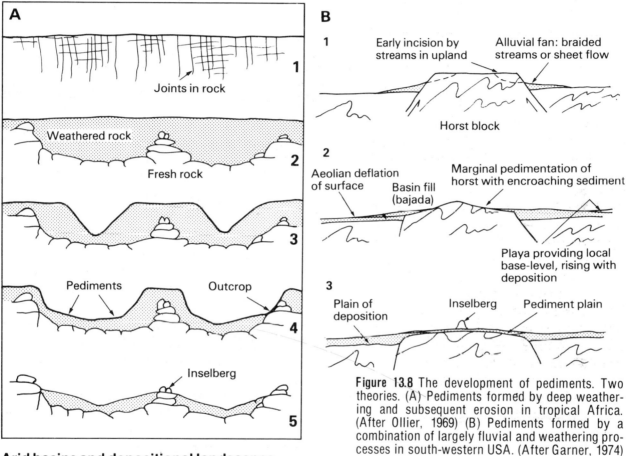

Figure 13.8 The development of pediments. Two theories. (A) Pediments formed by deep weathering and subsequent erosion in tropical Africa. (After Ollier, 1969) (B) Pediments formed by a combination of largely fluvial and weathering processes in south-western USA. (After Garner, 1974)

Arid basins and depositional landscapes

Desert basin landscapes are essentially areas of deposition, and aeolian processes may become particularly active. The resulting landforms in loose, surface sand range in scale from small ripples and ridges to the large-scale fixed and mobile dunes, deflation hollows and sand seas. These are superimposed on the landscape of stream-laid bahada and lake-precipitated salt deposits, and essentially re-work these materials.

An excess of evaporation over run-off produces a lack of permanent surface streams and drainage is often strictly internal. Stream flow at intermittent periods brings debris and matter in solution to the basin, but its flow weakens with evaporation and percolation, and the water increases in salt concentration. This water flows into a central lake, but this is a temporary feature and dries up when the flow of water ceases, giving rise to playa deposits.

Arid landscapes — case studies

The world's arid regions present contrasts within their boundaries, which necessitate more detailed studies, and three have been chosen to demonstrate some of the variations. These include the hot tropics with stable and unstable tectonic conditions, varying rainfall totals, and the world's coldest regions.

The Antarctic

Despite the apparent paradox of its being created almost entirely out of ice, the Antarctic (which is nearly twice the size of the USA) must rank alongside the Gobi, the Sahara and the Arabian Peninsula as one of the world's great deserts. All save save 5 per cent of its surface of 1.75 million km² is covered with ice. Unlike the Arctic, where the average thickness of ice is only 2 m, floating on the waters of the Arctic Ocean, that of the Antarctic is currently estimated to have an average thickness of 2500 m and a maximum of 4000 m. Only in two areas, Victoria Land and along the Antarctic Peninsula (Figure 13.9), do rocks protrude to a significant extent and form nunataks. For the majority of its area the rocks are downwarped beneath the weight of ice to a level approximately equal to the present sea-level. The maximum height of the ice is 4000 m near the Soviet Vostok station. It is estimated that the ice was another 300 m thick during the Quaternary Ice Age.

Precipitation over Antarctica as a whole is thought to average 110 mm of water equivalent per year. With the exception of very small quantities of rain around the edges during the warm season, all of this falls as dry snow crystals, since the temperatures are always below freezing except for very short periods on the coast. Considerably higher precipitation (400 mm water equivalent) occurs on the coasts, but the South Pole has only 60 mm. The higher precipitation on the coast is due to the stormy conditions which result from the juxtaposition of the high pressure over the ice which leads to winds blowing outwards from the Pole. These winds tend to create a separate circumpolar storm zone, isolated from the warmer areas to the north (Figure 13.10). Precipitation figures alone are sufficient to place the Antarctic and the deserts of middle latitudes in the same category.

While precipitation is low the wind strength is greater than anywhere else on Earth. Winds up to 240 km per hour have been recorded at Commonwealth Bay, while average figures of 65 km per hour and 335 days of gales in one year have been recorded in coastal areas. Averages and maximum figures can both be misleading, but the world's record for the coldest temperature is also found in the Antarctic (–88.3° C was recorded at Vostok in 1960 at an

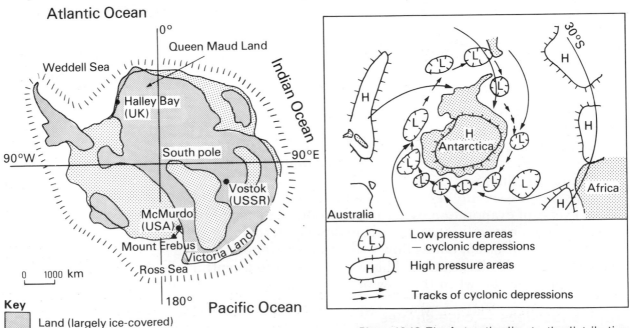

Figure 13.9 Antarctica: major features.

Figure 13.10 The Antarctic climate: the distribution of cyclonic depressions and high pressure areas at 6 a.m., 17.2.58. The arrows show the regular paths taken by depressions in that year.

altitude of over 3000 m). The highest areas of Antarctica experience annual average temperatures of below –55° C with low diurnal ranges. These extremes of cold, high winds and low precipitation render Antarctica a truely arid region. Driven snow, very dry because of the low temperatures, rapidly masks any obstruction in a fashion similar to wind-driven sand, creating a wide variety of dune-like drifts. Blizzards remove vast quantities of unconsolidated snow, leaving behind a wind-scoured plateau of ice. During a blizzard at Terra Adelie 240 tonnes of snow were driven over a 1 m section of coastline. The average for such snow movement over the continent as a whole would be nearer 8 tonnes, but, remembering the lightness of such snow, this is a large quantity to be moved in one day. Wind-shaped structures in ice, known as sastrugi, show the direction of the prevailing winds by the orientation of their long axis. Elongated and up to 2 m in height, but more often less than 1 m, they give a dissected appearance to much of the surface, but vast areas remain as featureless deserts of ice with the occasional drift of snow. Crevasses add variety to the monotonous landscapes only where ice is moving relatively quickly in glaciers, ice streams and the like.

The Central Desert of Iran

Iran occupies a transitional position between the arid, subtropical zone of the Northern hemisphere and those arid lands of temperate latitudes. The more northerly areas, too far from maritime influences to receive much rain, stretch from the eastern shores of the Caspian and along the northern flanks of the Himalayas into Mongolia, the Kara Kum and Kyzyl Kum, the Takla Makar and the Gobi. Iran is a mountainous country consisting of two main mountain ranges of late Alpine origin, the Zagros and the Elburz, enclosing an intermontane basin of some 750 000 km² at an average altitude of 1000 m (Figure 13.11). The two ranges

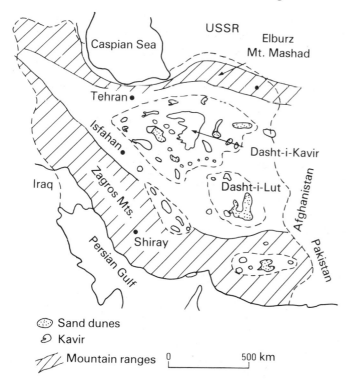

Figure 13.11 Sand dunes and kavirs in Iran: they are situated in high-level basins within the mountain ranges. (After NIOC, 1957, in Fisher, 1968)

coalesce in the north-western part of the country in the geological confusion of the Caucasus and the Anatolian Plateau. Mount Ararat (5121 m) marks the focal point of the ranges. The Elburz ranges are the higher group, with Mount Damavand (5790 m) standing to the north of Tehran from where the range declines in height towards Afghanistan. The Zagros ranges reach a highest point of 4520 m (Zardeh Kuh) and also decrease in height to the east. The fact that these two ranges lie across the paths of easterly-moving depressions reduces the amount

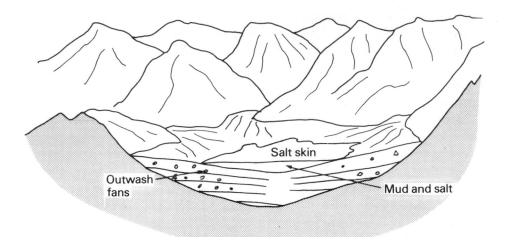

Figure 13.12 Generalised features of the kavirs of Iran.

of precipitation to under 300 mm in the Central Plateau, where summer temperatures reach 50° C and diurnal ranges are up to 25° C.

Much of the Central Plateau has endoreic drainage. Only in the east near Mashad is the drainage to the Caspian. Virtually all the rainfall on the inland flanks of the mountains drains towards a series of central basins forming the unique Kavir systems (Figure 13.12). These are fed by either surface seasonal streams or shatts — subsurface water or mud channels whose position is difficult to detect until the ground gives way under a person's weight. It is impossible to understand fully the Kavir system in terms of present climatic conditions. Various theories concerning the pluvial conditions of the Quaternary have been postulated. Some of these involve large inland seas, or a series of lakes, stretching across into Afghanistan; others are less ambitious and suggest simply a contraction of the most arid zones. What is certain, however, is that the wetter periods, accompanied by glaciation in the mountains, gave increased erosional and transportational capacity to the streams. This resulted both in the deepening of the valleys and in the spreading out of vast fans of sand, gravels and boulders away from the mountains. Even if higher precipitation was limited only to the mountains during the Quaternary period the effects would have been widespread. During pluvial phases there was additional outwash from the mountains, and the increased power of the rivers led to accumulations of fluviatile material in the residual lakes within the desert basins. Whilst this was deposited with a force which is still partially observable today, there also occurred the removal of deposits that had formed on the valley bottoms along the upper courses of the streams. These deposits were a feature of drier phases, when the river's transportational power was insufficient to carry outwash far into the foreland. Consequently, during arid intervals of the Quaternary, the accumulation of rock debris and pebbles was restricted to the bordering mountain ring and its valley floors, whilst deflational removal of lighter deposits, such as silt and sand, by wind action, was then prevalent in the lower desert regions.

It is not possible to understand the features of the Central Basin without an appreciation of these earlier conditions. Together with tectonic influences they have produced the relief on which present processes operate. The landscape in its present detail is the result of the decisive influences of seasonal water action and virtually continuous wind action.

During the earliest months of the year many parts of the inner basins are under water, mainly as a consequence of run-off from winter precipitation occurring over the border ranges that define inner Iran. Most of the north-eastern segment of the Dasht-i-Kavir, together with recently formed talus that is now covered by rainwater funnels or worked over by drainage furrows, point unmistakeably to the valleys' recent formation as the result of episodic falls of rain, which occur almost always as instability showers. Although denudation by the wind plays a subordinate part, the transport of wind-borne material must in no way be overlooked or underestimated. This view, which emphatically contradicts opinions held at present, is fully substantiated by, amongst other things, the existence of dunes on the southern margins of Dasht-i-Kavir; and in the interior of the Shahdad basin (southern Lut), morphological features produced by water action become fewer and fewer, whilst aeolian forces occupy unmistakably first place among the influences which have moulded surface land forms.

Large areas of sand are also to be found in the province of Sistan in south-east Iran. Here the 'wind of a hundred and twenty days' hollows out temporary basins arranged in rows and valleys with no outlets.

The Sahara

This desert is larger than the USA, and thus may be expected to exhibit a wide diversity of landforms, related to different geological arrangements and to the varieties of aridity within its borders. It is also a region which has an extensive core, or refuge, which has remained arid over long periods of time, and has constantly migrating boundaries — related to a combination of climatic changes and man's intervention which is difficult to unravel.

The size of the Sahara means that its aridity is related to a variety of atmospheric conditions. The west coast is affected by the cold Canaries Current, where cold water wells up and the winds blow virtually parallel to the coast (Figure 13.13). This is one of the most arid sectors of the Sahara. Inland the southern margins of the Sahara form an uncertain zone of competition between dry north-easterly winds and moist south-westerly winds, whilst in the north there are invasions by midlatitude depressions. Much of the atmospheric circulation near the surface, however, consists of winds blowing out from the land, and over shallow

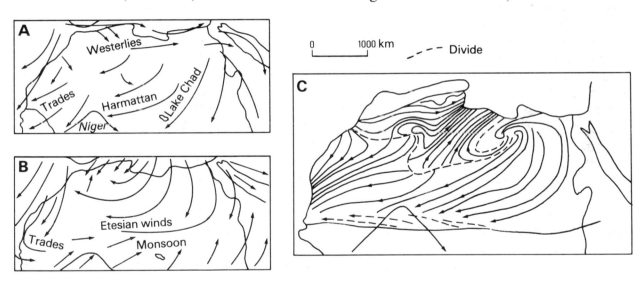

Figure 13.13 The Sahara: winds and dune alignments.
(A) January wind patterns; (B) July wind patterns; (C) Sandflow lines. ((A) and (B) after Dubief, 1953; (C) after Wilson, 1971; both in Cooke and Warren, 1973)

convectional movements beneath a major inversion which prevents most upward movements of the air. Disturbances likely to result in precipitation occur in the middle troposphere, and need to coincide with the right group of conditions for moisture to be precipitated. Showers are concentrated on upland areas such as the Tibesti and Ahaggar massifs. Large areas of the central Sahara receive little, if any, rain. The Saharan climate is noted for its extremes: recorded temperatures range from $-7°$ C in the Tibesti Mountains to $57°$ C in Algeria; annual precipitation has varied from 6.4 mm to 159 mm at Tamanrasset in Algeria; evaporation rates are high throughout the area; and larger storms commonly precipitate 1 mm rain/minute. Flooding occurs after even small storms: one observed in 1960 produced 16 mm in a nearby

314

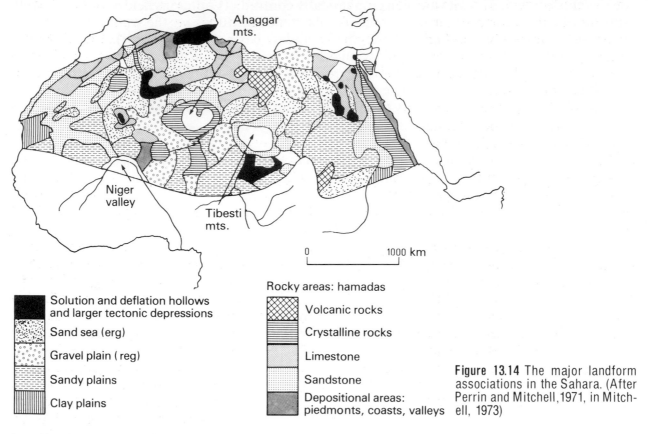

Rocky areas: hamadas

Solution and deflation hollows and larger tectonic depressions

Sand sea (erg)

Gravel plain (reg)

Sandy plains

Clay plains

Volcanic rocks

Crystalline rocks

Limestone

Sandstone

Depositional areas: piedmonts, coasts, valleys

Figure 13.14 The major landform associations in the Sahara. (After Perrin and Mitchell, 1971, in Mitchell, 1973)

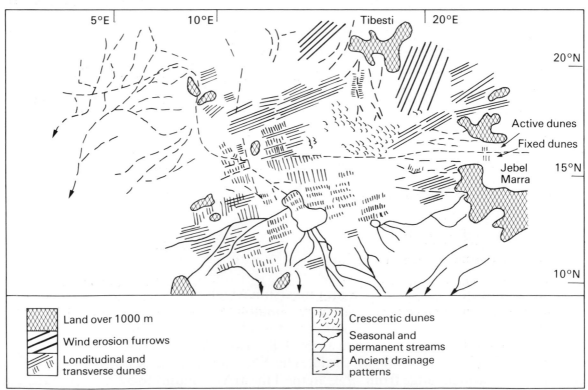

Land over 1000 m

Wind erosion furrows

Londitudinal and transverse dunes

Crescentic dunes

Seasonal and permanent streams

Ancient drainage patterns

Figure 13.15 The southern margin of the Sahara. Many of the dunes south of the active/fixed boundary have since begun to migrate again. Compare the dune orientations with Figure 13.13. (After Grove and Warren, 1967, in Cooke and Warren, 1973)

upland rain gauge (cf. mean annual precipitation of area = 18 mm), but a flood of 120 m³/s occurred in the fourth order stream, probably reaching a maximum of 1600 m³/s, whilst suspended sediment was as high as 1712 g/1. This contrasts with a high runoff value of 58 m³/s in the river Plym of south Devon following precipitation of 30-40 mm rain on land which was already sodden after several weeks of heavy rains, and emphasises the effectiveness of surface runoff in arid regions.

The landforms of the Sahara are dominated by massive sand seas (ergs) and by extensive areas of flat, bare rock (hamada), gravel-covered plain (reg), or of plains with a thin sand cover (Figure 13.14). There are also landscapes where stream-carved landforms dominate, and these are commonest in upland areas. The erg landscapes are particularly distinctive, and new studies of these from satellite images are adding much to the knowledge of dune types and formation. But it is the horizontal landforms which are most extensive: some are formed on bare rock; others have a thin covering of gravel; and others have sand swept into small local dunes.

The Sahara margins are zones of changing conditions. The southern margin has plentiful evidence of such changes (Figure 13.15) having shrunken remnants of former inland seas, clear lines of ancient drainage focusing on upland areas, and dunes fixed by vegetation. The boundary of the desert is difficult to define, and a few successive dry years are sufficient to remove it southwards, exposing dune sands to renewed wind-induced movement, and causing lake levels to fall. The northern margins now bring desert features and unvegetated landscapes to the shores of the Mediterranean in Libya and Egypt. In Roman times, however, this was a well-endowed farming region producing grain for export to Rome. Some have suggested that the subsequent deterioration was due to a climatic change, but others have blamed either the effects of grain cultivation (resulting in a dust bowl), or the nomadic peoples who followed the Romans. It is clear that man's activities on the delicately-balanced margins of a desert have to be considered carefully in the light of a need to conserve the small quantities of moisture available.

An aridity index

Several attempts have been made to devise a quantitative index expressing the precipitation–temperature–evaporation inter-relationship which determines aridity.

a) Thornthwaite (1931): the Precipitation Effectiveness Index. This is essentially a ratio of precipitation (P) and evaporation (E):

$$\frac{P}{E} = \sum_{1}^{12} \frac{Pm}{Em} \times 10$$

Pm = mean monthly precipitation; Em = mean monthly evaporation rate; both measured in the same units.

b) If the evaporation measurement is not available, an alternative is suggested:

$$\frac{P}{E} = \sum_{1}^{12} 115 \left\{ \frac{Pm}{Tm-10} \right\} \frac{9}{10}$$

(Tm = mean monthly temperature)

NB. The P/E ratio = 31 for semi-arid/humid margin; 16 for the arid/semi-arid margin.

c) Another index was designed by Thornthwaite in 1948, which included transpiration by plants. This is the potential evapotranspiration index. A simplifed version was proposed by Khosla in 1949:

$$Lm = \frac{Tm - 32}{9.5}$$

(Lm = mean monthly water loss in inches)

N.B. An Lm of 40 marks the semi-arid/humid boundary; 20 marks the arid/semi-arid boundary.

14

Geomorphology

Geomorphology, the science of landform study

The study of landforms is known as geomorphology — a science which has developed with the aid of geologists, geographers and engineers. Because there has been such a wide range of interests involved, there have been many different definitions of geomorphology based on personal points of view. Thus there are those who define it as 'the study of the origin and evolution of landforms', being particularly interested in using studies of landforms to elucidate the most recent geological history of an area. Unfortunately much of this work has been based on concepts and models which are too simple, and have come too early in the development of the science. The fallibility of this approach led geomorphologists in the 1950s and 1960s to see their study as 'the science of landforms and land-forming processes': they paid increasing attention to the measurement of the landforms and the ways in which processes may affect them. The main drawback discovered in this approach has been that many landforms cannot be explained purely by reference to the processes which are acting on them today. Simple associations are now regarded as unlikely, since a variety of complexly interacting factors are, and have been, at work to produce the range of landforms in an area. This much has been made clear throughout this book, and particularly in relation to the discussion of arid landforms (chapter 13). It is important, too, to realise just how much man's activities have affected the relationships which have been established in the natural order. The result of such discoveries has been to give a wider definition to gemorphology. It can be seen as studying the interactions taking place between the solid Earth surface and the atmospheric gases and oceanic waters. This involves a continuously changing situation of balance between natural and man-induced forces: the form of the land surface is thus subject to changes as the balance between the forces is modified. Geomorphology is the study of the Earth's changing surface. This view also has the virtue of releasing the subject from too rigid adherance to one model (e.g. the Davisian cycle), and thus makes it of greater value to those geologists and geographers who relate their studies to the landforms and processes operating at the Earth's surface. Studies of natural hazards, terrain evaluation and sedimentation in particular benefit from this view.

The changing definition of geomorphology has been accompanied by modifications in the methods adopted by geomorphologists in their investigations. In particular there has been a move from simple, descriptive writings, often clearly inspired by the beauty of the natural landscape, to more scientific approaches. This is due partly to engineers and geologists, who have devised techniques of measurement and rigorous procedures of dealing with geomorphological problems. Hydrologists and soil scientists have made particular contributions, too, so that the geomorphologist increasingly considers his field in the light of basic chemical and physical principles. This has been accompanied by a growing degree of quantified measurement of both form and process, and relationships have been refined to the point at which they can be expressed in mathematical terms.

The 'scientific' and 'quantitative' emphases have contributed much to the development of geomorphology, but it will probably never be possible to reduce the entire natural system to a set of mathematical models which sum the total of interacting factors and are able to predict results from certain combinations of these factors. The complexity of natural systems and the irregularity of landform-changing events, along with the changing patterns through time, makes a combination of quantitative and qualitative observations necessary. It is often extremely difficult to take the necessary measurements, or to define which are the significant measurements to be made amongst the wealth of variables. Physicists, used to laboratory conditions in which individual variables can be isolated and their inter-relationships examined, are often frightened away by the multivariate nature of geomorphological problems. Size and inaccessibility also create difficulties for field operations. Even where measurements are available they may be too few at too infrequent intervals, and commonly relate to a different purpose from the one in hand, if older results are used. The reduction of measurement and relationships in geomorphology to mathematical form carries an inherent danger that such expressions attract an aura of precision and finality which is not justified by their bases.

The major contribution of such developments in geomorphology has been to force the geomorphologist to go back to the basic field evidence. There is much to challenge the scientist here, since new methods of measurement often have to be devised, and it is better if continuous records can be kept instead of occasional, random visits: unusual events leading to landform changes tend to take place at night, or in the middle of winter. The parallel attempt to bring the field situation into the laboratory has not been so successful, although many concepts in stream and sediment flow can be illustrated in this way. The reduction of natural systems to laboratory-size wave tanks and flumes meets the problem of scale in time and space. It has proved impossible, for instance, to produce a muddy sediment of a particle size which acts like real mud in a scaled-down model of the river Thames estuary.

Geomorphology has been seen sometimes as part of the sedimentologists' field; in the United Kingdom it has formed a part of geographical training, and wider regional divisions were for long based on physiographic characteristics; the field of applied geomorphology (e.g. applying the understanding of natural processes to the prediction of what gradient is needed in a road-cutting, or of rates of silting in harbours) is largely in the hands of civil engineers. The science has thus always had an essentially interdisciplinary basis and relevance, but recent developments in approach are drawing together the geological aspects with the geographical study of man's effect on the landscape, and with categories of data of interest to engineer, agriculturalist and soldier. The study of landforms is being related to terrain evaluation and the conservation of land resources as man fills his world and wishes to use its surface to the full. Team projects, like the Morecambe Bay Barrage scheme, or the Sabrina Project of the lower river Severn basin, rely on the cross-fertilisation of ideas which is now increasingly possible.

The development of ideas in geomorphology

Ideas have always been easy to come by: the hard work is involved with relating these to real and useful situations. The history of the development of geomorphology illustrates how the increasing facilities open to man in the examination of his environment have led to more rigorous testing of his speculative fantasies.

In the years before the period AD 1700–1800 men lacked the means for extensive travel, and for the methods of testing ideas. Many accounts of natural phenomena were necessarily based on local observation and were explained by reference to conventional authorities, including the scriptures and an accepted lore which had grown up around natural features. The idea that the water issuing into streams from springs has been drawn up from the oceans through rocks was repeated in many places: this was a world before the full implications of gravitational force could be applied to such situations, and where most explanations were of the sort that could not be tested by direct observations. There were people like Leonardo Da Vinci (1452–1519), who suggested that rivers eroded valleys, but his notebooks were not published for 200 years after his death. Communication of ideas was still slow and laborious at this stage. Although Bernard Palissy wrote in 1580 that stream water comes from rain, few took any notice of his pronouncement. Some of the first quantitative studies were recorded by Pierre Parrault in 1674. He measured the area of the river Seine basin above Aignay-le-Duc as

121.5 km², and calculated that the mean precipitation of that area from 1668 to 1670 was 500 mm. A total of 60 750 000 m³ of water were received each year, and this was more than enough to account for the discharge at Aignay-le-Duc (10 000 000 m³). During the eighteenth century engineers employed to build roads, canals and harbours had to contend with landforms and surface processes, and made reports on their observations and the problems overcome. This focused attention on small-scale effects and fairly rapid changes in stream banks, valley side slopes and beaches. At this time the scientific approach of observation leading to hypothesis and hence to further testing was being evolved in practical ways by such 'applied scientists'.

These developments, however, were overtaken and eclipsed by the growth of geological science in the early nineteenth century. The observations of James Hutton caused an increasing group to see time as having a vast extent, compared to the 6000 years of Earth history envisaged by some Bible commentators. He argued that the past could be understood from studies in the rock record by reference to processes operating today. When Charles Lyell based his 1830 textbook on this principle he was castigated as a 'uniformitarian', but the idea could be seen to work in the explanation of so much. It had two important results for geomorphological studies. One was that a series of 'former worlds' could be suggested in which phases of mountain uplift were separated by periods of erosion and the formation of new rocks, which were then often uplifted again. The second was an emphasis on the larger structural units of study and a corresponding neglect of the mechanisms by which the surface processes acted. Whilst the dramatic effect of volcanoes, and the pounding of the sea coast by waves, could be seen to be effective in moulding the land surface, little attention was paid to their modes of operation. More time was spent in acrimonious public debate as to whether all rocks were formed in the sea, or whether a large proportion were due to volcanic action.

At the end of the nineteenth century observations had increased to the extent that a number of simple conclusions were being drawn. Unfortunately the simple conclusions, instead of being tested further, became grandiose schemes, which dominated the subject and held back development. In the USA W.M.Davis followed up the work of the 1870-80 explorers of the western areas, like Powell, Dutton and Gilbert. They had introduced such terms as 'grade', 'base-level'; 'consequent', 'antecedent' and 'superimposed' drainage; and Davis combined these in the concept of the 'cycle of erosion' (Figure 14.1). This could be used to explain the

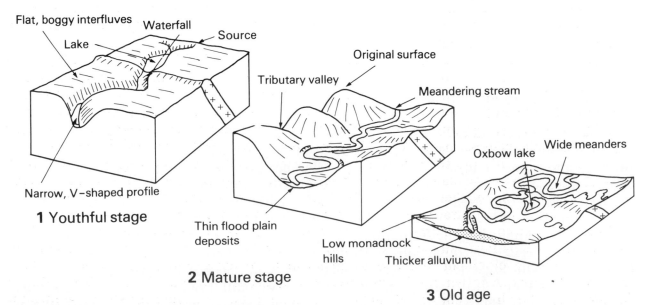

Figure 14.1 The 'normal' Davisian cycle of river erosion. Notice how this scheme suggests that irregularities are removed from the course and hills are worn down after initial incision by streams. The final stage of low-lying relief is known as a peneplain.

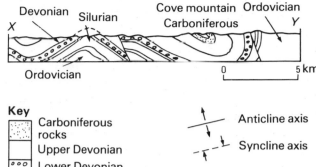

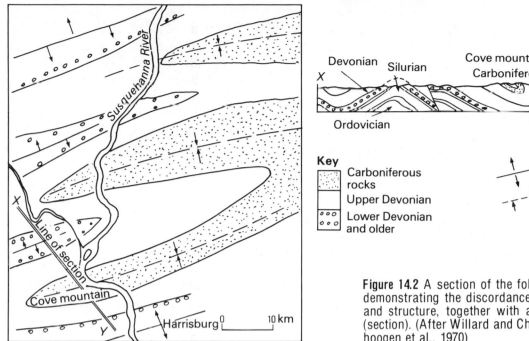

Figure 14.2 A section of the folded Appalachians, demonstrating the discordance between drainage and structure, together with accordant summits (section). (After Willard and Charles, 1938, in Verhoogen et al., 1970)

accordant summit levels and discordant drainage patterns of the Appalachians (Figure 14.2) and became widely accepted by geologists in relation to unconformities, which were seen to be ancient peneplains. It was also taken up by teachers of geography in schools, where simplicity of the model commended itself: it is still difficult to eradicate the idea from the minds of students once they have it fixed there!

The cycle of erosion concept has several major failings. It is difficult to apply it to actual situations, and even where a particular valley form has been designated 'youthful', for instance, little has really been added to the knowledge of valleys, and the designation does little to encourage further study of them. The difficulties of application are related to the facts that (a) it is now regarded as improbable that mountain building took place in the short-lived spasms envisaged by Davis, and (b) the changes of climatic conditions and base-levels take place too rapidly to allow a complete Davis-style cycle to run its course. This has meant that landform development can no longer be regarded as cyclic in sequence, but rather as a series of oscillations between changing materials, changing life forms and a variety of other changing conditions (e.g. climate and even the rotational speed of the Earth).

At the same period studies in the Sonoran Desert of Arizona by McGee, Page and Bryan led to conflicting ideas including parallel slope retreat and pediment formation. These were taken up and applied to humid landscapes by Penck in Germany in the 1920s. His ideas on slope evolution (chapter 6) were related to the interaction of tectonic and surface processes. Uplift would begin slowly, resulting in gentle slopes; it would then increase in intensity to give steeper slopes with convex upper parts. An abrupt lowering of base-level would result in downcutting at the stream mouth and the formation of knick-points migrating upstream. Slopes would retreat parallel to former states, leaving remnant inselbergs above low-angle plains. L.C. King of South Africa applied these ideas in a more general way to the interpretation of pediments in South Africa, and then traced them in most parts of the world in the 1950s.

Both the cycle concept of Davis and the parallel retreat of slopes were related to periods of uplift followed by fairly stable tectonic and base-level conditions. Both suggest a series of events in the development of the relief in an area, relating to alternations of uplift and erosional phases. Another model suggests that landscapes do not show evidence of past history once they have achieved a state of dynamic equilibrium, since the stream gradient is then adjusted to the quantity of water and sediment load and to their reactions with the materials in the stream bed. The concave long profile of the stream reflects the downstream

Plate 208 A lunar view from Apollo 15 (1971). The crater Aristarchus (35 km diameter) to the left of the meandering rille known as Schroeter's Valley. The rille has its head in the feature known as Cobra's Head, and beyond that is the crater Herodotus. Rilles such as this may have been formed by the eruption of hot gas and lava droplets melting their way through the surface rocks. (USGS)

Plate 209 Apollo 11 (1969) oblique view of the lunar far side. The largest crater has a diameter of 35 km. (USGS)

Plate 210 The lunar surface: part of a rock-strewn crater with the Apennine Front (left) and Hadley Delta (right) in the background. (USGS)

Plate 211 Meteor Crater, Arizona, was formed in a similar way to many Moon craters. (USGS)

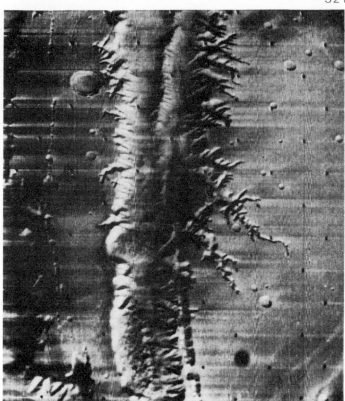

Plate 212 The north wall of Ganges Canyon, one of the branch canyons near the equator of Mars. Avalanches can be seen along the walls; there are sand dunes; and meteorite impact craters have affected the flatter lava flows to the right. This picture was obtained by Viking I on 1 July 1976.

Plate 213 The major canyon at the Martian equator, some 70 km wide and over 6 km deep. It is thought to be a relict of a giant fracture of the Martian crust, opened up by erosional processes. Picture taken by Mariner 9 on 12 January 1972.

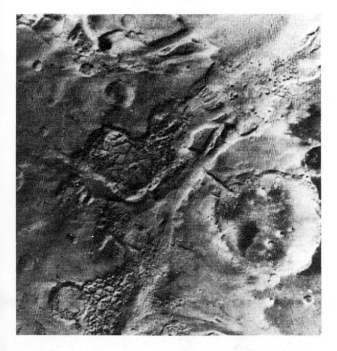

Plate 214 These massive flood channels, up to 40 km wide, were seen by Mariner 9. They must have been formed by sudden discharges of water or the melting of dry ice, leading to the collapse of their headwater areas.

Plate 215 The largest known volcano, Nix Olympica on Mars. It rises 24 km above the surface of the planet (cf. Mount Everest is 10 km high), and is 600 km across at the base. Masses of lava must have been poured out at the same place over many millions of years (cf. the Hawaiian islands and plate tectonics, Figure 2.14)

increase in discharge, and watershed in an area of uniform rock will tend to be at the same height. This produces a landscape where accordance of summit levels is not related to an event in the past. Other time-independent landforms may be produced if uplift of an area is at the same rate as the denudation of the area.

Further ideas also tend to confirm warnings that a simple model like the Davisian cycle cannot provide a far-reaching basis for landform studies. Most of the grand models so far devised have paid little attention to weathering processes, and there is often the implication that rock outcrops are soon covered by weathering products. Gilbert first noticed that whilst the rapid and partial removal of regolith accelerates rock decay, the complete removal often slows it down. Weathering is most active in the zone between the water table and the surface, due to the constant passage of water: this assumes rock permeability, but once joints are opened out weathering can attack from all sides, whereas a surface outcrop of bare rock, particularly in the tropics, may be weathered slowly due to its compactness. For this reason tropical rivers are often characterised by rapids which may grow into waterfalls, instead of being reduced by stream action. Benches may also become more prominent on hillsides once rock is exposed at the surface. Both of these events would be in direct conflict with a cycle of erosion which involved increasingly subdued relief and the elimination of irregularities.

The work of W.M. Davis also left the idea of 'normal erosion' to posterity: he regarded the work of streams as being in this category. Increasing knowledge of worldwide landforms and surface process operation has led in the twentieth century to the development of climatic geomorphology. Distinctive groupings of landforms are recorded in equatorial, savanna, desert, semi-desert, humid temperate, periglacial and glacial areas, and can be related to processes operating in the present and recent past (Figure 14.3). There must still be qualifications to such an approach, however, for the reasons advanced in chapter 5: inter-relationships

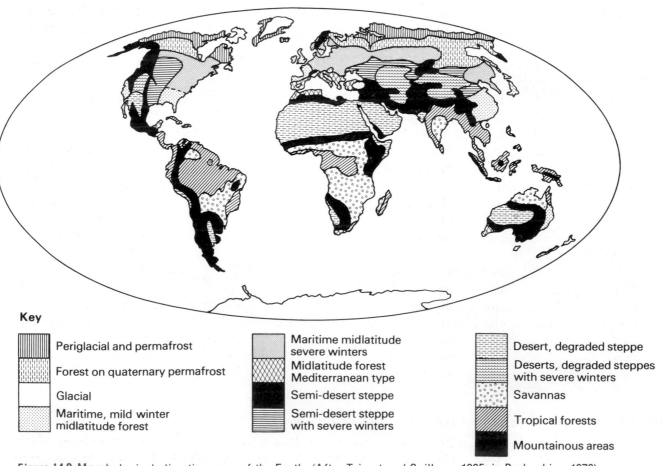

Key

▦	Periglacial and permafrost
▨	Forest on quaternary permafrost
☐	Glacial
⣿	Maritime, mild winter midlatitude forest
⣿	Maritime midlatitude severe winters
▩	Midlatitude forest Mediterranean type
■	Semi-desert steppe
▤	Semi-desert steppe with severe winters
▥	Desert, degraded steppe
▦	Deserts, degraded steppes with severe winters
⦂	Savannas
▨	Tropical forests
■	Mountainous areas

Figure 14.3 Morphological-climatic zones of the Earth. (After Tricart and Cailleux, 1965, in Derbyshire, 1973)

between form and process are poorly established, and the influence of man is playing an increasing part in interpretations.

The history of fashions in geomorphological ideas has led from the days of acrimonious discussion over rival, but largely unsupported, ideas to a situation where grand theories and schemes are unpopular. The stage has been reached when too many of these have been shown to be inadequate, and where it is realised that much basic observation and measurement is necessary before further grand models can be of use. It will be some time before a general theory of climatic geomorphology is acceptable, or truely scientific and quantitative approaches are viable. The science is now in a state of research, less fettered by preconceptions than for many years, and aided by increasingly sophisticated instruments. It is to be expected that this will make it more useful to a wider circle of people.

Principles of geomorphology

If grand models of landform development have been dismissed for the time being, certain basic concepts remain and are accepted widely. Many are still open to a degree of debate, and it is profitable to discuss them in this light, based on the studies which have been carried out in this book in relation to the processes and their interactions with the Earth's surface.

The stability of the Earth's surface

A major distinction was drawn in chapter 1 between the areas of the world which are tectonically active, experiencing earthquakes and volcanoes, and other areas which are not. The active areas have been correlated with plate margins and the formation of fold mountain ranges, whilst the stable areas between have been characterised as passive zones carried on the slowly moving plate foundations. The movements taking place at the plate boundaries are possibly as violent as any which have taken place during geological time: in short, the tectonic processes giving rise to major relief features act relatively slowly over millions of years, and are not sudden events.

At the same time the 'stable' regions can be subjected to important changes in level. Loading and unloading by ice accumulation or melting, sedimentation or erosion, or by the accumulation of volcanic materials, may lead to downwarping, or upwarping of the continental or ocean-floor surface. Such changes in patterns of ice storage and of ocean basin shape, have resulted in variations in the water levels in oceans. It is often difficult to separate such different effects, but measurements show that the railway line across the Caucasus Mountains rises by up to 7.3 mm per year in some places, and sinks by up to 6.4 mm elsewhere; figures have been given for the isostatic recovery in the Baltic Sea area; Quaternary sediments, formed in the last 2 million years, show that the Central Valley of Chile has subsided by 2000 m in this time. Even the more stable areas show evidence of considerable changes over the most recent phases of geological time (e.g. the west coast of France is rising by 0.9–2.8 mm per year, or 280 cm in 1000 years).

This picture of movement must be seen in the context of interpretations which suggest that uplift of land masses has been a short-lived process, followed by long periods of erosion related to the seldom changing base-level of the ocean surface. The interpretations of Quaternary sea-levels (Figure 8.21) provide a necessary consideration in landform studies, but must be seen as tentative, and in relation to local conditions of stability. Each area has developed according to a pattern of interactions between uplift and displacement of rocks, changing surface processes, and variations in base-levels, each aspect in this list being determined independently.

Uniformitarianism and catastrophic events

The basis of modern geological investigations is the concept of uniformitarianism, by which the results of past process activity is interpreted in relation to present understanding of ways in

Process	Volume	
	Karkevagge m³	Upper Rhine m³ x 10⁶
Rockfalls	50	0.016
Avalanches	88	0.25
Earthslides, including sheet slides, mudflows, schist slides	580	1400 plus
Creep: talus and solifluction	850 000	41.6
creep	—	1200
Running water: dissolved salts	150	0.224
fluviatile erosion	—	4

Figure 14.4 Processes in two contrasting environments. Karkevagge is in northern Scandinavia, where the slopes are not so great as in the Alpine zone of the upper Rhine valley, but where tundra conditions are prevalent. The upper Rhine valley has many slopes which are unstable because they are so steep and because the rocks are schists, which collapse easily. Compare the effects of rapid mass movement, slow mass movement and running water in the denudation of these two areas. Notice the differences in scale of the volumes. (After Rapp, 1960 and Jackli, 1957, in Leopold, Wolman and Miller, 1964)

which processes work today. This idea has been much misunderstood from the time of its original conception, and its implications of a slow, steady progress of landform development has always seemed to be in conflict with the 'catastrophist' approach which emphasises the role of sudden and short-lived events. The contrast has seemed to be most clear in places like the United Kingdom, where natural events are not so intense as in many parts of the world.

The studies of landform processes showed that a large proportion of the work accomplished by geomorphological agents is carried out after floods, avalanches, glacial surges, volcanic eruptions, storm-wave activity and where man carries out constructional activities. The relationship of such catastrophic events to the more uniform activity of processes varies with the local environment (Figure 14.4). It is also clear that much of the erosion and transport of rock debris carried out in short-lived events follows long periods of uniform weathering and slow movement, leading to an unstable position which is adjusted by a rapid movement. The long period of geological time puts the seeming conflict into its true perspective. Thus the short-lived, catastrophic events, occurring every 10, 100 or 1000 years, are seen in relation to periods of several million years during which the whole landscape is modified by a variety of factors. The time element is thus of great importance in an analysis of landform evolution.

Climatic and geological factors in landform genesis

The primacy of surface processes, or of the underlying geology (i.e. major structures together with details of rock lithology), in the formation of landforms and of their distinctive distributions and associations, has been debated widely. A geologist may stress the effects of rock-type or structural lineations on landforms, whilst a geographer may place emphasis on worldwide differences in distributions in an attempt to relate these to climatic regimes and the varying effects of atmospheric processes.

Ideas of climate-related origins for landforms began with W.M. Davis's adaptation of his cycle confept to arid and glacial areas. This approach was limited in its usefulness by vagueness of observation and by the need to fit events and landforms into a progression. He suggested, for instance, that glacial valleys become graded with time — although no smooth profiles akin to the suggested smooth concave stream profiles had been observed. Such views were perpetuated, however, and the cycle concept was extended to savanna and periglacial regions. Once again this produced little more than a series of imprecise and debateable statements.

In Europe a different approach led to a more profitable result. Penck, whose ideas on slope development have also been discussed (chapter 6), suggested in 1910 that climatic zones could be defined from a study of landforms and the processes affecting them. This was intended to contribute to the idea of 'natural regions' combining climatic, soil and vegetation zones, which was then popular. E. de Martonne, a French geographer, really set the future pattern in 1913 by emphasising the varying character of fluvial processes in different climatic zones; by distinguishing climatic and structural factors on a scale basis; and by including the factor of

'Facies'	Processes at work
Warm humid	Rock decomposition rapid and chemical processes dominant, producing great thicknesses of residual soils. Few rock outcrops or escarpments: slopes gentle and large alluvial cones. *Examples.* equatorial areas of Africa and South America
Humid temperate	Similar in many ways, but less developed: thinner soils and steeper slopes. *Examples:* west and central Europe, eastern USA.
Dry season	Mechanical weathering more common, producing a variable weathering mantle with coarse fragments. More rocky escarpments and higher slope grades. Streamflow irregular and violent, producing badlands. *Examples:* steppes and savannas, Mediterranean areas.
Desert	Mechanical decomposition produces coarse, unequally-broken debris accumulating in place: attacked by wind to produce stony hammada, saline basins, sandy wadis and dunes.
Cold dry	Polar regions and high mountains. Dry air and frozen soil with seasonal thaw. Mechanical weathering processes give rise to vast sheets of coarse debris.
Cold humid (glacial)	Areas of perpetual snow acting to wear away and transport rock as ice masses.

Figure 14.5 The 'geographic facies' concept of E. de Martonne, put forward in 1913. How far do his 'facies' relate to modern ideas of morphological-climatic zones, and in what ways would you quarrel with the ideas set out here on the basis of modern knowledge? (After de Martonne, 1913, in Derbyshire, 1973)

climatic change. He concluded by recognising a group of 'geographic facies' (Figure 14.5). In 1926 there was a conference in Germany on *'The morphology of the climatic zones'*, and this looked forward to many of the ideas being taken up today.

In more recent years the simple climate-landform relationship has been regarded as an insufficient, and perhaps misleading, correlation. Emphasis has switched from temperature and precipitation controls of processes to the inclusion of complex physical and biochemical reactions: vegetation and soil processes serve as intermediaries between the atmospheric processes and landforms in many parts of the world. There is a realisation, too, that the results of ancient climatic changes may affect the present features overwhelmingly, and that general statements are too vague at this stage to mean much. The climatic geomorphologist, like any other, must examine the full range of interactions in the environment, and must attempt quantitative and long-term experiments to determine adequate correlations.

A recent contribution to this debate (Thomas, 1974) examines tropical processes and landforms essentially because they show the results of weathering, mass movement and streamflow and exclude frost action. The processes may not produce landforms restricted to a particular climatic zone, but the tropical conditions do restrict the forces operating. Another important consideration in tropical areas is the fact that the Quaternary climatic changes have been less drastic in the tropics than elsewhere, so that the effects of the processes over longer periods can be assessed. An attempt is made to cut through the one-sided picture of earlier studies in tropical areas, which studied mainly areas of ancient Precambrian shield and crystalline silicate rocks: there are also highland areas in the tropics which have received less attention. The main message of this study is, however, to place the emphasis on the full study of atmosphere-soil-plant-animal-landform-tectonic interactions over time. Consideration of any pair of these factors on its own can be expected to result in a false picture.

Dealing with natural complexity

Geomorphology is a study of interdependent forces at work in nature over long periods of time. It is thus a complex study, and the first temptation is, perhaps, to generalise and simplify. It is clear, however, from the progress of the subject, that all generalisations concerning relationships between form and process must be examined carefully — and that includes those made in this book! To be useful any generalisation must be based on the study of reality, rather than of abstract speculation or views from the tops of prominent hills. If a generalised statement or model assists further enquiry, or enables a student to understand a particular concept more easily, it is worth pursuing. All such generalisations must be seen as a temporary view of a state of knowledge, and must not be allowed to determine future developments. Thus the study of a new area of topic should not be conditioned by trying to fit a particular model to the situation before full investigations have been carried out. Models should be seen as a form of hypothesis to be tested or modified in the light of new measurement and observations. The value of a model lies more in its 'fruitfulness' as a source of further discovery, than in its 'correctness'.

Natural complexity also makes it difficult to decide which measurements will be the most significant. It is easiest to deal with the smallest number of variables, but such an approach looks only at a part of the whole problem, since virtually all natural situations of the type facing the geomorphologist are multivariate. A promising approach to this difficulty is through systems analysis, the principles of which are outlined in chapter 7. There is a temptation to become involved with systems thinking and terminology for its own sake. This form of analysis is a tool for the geomorphologist to use and apply, and allows him to reduce situations to their basic parts, which can still be seen to interact. The systems approach also tends to emphasise the concept of 'steady state' (i.e. input = output), or of 'dynamic equilibrium' in larger-scale situations where conditions are more variable. These may be applied in landform and process studies, but it must be realised that landforms evolve over time and as the external forces change in emphasis.

Applying geomorphology

It has been suggested in the past that the geomorphologist has little to contribute to other fields, and particularly to practical applications of his subject. In this view the value of studies in geomorphology is as an intellectual activity involved with controversies over rival theories, and having the increasing challenge of demands for scientific rigour of observation to be combined with reasoning and imagination. This, however, reflects the approach to geomorphology which was encouraged by Davisian thinking, and the results of that approach which made the subject of so little use. Civil engineers had to solve the practical problems in their own ways because geomorphological ideas were so conditioned by the theoretical models that they could not be applied to real situations in a meaningful way.

New approaches to the subject encourage the view that it has much to contribute today in a variety of fields, including both the academic and applied studies.
1) Geomorphological studies become more valuable to the geographer, since they take an increasing account of man's part in modifying natural processes and landforms (chapter 11).
2) Geomorphological studies become more useful to the geologist, and in a new way. In the past they have suggested ideas to help in the interpretation of the past — peneplains and unconformities were compared, and buried landscapes related to modern features. More recent quantitative studies of streamflow and sediment movement, together with the field observations of glaciers and wind activity, have aided a fuller understanding of sedimentary processes — although it must be emphasised that this has often come from the researches of geologists themselves.
3) Terrain analysis, with its correlation of landform, soil and surface land use, has grown from the wider view in which modern geomorphology is finding its place. This field has connotations of environmental conservation, and also has military applications (Figure 14.6).
4) Geomorphological studies are also relating more closely to the experience of the engineer,

Surface composition

- ☐ Consolidated rock: outcrops of granite, gneiss
- ◻ Non-consolidated material
- ⊙ Mineral soil — poorly graded sands and silty sands
- ⊛ Organic soil — fine and coarse fibrous muskeg
- ○ Water bodies more then 1 m deep and 0.5 hectare in area

Surface morphology: Macromorphology

Slope steepness		Slope form	
△1 0-6°	0-10%	1 Convex, smooth	
△2 6-14°	10-25%	2 Planar, smooth	
△3 14-26½°	25-50%	3 Concave, smooth	
△4 26½-45°	50-100%	4 Convex, rough	
△5 Over 45°	Over 100%	5 Planar, rough	
△6 Classes I, II		6 Concave, rough	
△7 Classes II, III		7 Classes 1, 3	
△8 Classes III, IV		8 Classes 1, 2, 3	
△9 Classes I, II, III		9 Classes 4, 5, 6	

Surface morphology: Micromorphology

- ◣ Positive features of mineral soil with a random linear pattern.
  Slopes 6, 7, lengths 400-1800 ft (125-550 m), width-length ratio 1:4-1:20, amplitude 1-30 ft (3-10 m), spacing 10 per mile (6 per km), non-symmetric sigmoid in section.
  Aeolian — fixed sand dunes.

- ◣ Positive features of mineral soil in a random linear pattern.
  Slopes 6, 7, lengths 50-400 ft (15-125 m), width-length ratio 1:2-1:8, amplitude less than 10 ft (3 m), spacing 18 per mile (12 per km), irregular sigmoid in cross-section.
  Aeolian — sand sheets and ripples.

- ⊖ Negative features in mineral soil in a clustered, non-linear, overlapping pattern.
  Slopes 7, 9, lengths 10-200 ft (3-65 m), width-length ratio 1:1-1:2, amplitude 10-50 ft (3-15 m), spacing 15 per mile (10 per km), irregular cardioid in section.
  Glaciofluvial — kettle holes.

- ● Positive features of consolidated rock in a random, non-linear pattern. Slopes 5, 5, lengths 20-100 ft (6-30 m), width-length ratio 1:1-1:2, amplitude 10-30 ft (3-10m), spacing calculation not possible, irregular rectilinear in section.
  Glacial — rock outcrops and erratics.

Surface cover: Vegetation structure and spacing

Height	Stem type	Form	Spacing	
↑ Over 25 ft (8 m)	Woody	Trees	① Over 250 ft (80 m)	⑥ 25-40 ft (8-12 m)
⇞ 5-25 ft (2-8 m)	Woody	Young trees	② 140-250 ft (45-80 m)	⑦ 15-25 ft (5-8 m)
▼ 2-5 ft (1-2 m)	Woody	Tall shrubs	③ 90-140 ft (30-45 m)	⑧ 10-15 ft (3-5 m)
▼ Under 2 ft (1 m)	Woody and non-woody	Low shrubs, grasses,	④ 60-90 ft (20-30 m)	⑨ 0-10 ft (0-3 m)
			⑤ 40-60 ft (12-20 m)	

Figure 14.6 Part of the system of terrain mapping used by the Canadian army. Suggest ways in which this can be used for military purposes (e.g. in the passage of an army through an area, including heavy armour), and apply the scheme to an area you know. (After Parry et al., 1968, in Mitchell, 1973)

as several of the studies in this book emphasise. Studies of mass movement, fluvial and marine processes are most vital in this sphere, and it is often the work of engineers which has re-vitalised basic geomorphological investigations.

In short, geomorphological studies are coming of age, and in future will make increasingly important contributions to man's use of his living space.

Lunar landforms and processes

The increasing knowledge of surface features on the Moon and the planets near to the Earth (Mars, Venus, Mercury) prompts a degree of comparison to be made. These bodies experience different surface conditions, due to their sizes, distances from the Sun and rotational characteristics (Figure 4.22). The science of these bodies is sometimes called 'planetology', but the geological and geomorphological aspects have been referred to as 'planetary geoscience'.

The Moon, the closest body to the Earth, has been known in some detail from telescope observations for many years. The resolution of these observations before 1959 was only 500 m. Photographs taken by Moon-orbiting spacecraft in the 1960s, and then the observations made by the landings from 1969 onwards, have added immensely to knowledge of the surface features and conditions and to an understanding of the origin and history of the Moon. The studies of rocks, landforms, radiometric dates, earthquake activity and magnetic field in particular have enabled scientists to work out a more accurate idea of Moon evolution.

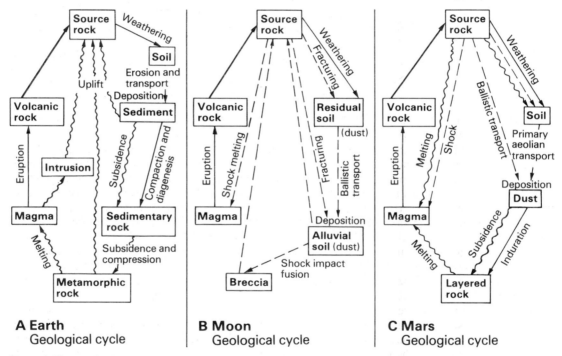

A Earth Geological cycle

B Moon Geological cycle

C Mars Geological cycle

Figure 1 The geological cycles of the Earth, the Moon and Mars compared. Look out for further information for these and other planets. What are the common features and differences? (After Wilson, Pickard and Harp in 'Geology', 1974)

The Moon has no atmosphere, the main processes affecting the surface being a combination of volcanic activity (internal energy) and impact cratering due to meteorites (Figure 1). Both processes also occur on the Earth, but meteorite impact is a relatively rare type of landform origin, since most meteorites burn up in the atmosphere and the craters of those which do reach the surface are destroyed relatively soon by erosion. Although some volcanic activity may still occur on the Moon (as suggested by smoke seen through telescopes), it is now restricted to gas emission, and most of the Moon rocks belong to two time groups: one phase of volcanic eruption took place over 4000 million years ago, and the second between 3700 and 3400 million years ago according to the radiometric dates of the samples brought back to Earth. These two periods of volcanic rock formation have given rise to the major twofold distinction in the Moon's surface features — the 'highlands' and the 'maria' (i.e. the 'seas'). The highlands consist of the older rocks, which were intensively cratered by meteoritic impacts before the maria were formed as lava flowed into many of the craters formed by this process (Figure 2). A major distinction between these zones is in the degree of cratering, since the maria have been affected less over the somewhat shorter period available. Both groups of rocks are extremely old by comparison with the Earth's surface, where

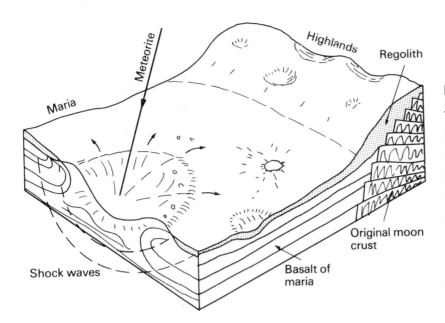

Figure 2 The origin of lunar surface features. The original crust was shattered by meteorite impact and the eruption of lavas which filled the maria: this still forms the foundation of the high-land areas. Much of the surface modification is now carried out by tiny meteorites: larger meteorites form a crater as shown. (After Aglinton, Maxwell and Pilinger, 1973, in Press and Siever, 1974)

the oldest-known rocks are 3900 million years old, but only 20 per cent of the land surface (i.e. 6 per cent of the whole Earth) has rocks older than 600 million years.

The difference between the relief features and the ages of the Moon and Earth surfaces is due to the combination of the Moon having a smaller mass and less internal energy with the absence of a lunar atmosphere. Most of the Moon's internal energy was used up at an early stage in its evolution, and it now has a thick crustal layer and little internal melting: volcanic activity of more recent times is related to the heat generated by meteoritic impacts which may result in local melting. The rilles, for instance, may be due to the effect of lava flowing across the dusty surface, and many show features akin to lava channels on the Earth. The absence of an atmosphere means that the Moon has no sedimentary rocks: weathering, streamflow, glaciers and wind are absent. It seems, then, that the Moon surface may represent an early stage in the development of the Earth, but one which has been frozen for the remainder of its existence by the depletion of internal energy sources and the lack of external interactions apart from meteorite impacting.

Mars has also been surveyed photographically by orbiting satellites, and craft have landed on the surface. Its surface bears many resemblances to that of the Moon, since it is also heavily cratered, but the planet does have a thin atmosphere (giving a surface pressure of approximately one-twentieth of that at the Earth's surface) and ice caps accumulate seasonally at the poles. The atmosphere is 90 per cent carbon dioxide and temperatures are generally low, since Mars is farther from the Sun than the Earth, and it seems that wind activity and the movement of frozen carbon dioxide may be effective agents of surface change, as well as the volcanic and meteoritic activity found on the Moon. Enormous volcanoes and canyons suggest that plate tectonics may have scarcely started before major activity ceased.

Venus is more akin to the Earth in size and mass, but it is nearer to the Sun. A Russian probe has suggested that surface atmospheric pressures are 90 times greater than those on the Earth, and that temperatures there reach 500° C. Little is known of the surface features due to the dense cloud cover, but it is thought that volcanic activity is more common than it is on the Earth today. The Venusian atmosphere is composed of sulphuric acid and perhaps hydrofluoric acid droplets, and these must result in very rapid weathering of the rocks at the surface.

Mercury is smaller than the Earth and is the planet closest to the Sun. Mariner 10 flew within 720 km of the surface in March 1974. Mercury's surface is pock-marked like the Moon. It has a very thin atmosphere of gases like argon, neon and helium, and temperatures up to 510° C (enough to melt lead), but down to −210°C on the dark side.

The more that is known about other planets, the more it seems that the Earth's surface is unique in the Solar System. Some of the processes occurring on the Earth are mirrored elsewhere, but there are many differences in the processes acting and in their emphases.

Bibliography

This is partly a record of the authorities consulted in the writing of this book, but also a basis for further study. The list has thus been subdivided into four categories:

Section A: texts and readers which can be used in Sixth Form courses, of an approximately parallel standard to this book.

Section B: advanced texts, dealing with specialist areas within geomorphology, or with the whole in greater detail.

Section C: individual papers and selections of scientific papers, mainly of a review nature. These are useful for background reading.

Section A

Bloom, A. L. (1969). *The surface of the Earth.* Prentice-Hall.

Bradshaw, M. J. (1977). *Earth, the living planet.* Hodder & Stoughton Educational.

Brunsden, D. & Doornkamp, J., Editors (1974). *The unquiet landscape.* David and Charles.

Calder, N. (1972). *Restless Earth.* BBC Publications.

Calder, N. (1974). *The weather machine and the threat of ice.* BBC Publications.

Curran, H. A., Justus, P. S., Perdew, E. L., & Prothero, M. B. (1974). *Atlas of landforms,* 2nd Edition. Wiley.

Francis, P. (1976). *Volcanoes.* Penguin.

Hallam, A. (1973). *A revolution in the Earth Sciences.* Clarendon Press.

Hamilton, W. R., Wooley, A. R., & Bishop, A. C. (1974). *Minerals, rocks and fossils.* Hamlyn.

Hoskins, W. G. (1955). *The making of the English landscape.* Hodder and Stoughton.

Laporte, L. F. (1972). *The Earth and human affairs.* Harper and Row.

Leopold, L. B. (1974). *Water: a primer.* W. H. Freeman.

Newson, M. D. (1975). *Flooding and flood hazard in the United Kingdom.* Oxford University.

Oakeshott, G. B. (1976). *Volcanoes and earthquakes: geologic violence.* McGraw-Hill.

Open University Environment Course Team (1972). *Coastal Environments.* Open University Press.

Press, F. & Siever, R. (1974). *Earth.* W. H. Freeman.

Small, R. J. (1972). *The study of landforms.* Cambridge University Press.

Sparks, B. W. (1960). *Geomorphology.* Longmans.

Strahler, A. N. (1969). *Physical geography,* 3rd Edition. Wiley.

Strahler, A. N., & Strahler, A. H. (1973). *Environmental geoscience.* Wiley.

Twidale, C. R. (1968). *Geomorphology.* Nelson.

Weyman, D. R. (1975). *Runoff processes and streamflow modelling.* Oxford University.

Whalley, W. B. (1976). *Properties of materials and geomorphological explanation.* Oxford University.

Section B

Allen, J. R. L. (1970). *Physical processes of sedimentation.* George Allen and Unwin.

Bagnold, R. A. (1941). *The physics of blown sand and desert dunes.* Methuen.

Blatt, H., Middleton, G., & Murray, R. (1972). *Origin of sedimentary rocks.* Prentice-Hall.

Charlesworth, J. K. (1957). *The Quaternary era.* Arnold.
Chorley, R. J., Editor (1969). *Water, Earth and Man.* Methuen.
Cooke, R. U. & Doornkamp, J. C. (1974). *Geomorphology in environmental management.* Clarendon Press.
Cooke, R. U. & Warren, A. (1973). *Geomorphology in deserts.* Blandford.
Davies, J. L. (1972). *Geographical variation in coastal development.* Oliver and Boyd.
Embleton, G. & King, C. A. M. (1975). *Glacial and periglacial geomorphology,* 2nd Edition. Arnold.
Estes, J. E. & Senger, L. W. (1974). *Remote sensing: techniques for environmental analysis.* Hamilton Publishing Co./Wiley.
Fisher, W. B. (1968). *The Cambridge history of Iran: Volume 1, the land of Iran.* Cambridge University Press.
French, H. M. (1976). *The periglacial environment.* Longman.
Garner, H. F. (1974). *The origin of landscapes: a synethsis of geomorphology.* Oxford University Press.
Garrels, R. M. & Mackenzie, F. T. (1971). *Evolution of sedimentary rocks.* Norton, New York.
Gross, M. G. (1972). *Oceanography.* Prentice-Hall.
Holmes, A. (1965). *Principles of physical geology,* 2nd Edition. Nelson.
King, C. A. M. (1972). *Beaches and coasts,* 2nd Edition. Arnold.
King, C. A. M. (1976). *The geomorphology of the British Isles: northern England.* Methuen.
Leopold, L. B., Wolman, M. G., & Miller, J. P. (1964). *Fluvial processes in geomorphology.* Freeman.
Ollier, C. D. (1969). *Weathering.* Oliver and Boyd.
Open University Science Foundation Course Team (1971). *Earth History I, II.* Open University Press.
Pitty, A. F. (1971). *Introduction to geomorphology.* Methuen.
Price, R. J. (1972). *Glacial and fluvioglacial landforms.* Oliver and Boyd.
Serventy, V. (1968). *Landforms of Australia.* Angus and Robertson.
Short, N. M. (1975). *Planetary geoscience.* Prentice-Hall.
Sissons, J. B. (1967). *The evolution of Scotland's scenery.* Oliver and Boyd.
Sissons, J. B. (1976). *The geomorphology of the British Isles: Scotland.* Methuen.
Smailes, A. E. (1960). *North England.* Nelson.
Spate, O. H. K. (1968). *Australia.* E. Benn.
Stoddart, D. R., & Yonge, Sir M. (1971). *Regional variation in Indian Ocean coral reefs.* Academic Press.
Sugden, D. E., & John, B. S. (1976). *Glaciers and landscape.* Edward Arnold.
Sweeting, M. M. (1972). *Karst landforms.* Macmillan.
Taylor, G. (1958). *Australia.* Methuen.
Thomas, M. F. (1974). *Tropical geomorphology.* Macmillan.
Thornbury, W. D. (1954). *Principles of geomorphology.* Wiley.
Tricart, J. (1963). *Geomorphology of cold environments.* (English translation 1969) Macmillan.
Tricart, J., & Caileux, A. (1965). *Introduction to climatic geomorphology.* (English translation 1972) Longman.
Tricart, J. (1968). *Structural geomorphology.* (English translation 1974) Longman.
Ward, R. C. (1975). *Principles of hydrology,* 2nd Edition. McGraw-Hill.
Washburn, A. L. (1973). *Periglacial processes and environments.* Arnold.
Wooldridge, S. W., & Linton, D. L. (1955). *Structure, surface and drainage in south-east England.* Philip.
Young, A. (1972). *Slopes.* Oliver and Boyd.
Zenkovich, V. P. (1962). *Processes of coastal development.* (English translation 1967) Oliver and Boyd.

Section C

Brown, E. H., & Waters, R. S., Editors (1974). *Progress in geomorphology: papers in honour of David L Linton.* Institute of British Geographers, Special Publication 7.

Brunsden, D., & Doornkamp, J., Editors (1972–73). *The unquiet landscape: a series from the Geographical Magazine.* David and Charles.

Coates, D. R., Editor (1972). *Environmental geomorphology and landscape conservation.* Vol I: *Prior to 1900;* (1974) Vol II: *Urban areas;* (1973) Vol III: *Non-urban regions.* Dowden, Hutchinson and Ross.

Cox, A., Editor (1973). *Plate tectonics and geomagnetic reversals.* Freeman.

Derbyshire, E. (1972). Tors, rock weathering and climate in southern Victoria Land, Antarctica, in Price, R. J., and Sugden, D. E.: *Polar geomorphology.* I.B.G. Special Publication 4.

Derbyshire, E., Editor (1973). *Geographical readings: climatic geomorphology.* Macmillan.

Dury, G. H., Editor (1966). *Essays in geomorphology.* Heinemann.

Dury, G. H., Editor (1970). *Geographical readings: rivers and river terraces.* Macmillan.

Embleton, C., Editor (1972). *Geographical readings: glaciers and glacial erosion.* Macmillan.

Gass, I. G., Smith, P. J., & Wilson, R. C. L. (1972). *Understanding the Earth,* 2nd Edition. Open University Press.

Goldthwait, R. P., Editor (1975). *Glacial deposits.* Dowden, Hutchinson and Ross.

King, C. A. M., Editor (1976). *Landforms and geomorphology.* Dowden, Hutchinson and Ross.

King, C. A. M., Editor (1976). *Periglacial processes.* Dowden, Hutchinson and Ross.

Lamb, H. H. (1966). *The changing climate.* Methuen.

Moor, J. R., Editor (1971). *Oceanography: readings from Scientific American.* Freeman.

National Oceanic and Atmospheric Administration (1972). *Collected reprints: Atlantic Oceanographic and Meteorological Laboratories.*

Oxburgh, E. R. (1974). The plain man's guide to plate tectonics. *Proceedings of the Geologists' Association,* **85,** 299–359.

Press, F., & Siever, R., Editors (1974). *Planet Earth: readings from Scientific American.* Freeman.

Schumm, S. A., Editor (1972). *River morphology.* Dowden, Hutchinson and Ross.

Steers, J. A., Editor (1971). *Geographical readings: introduction to coastline development.* Macmillan.

Steers, J. A., Editor (1971). *Geographical readings: applied coastal geomorphology.* Macmillan.

Tank, R. W. (1973). *Focus on environmental geology.* Oxford University Press.

Wilson, J. T., Editor (1973). *Continents adrift: readings from Scientific Amercian.* Freeman.

Stereo glasses

These may be obtained in the United Kingdom from Casella Ltd., who market plastic- and metal-framed glasses. The latter, though more expensive, are much more robust and worth the extra money. In the USA metal-framed viewers may be obtained from Air Photo Supply Corporation, 158-PE, South Station, Yonkers, New York 10705.

Index

Ablation 188–9
Abyssal plain 21, 22–4
Alluvial fan 139–40, 147
Ancient climates 16
Animals and weathering 260–2
Antarctic ice sheet 310–11
Antecedent drainage 282–3
Arid basin 309
Arid erosion 298–301
Aridity 297–8, 315
Arid upland landscape 305, 306–7
Asthenosphere 14
Atlantic Ocean history 59–60
Avalanche 190

Basalt 28, 30, 32, 51, 162, 329
Base-level changes 149–53
Beaches, bars, spits 171, 172, 247, 249–55, 289
Beach sediment 250
Block mountains 35, 36, 38–41
Braided stream 121, 165
Broads, The 288–9

Calcrete 174
Caves 178–80, 183
Cirque 168, 169, 191, 198–9, 200
Cliffed coasts 171, 172, 248–9
Cliffs 242, 246, 278, 279, 281, 284
Climate and landforms 275–6
Climate and weathering 87–9, 91
Climatic change and streamflow 125
Climatic geomorphology 324–5
Coastal system 240
Coastal types 243–7
Coastal zone 230
Coast classification 256–8
Coasts and climate 245, 249, 258
Coasts and plate theory 243–4
Continent evolution 58–61
Continental drift 15–17

Continental rise, shelf and terrace 22, 34, 35, 44–6, 53, 162
Continents 20, 34
Cordillera 54–5, 163
Core 71–2
Creep 99
Crevasse 192–3
Crust 74–6
Currents (ocean) 239
Cycle concept 6

Davis, W.M. 6, 144–5, 318, 319, 322
Deflation 170, 225, 226–7
Delta 167, 255–6
Deltaic deposits 140–1, 146
Denudation 81
Desert deposition 301 2
Desert landforms 170, 174, 175, 176, 301
Dolines 160, 182
Drainage basin 115–6, 148–9
Drainage basins and climatic change 153–5
Drainage patterns 147
Drumlin 195–6, 210, 211
Dry valley 177, 181, 184, 281
Dune 170, 175, 176, 215, 218–225, 285
Dust Bowl 170, 226

Earth age 63–5
Earth interior 68–9, 71–6
Earth interior temperature 29
Earth origin 79
Earthquake activity 10, 46–51
Earthquake prediction and control 69–70
Earthquake records 65–9
Earth's gravitational field 71, 75
Earth shape 62–4
Earth's magnetic field 71, 72, 73, 74
Electromagnetic spectrum 8
Erosion 81
Erratic boulder 169, 173, 194, 285, 288

Escarpment (cuesta) 43, 278, 280, 292
Esker 142–3, 168
Estuary 256
Exfoliation 92, 164

Faults 38, 42, 284
Fjord 168, 201–2, 296
Floods 119–124, 146, 182
Flood-plain 138–40, 147, 166
Fluvial, aeolian processes in deserts 302–4, 306–7
Fluvioglacial deposit 141–4, 146, 173
Fold 37, 38, 39, 42, 43, 152, 156, 163, 173, 280, 284
Fold mountain 36–8, 51–59, 276
Frost action 100

Geomorphology (definition) 316–7
Geomorphology, applied 326–7
Geomorphology and development of ideas 317–23
Geomorphology—principles 323–26
Geosyncline 40–1
Glacial deposits 193–7, 287–8
Glacial drainage diversion 212, 293–4
Glacial erosion 197–202
Glacial trough 169, 199–202, 291, 295–6
Glacier 168, 186–7, 190, 191, 210
Glacier regimes 187–8
Gorge 166
Graded profile 145

Head 285
Himalayas 56–7
Hydrological cycle 81

Ice accumulation 185–6
Ice Ages 203–7
Ice decay 188–9
Ice movement 189, 192–3
Ice sheet 168, 186–7, 306
Ice thermal conditions 197
Ice transport 193
Igneous rocks 28, 32, 50
Impact crater, 320–1
Incised valley 286

Iran: central desert 311–13
Island arc 14, 32, 52–4, 56
Isostasy 75–6

Kame 143
Karren 159–60
Karst landforms 95, 156–161, 167, 177–84
Karst landforms and climate 180–1
Kettle hole 143

Land capability classification 149
Lava plateau 51
Limestone 156–7
Lithosphere 14
Loess 223, 224
Longshore drift 234–5
Lowland plains, basins 43–4
Lunar landforms and processes 328–9

Magnetic field readings 25, 26, 27
Man and land forming processes 268–75
Man and landforms 263–5, 266–7
Mantle 68, 72–4
Marine erosion 239–41
Mars 321, 329
Mass movement processes 96, 97, 98–101
Meander 121, 138, 146, 147, 154, 155, 165–6
Meltwater inputs to streams 125–6
Mercury 329
Minerals and weathering 85
Moon 80, 320, 328–9
Moraine 193–7, 210, 211

Nappe 38
Nodule 33

Ocean basin 20–33, 34
Ocean basin evolution 32, 163
Ocean–floor minerals 33
Ocean–floor spreading rates 26
Ocean ridge 21–2, 24–7, 28–9, 32
Ocean trench 21, 22, 25, 27
Oil and gas resources 53
Ooze 22, 23
Overflow channel 196, 212

Pediment landscape 174, 305, 307, 308–9
Permeability 154–5
Permafrost 111–114
Periglacial landforms 112–3, 288
Periglacial processes 99–100, 111–2, 285
pH value 87
Phreatic water 158
Pingo 100, 113
Plants and landforms 266
Plants and sedimentation 262
Plants and weathering 260–2
Plateau 283–4, 286
Plate boundary, constructive 13, 14
Plate boundary, destructive 13, 14
Plate movement 13, 162
Plates and fold mountains 51–61
Plates and mineral resources 52–3
Plate, subducted 13, 14
Plates, world distribution 12
Plate tectonics 10–15
Platforms for sensors 9
Polje 177–8
Polygonal ground 113, 114
Porosity 154–5
Precambrian continents 61
Precambrian shield 35, 41–3, 52

Quaternary glaciation in the British Isles
    208–14
Quaternary (Pleistocene) Ice Age 204, 205,
    206, 207–214

Raindrop impact 120
Rainfall inputs to streams 117–9
Raised beach 241–2, 246
Red clay 23
Reef 172, 262–4
Regolith 83
Relict landforms 275–6, 281
Remote sensing 8–9
Rift valley 40, 51
River terraces 153
Rock cycle 35
Rocks and weathering 86, 90

Sahara 313–5
San Andreas fault 10–2, 18

Sand 229
Sandur 143–4
Satellite images 161–76, 190, 278, 279, 290
Satellites 8–9
Scale of landforms 7
Sea–level changes 241–2, 149–53, 208–14
Seamounts (guyots) 27
Seawater 239
Sediment inputs to streams 125–6
Slope evolution 108–111
Slope failure 101
Slope forms 165
Slope processes 102–3, 106–7
Slope profile and plan 103–5
Slope regolith 105
Slope study 102
Snow line 185
Solar system 76–8, 329
Solifluction 100
Springs 156
Stream channels—
    anastomosing 129
    braided 129
    meandering 128–9
    straight 128
Stream deposition 138–40, 301–2
Stream discharge 130–1
Stream erosion 137–8
Stream flow 146
Stream hydrograph 124
Stream load 131–7
Stream long profile 144–5
Stream network 148–50, 151
Streams in deserts 300–1
Streams in urban areas 274
Striations 210
Subduction (Benioff) zone 31
Subsidence 101
Superimposed drainage 281–2
Swallow holes (swallets) 160
Systems 115–7

Terrain mapping 327
Thalweg 128
Thermokarst 112
Tidal environments 237–8, 247
Tidal range 237–8
Tidal surge 236–7
Till 142, 193–4, 291

Tors 92–4
Transform faults 13
Tsunami 234–5

Underfit stream 154, 284–5
Underground water 154–61, 177–84

Venus 329
Volcanic activity 10, 19, 28–31, 32, 46–51
Volcanic peak 21, 22, 27, 29–30, 32
Volcano 163, 321
Volcanoes and man 50

Water in channels 130
Water in ocean basins 33
Water table 85, 156
Wave action 233–4
Wave environment 230–3
Waves 171, 172
Weathered zone 83–4
Weathering 81, 83–97, 164
Weathering and time 95
Weathering in deserts 299
Weathering of buildings 89
Weathering of limestone 157
Weathering on coasts 94, 259
Weathering processes 84
Wind deposition 301
Wind erosion 226–8, 299–300
Wind transport 216–9

Young fold mountains 35, 36–8, 46–7, 50